Industrial Inorganic Pigments

Edited by
Gunter Buxbaum

Other Titles of Interest:

W. Herbst, K. Hunger
Industrial Organic Pigments
Second, Completely Revised Edition 1997
ISBN 3-527-28836-8

Paints, Coatings, and Solvents
Edited by Dieter Stoye and Werner Freitag
Second, Completely Revised Edition 1998
ISBN 3-527-28863-5

Automotive Paints and Coatings
Edited by Gordon Fettis
First Edition 1995
ISBN 3-527-28637-3

Hans G. Völz
Industrial Color Testing
First Edition 1995
ISBN 3-527-28643-8

Heinrich Zollinger
Color Chemistry
Second, Revised Edition 1991
ISBN 3-527-28352-8

Industrial Inorganic Pigments

Second, Completely Revised Edition

Edited by Gunter Buxbaum

WILEY-VCH

Weinheim · New York · Chichester · Brisbane · Singapore · Toronto

Dr. Gunter Buxbaum
Bayer AG
Anorganische Industriechemikalien
F & T
-R 86-
47812 Krefeld
Federal Republic of Germany

First Edition 1993
Second, Completely Revised Edition 1998

Library of Congress Card No. applied for.

British Library Cataloguing-in-Publication Data: A catalogue record for this book is available from the British Library.

Deutsche Bibliothek Cataloguing-in-Publication Data:

Industrial inorganic pigments / ed. by Gunter Buxbaum. – 2., completely rev. ed. – Weinheim ; New York ,
Chichester ; Brisbane ; Singapore ; Toronto : Wiley-VCH, 1998
ISBN 3-527-28878-3

Printed on acid-free and chlorine-free paper.

Composition: Graphischer Betrieb Konrad Triltsch, D-97070 Würzburg
Printing: Betz-Druck, D-64291 Darmstadt
Bookbinding: J. Schäffer GmbH & Co. KG, D-67269 Grünstadt

Preface to the Second Edition

The fact that after only five years a second edition of this book is necessary demonstrates its success. This second edition is not a mere reprint but we have used the opportunity to review all the chapters and the commercial data. Some attention has been given to company mergers in the pigment industry, but this is something that is continually changing. The lists of the standards have been updated as well as the references. New trends in the field of inorganic pigments are described, e.g. the growing importance of luster pigments has led to the inclusion of a more detailed description of them. Sections on pigments whose importance has decreased have been shortened.

Nearly every chapter has been rewritten. Some authors of the first edition are now retired; their contributions have been revised by younger colleagues of known excellence. We express our special thanks to the readers of the first edition who made contributions or gave us valuable hints for this new edition.

Finally we thank the publisher for patience and support.

Preface to the First Edition

Inorganic pigments have a long history. Their chemistry is manifold and the information is spread over a vast number of books and articles with varying degrees of actuality. "Industrial Inorganic Pigments" covers the whole field and is written by experts in the field for all those dealing with the application of inorganic pigments.

Inorganic pigments significantly change our ambient; they are irreplaceable for the coloring of construction materials. They show good light and weather resistance and they withstand the attack of heat and chemicals. Their applications range from concrete to artist's colors, from industrial paints to toners in photocopiers, from coloring foodstuffs to their use as raw materials for catalysts.

The application properties of pigments depend not only on their chemistry but also on their physical appearance and to a greater extent on the manufacturing process. Therefore, the book places much emphasis on the description of industrial production processes. The inclusion of extensive descriptions of applications means that this book is far more than a mere list of pigments and their properties.

Since color is the most important aspect, the book opens with a basic chapter dealing with color and its measurement, incorporating the latest standards. The increasingly important environmental and health regulations are described for each separate class of pigments.

The large number of references (more than 800) will enable the reader to acquire further knowledge of this extensive field.

List of Authors

Heinrich Heine, Hans G. Völz, Bayer AG, Krefeld, Federal Republic of Germany (Chap. 1)

Jürgen Kischkewitz, Bayer AG, Krefeld, Federal Republic of Germany (Section 2.1)

Wolf-Dieter Griebler, Sachtleben Chemie GmbH, Duisburg, Federal Republic of Germany (Section 2.2)

Marcel De Liedekerke, Zinkwit Nederland bv, Eijsden, The Netherlands (Section 2.3)

Gunter Buxbaum, Helmut Printzen, Bayer AG, Krefeld, Federal Republic of Germany (Section 3.1.1)

Manfred Mansmann, Dieter Räde, Gerhard Trenczek, Bayer AG, Krefeld, Federal Republic of Germany (Section 3.1.2)

Manfred Mansmann, Dieter Räde, Bayer AG, Krefeld, Federal Republic of Germany; Volker Wilhelm, Bayer AG, Leverkusen, Federal Republic of Germany (Section 3.1.3)

Stefanie Schwarz, BASF Aktiengesellschaft, Produktsicherheit Pigmente, Ludwigshafen, Federal Republic of Germany (Section 3.2)

Henning Wienand, Jörg Adel, BASF Aktiengesellschaft, Ludwigshafen, Federal Republic of Germany (Section 3.3)

Karl Brandt, Gerhard Adrian, Dr. Hans Heubach GmbH & Co. KG, Langelsheim, Federal Republic of Germany (Section 3.4)

William B. Cork, Reckitt's Colours Ltd., Hull, United Kingdom (Section 3.5)

Heinrich Winkeler, Wilfried Mayer, Degussa AG, Hanau, Federal Republic of Germany; Klaus Schneider, Degussa AG, Frankfurt, Federal Republic of Germany (Section 3.6)

Peter Kleinschmit, Manfred Voll, Degussa AG, Hanau, Federal Republic of Germany (Chap. 4)

Lutz Leitner, Bayer AG, Leverkusen, Federål Republic of Germany; Hendrik Kathrein, Bayer AG, Krefeld, Federal Republic of Germany (Sections 5.1.1, 5.1.2, and 5.1.4)

Ronald Veitch, Helmut Jakusch, Ekkehard Schwab, Manfred Ohlinger, BASF Aktiengesellschaft, Ludwigshafen, Federal Republic of Germany (Section 5.1.3)

RONALD VEITCH, HELMUT JAKUSCH, EKKEHARD SCHWAB, BASF Aktiengesellschaft, Ludwigshafen, Federal Republic of Germany (Section 5.1.5)

GÜNTER ETZRODT, BASF Aktiengesellschaft, Anwendungsentwicklung Pigmente, Ludwigshafen, Federal Republic of Germany (Section 5.2)

GERHARD PFAFF, KLAUS-DIETER FRANZ, Merck KGaA, Darmstadt, Federal Republic of Germany; RALF EMMERT, Rona, Hawthorne, New York, United States; KATSUHISA NITTA, Merck Japan, Onahama, Japan (Section 5.3.1)

ROBERT BESOLD, Carl Schlenk AG, Nürnberg, Federal Republic of Germany (Section 5.3.2)

HARALD GAEDCKE, BASF Lacke + Farben AG, Besigheim, Federal Republic of Germany (Section 5.4)

KARL A. FRANZ, WOLFGANG G. KEHR, ALFRED SIGGEL, JÜRGEN WIECZORECK, Riedel-de Haën AG, Seelze, Federal Republic of Germany (Section 5.5)

Contents

1. Introduction

1.1. General Aspects

Definition. The word "pigment" is of Latin origin (pigmentum) and originally denoted a color in the sense of a coloring matter, but was later extended to indicate colored decoration (e.g., makeup). In the late Middle Ages, the word was also used for all kinds of plant and vegetable extracts, especially those used for coloring. The word pigment is still used in this sense in biological terminology; it is taken to mean dyestuffs of plant or animal organisms that occur as very small grains inside the cells or cell membranes, as deposits in tissues, or suspended in body fluids.

The modern meaning associated with the word pigment originated in this century. According to accepted standards (Table 1, "Coloring materials: Terms and definitions"), the word pigment means a substance consisting of small particles that is practically insoluble in the applied medium and is used on account of its coloring, protective, or magnetic properties. Both pigments and dyes are included in the general term "coloring materials", which denotes all materials used for their coloring properties. The characteristic that distinguishes pigments from soluble organic dyes is their low solubility in solvents and binders. Pigments can be characterized by their chemical composition, and by their optical or technical properties. In this introductory chapter, only inorganic pigments used as coloring materials are discussed.

Extenders (fillers) are substances in powder form that are practically insoluble in the medium in which they are applied. They are usually white or slightly colored, and are used on account of their physical or chemical properties. The distinction between an extender and a pigment lies in the purpose for which it is used. An extender is not a colorant, it is employed to modify the properties or increase the bulk (volume) of a given material. Extenders are beyond the scope of this book and will not be discussed in detail.

Historical. Natural inorganic pigments have been known since prehistoric times. Over 60000 years ago, natural ocher was used in the Ice Age as a coloring material. The cave paintings of the Pleistocene peoples of southern France, northern Spain, and northern Africa were made with charcoal, ocher, manganese brown, and clays, and must have been produced over 30000 years ago. About 2000 B.C., natural ocher was burnt, sometimes in mixtures with manganese ores, to produce red, violet, and black pigments for pottery. Arsenic sulfide and Naples yellow (a lead antimonate) were the first clear yellow pigments. Ultramarine (lapis lazuli) and artificial lapis lazuli (Egyptian blue and cobalt aluminum spinel) were the first blue pigments. Terra verte, malachite, and a synthetically prepared copper hydroxychloride were the first green pigments. Colored glazes for bricks (i.e., ceramic pigments) were widely used by the Chaldeans. Calcite, some phases of calcium sulfate, and kaolinite were the white pigments used at that time.

Table 1. Listing of standards for pigments

Key words	ISO	EN	ASTM	DIN
Acidity/alkalinity	787-4	ISO 787-4	D 1208	EN ISO 787-4
Aluminium pigments and pastes:				
Sampling and testing			D 480	55923
Specification	1247		D 962	55923
Barium chromate pigments:				
Specification	2068			
Bleeding	787-22		D 279	53775-3
Carbon black pigments (see also lampblack):				
Black value				55979
Solvent-extractable material			D 305	55968
Specification			D 561	55968
Cadmium pigments:				
Specification	4620			
Chalking degree:				
Adhesive tape method	4628-6			53223
KEMPF method			D 4214	53159
Change in Strength (see ease of dispersion and PVC)				
Chemical resistance	1812-1	ISO 2812-1		EN ISO 2812-1
Chlorides, water-soluble (see matter soluble)				
Chromium oxide pigments:				
Specification	4621		D 263	ISO 4261
Climates:				
Containing evaporated water				50017
Standardized	554			50014
Open air				50019-1
SO_2 atmosphere	6988	ISO 6988		EN ISO 6988
Coating materials:				
Terms and definitions	4618-1 to 4618-4	971-1		EN 971-1 55945
Color differences:				
CIELAB	7724-3		D 1729 D 2244 E 308	6174
Conditions/evaluation of measurements	7724-2			53236
Significance				55600
Color in full-shade systems:				
Black pigments	787-1		D 3022	55985-2
Colored pigments	787-1		D 3022	55985
White pigments	787-1		D 2805	55983
Coloration of building materials				53237
Colorimetry	7724-1 7724-2 7724-3		E 259 E 308	5033-1 to 5033-9 6174
Coloring materials:				
Classification				55944
Terms and definition	4618-1	971-1		55943 EN 971-1

Table 1. (continued)

Key words	ISO	EN	ASTM	DIN
Corrosion testing:				
NaCl	9227		B 117	50021
SO_2	6988	ISO 6988		EN ISO 6988
Density:				
Centrifuge method	787-23	ISO 787-23		EN ISO 787-23
Pycnometer method	787-10		D 153	EN ISO 787-10
Dusting behavior of pigments:				
Drop method				55992-2
Dusting value				55992-1
Ease of dispersion:				
Alkyd resin and alkyl-melamine system:				
Hardening by oxidation				53238-30
				53238-33
Stove type				53238-31
Automatic muller	8780-5	ISO 8780-5	D 387	EN ISO 8780-5
Bead mill	8780-4	ISO 8780-4		EN ISO 8780-4
Change in gloss	8781-3	ISO 8781-3		EN ISO 8781-3
Change in tinting strength	8781-1	ISO 8781-1		EN ISO 8781-1
Fineness of grind (see below)				
High speed impeller mill	8780-3	ISO 8780-3		EN ISO 8780-3
Introduction	8780-1	ISO 8780-1		EN ISO 8780-1
Oscillatory shaking machine	8780-2	ISO 8780-2		
Triple roll mill	8780-6	ISO 8780-6		EN ISO 8780-6
Fineness of grind	1524		D 1210	EN 21524
	8781-2	ISO 8781-2		EN ISO 8781-2
Heat stability (see also PVC)	787-21		D 2485	53774-5
Hiding power:				
Contrast ratio	6504-3			
Pigmented media	6504-1		D 2805	55987
Wedge-shaped layer	6504-5			55601
White and light gray media			D 2805	55984
Hue of near white specimens				55980
Hue relative of near white specimens				55981
Iron blue pigments:				
Methods of analysis	2495		D 1135	
Specification	2495		D 261	55906
Iron, manganese oxide pigments:				
Methods of analysis	1248		D 50	ISO 1248
Natural, specification	1248		D 3722	ISO 1248
Sienna, specification	1248		D 765	ISO 1248
Umber, specification	1248		D 763	ISO 1248
Iron oxide pigments:				
Black, specification	1248		D 769	ISO 1248
Brown, specification	1248		D 3722	ISO 1248
			D 3724	
FeO content			D 3872	
Methods of analysis	1248		D 50	55913-2
				ISO 1248

Table 1. (continued)

Key words	ISO	EN	ASTM	DIN
Red, specification	1248		D 3721	55913-1
				ISO 1248
Yellow, specification	1248		D 768	ISO 1248
Lampblack pigments:				
Specification			D 209	55968
Lead chromate pigments:				
Method of analysis	3711		D 126	ISO 3711
Specification	3711		D 211	ISO 3711
Lead chromate/phthalocyanine blue pigments:				
Methods of analysis			D 126	55972
Specification			D 212	55972
Lead chromate green pigments:				
Methods of analysis	3710		D 126	55973
Specification	3710			55973
Lead red (see red lead)				
Lead silicochromate pigments (basic):				
Methods of analysis			D 1844	
Specification			D 1648	
Lead white (see white lead)				
Light stability (see also resistance to light):				
Short test	2809			53231
Lightening power of white pigments	787-17		D 2745	55982
Lightness:				
White pigment powders				53163
Lithopone pigments:				
Specification	473		D 3280	55910
Matter soluble in HCl:				
Content of As, Ba, Cd, Co, Cr, Cr(VI),	3856-1		D 3718	53770-1
Cu, Hg, Mn, Ni, Pb, Sb, Se, Zn	to 3856-7		D 3618	to 53770-15
			D 3624	
			D 3717	
Preparation of extract	6713			52770-1
Matter soluble in water:				
Chlorides	787-13			ISO 787-13
Cold extraction	787-8	ISO 787-8	D 2448	EN ISO 787-8
Cr(VI) content				53780
Hot extraction	787-3	ISO 787-3	D 2448	EN ISO 787-3
Nitrates:				
Nessler reagent	787-13			ISO 787-13
Salicylic acid method	787-19	ISO 787-19		EN ISO 787-19
Sulfates	787-13			ISO 787-13
Matter volatile	787-2	ISO 787-2	D 280	EN ISO 787-2
Molybdenum orange pigments:				
Methods of analysis	3711		D 2218	ISO 3711
Nitrates, water soluble (see matter soluble)				
Oil absorption	787-5	ISO 787-5	D 281	EN ISO 787-5
			D 1483	
Opacity: paper, cardboard	2471			53146

Table 1. (continued)

Key words	ISO	EN	ASTM	DIN
Particle size analysis:				
Representation:			D 1366	53206-1
Basic terms	9276-1			66141
Logarithmic normal diagram				66144
Power function grid				66143
RRSB grid				66145
Sedimentation method:				
Balance method				66116-1
Basic standards			D 3360	66111
Pipette method				66115
pH value	787-9	ISO 787-9	D 1208	EN ISO 787-9
Phthalocyanine pigments:				
Methods of analysis			D 3256	
PVC, nonplasticized:				
Basic mixture				53774-1
Heat stability				53774-5
Test specimen preparation				53774-2
PVC, plasticized:				
Basic mixture				53775-1
Bleeding				53775-3
Change in strength				53775-7
Heat stability, in oven				53775-6
Heat stability, mill aging				53775-5
Test specimen preparation				53775-2
Red lead:				
Specification	510		D 49	55916
			D 83	
Reflectance factor; paper, cardboard:				
Fluorescent				53145-2
Nonfluorescent	2469			53145-1
Reflectometer (gloss assessment)	2813		E 430	67530
			D 523	
Residue on sieve:				
By water	787-7			53195
Mechanical method	787-18	ISO 787-18		EN ISO 787-18
Resistance to light	787-15	ISO 787-15		EN ISO 787-15
Resistivity, aqueous extract	787-14		D 2448	ISO 787-14
Sampling:				
Terms	842		D 3925	53242-1
Solid material	842		D 3925	53242-4
Scattering power, relative:				
Gray paste method	787-24	ISO 787-24		EN ISO 787-24
Black ground method				53164
Specific surface area:				
BET method				66131
N_2 adsorption				66132
Permeability techniques				66126-1

Table 1. (continued)

Key words	ISO	EN	ASTM	DIN
Standard depth of shade:				
Specimen adjustment				53235-2
Standards				53235-2
Strontium chromate pigments:				
Specification	2040		D 1845	55903
Sulfates, water-soluble (see matter soluble)				
SO_2 resistance	3231			53771
				ISO 3231
Tamped volume	787-11	ISO 787-11		EN ISO 787-11
Test evaluation:				
Scheme	4628-1			53230
Thermoplastics:				
Basic mixtures				53773-1
Heat stability				53772
Test specimen preparation				53773-2
Tinting strength, relative:				
Change in ~	8781-1	ISO 8781-1		EN ISO 8781-1
Photometric	787-24	ISO 787-24	D 387	55986
Visual	787-16	ISO 787-16		EN ISO 787-16
Titanium dioxide pigments:				
Methods of analysis	591		D 1394	55912-2
			D 3720	
			D 3946	
Specification	591		D 476	55912-1
Test methods	591		D 4563	55912-1
			D 4767	
			D 4797	
Transparency:				
Paper, cardboard	2469			53147
Pigmented/unpigmented systems				55988
Ultramarine pigments:				
Methods of analysis			D 1135	
Specification	788		D 262	55907
Viscosity	2884		D 2196	53229
Weathering in apparatus	4892			53231
	11341			53387
White lead:				
Methods of analysis			D 1301	
Specification			D 81	
Zinc chromate pigments:				
Specification	1249			55902
Zinc dust pigments:				
Methods of analysis	713			
	714			
	3549		D 521	55969
Specification	3549		D 520	55969

Table 1. (continued)

Key words	ISO	EN	ASTM	DIN
Zinc oxide pigments:				
Methods of analysis			D 3280	55908
Specification			D 79	
Zinc phosphate pigments:				
Methods of analysis	6745			ISO 6745
Specification	6745			ISO 6745

Painting, enamel, glass, and dyeing techniques reached an advanced state of development in Egypt and Babylon. A synthetic lapis lazuli (a silicate of copper and calcium) is still known as Egyptian blue. Antimony sulfide and galena (lead sulfide) were commonly used as black pigments, cinnabar as a red pigment, and ground cobalt glass and cobalt aluminum oxide as blue pigments. According to PLUTARCH, the Greeks and Romans did not regard the art of dyeing very highly, and made very little contribution to the development of new pigments. PLINY (23–79 A.D.) describes the pigments orpigment, realgar, massicot, red lead, white lead, verdigris, and pigments laked with alum, as well as the pigments already listed above. Certain types of chalk and clay were used as white pigments.

From the age of the migration of the peoples (fourth to sixth century A.D.) to the end of the late Middle Ages, there were no notable additions to the range of coloring materials. The reinvented pigment Naples yellow and certain dyestuffs for textiles from the orient were the only innovations. New developments in the field of pigments first occurred during the early Renaissance. Carmine was introduced from Mexico by the Spanish. Smalt, safflore, and cobalt-containing blue glasses were developed in Europe.

The pigment industry started in the 18th century with products such as Berlin blue (1704), cobalt blue (1777), Scheele's green, and chrome yellow (1778).

In the 19th century, ultramarine, Guignet's green, cobalt pigments, iron oxide pigments, and cadmium pigments were developed in quick succession.

In the 20th century, pigments increasingly became a subject of scientific investigation. In the past few decades, the synthetic colored pigments cadmium red, manganese blue, molybdenum red, and mixed oxides with bismuth came onto the market. Titanium dioxide with anatase or rutile structures, and acicular zinc oxide were introduced as new synthetic white pigments and extenders, respectively. Luster pigments (metal effect, nacreous, and interference pigments) have assumed increasing importance.

Economic Aspects. World production of pigments in 1995 was approx. 5×10^6 t. Inorganic pigments accounted for ca. 97% of this.

About one-third of this total is supplied by the United States, one-third by the European Community, and one-third by all the remaining countries. The German pigment industry supplied about 40% of the world consumption of inorganic colored pigments, including about 50% of the iron oxides. Estimated world consumption of inorganic pigments in 1995 can be broken down as follows [1.1]:

Titanium dioxide	66%
Iron oxides (natural and synthetic)	14%
Carbon black pigments	10%
Lithopone (incl. ZnS)	4%
Chromates	3%

Chromium oxide	1%
Zinc oxide	< 1%
Molybdates/lead chromates	< 1%
Luster pigments	< 1%
Mixed metal oxide pigments	< 0.5%
Iron blue pigments	< 0.5%
Ultramarine	< 0.5%

Pigment production is a still growth industry, but in the future the growth rate will decrease. The sales of inorganic pigments in 1995 amounted to ca. $\$13 \times 10^9$. The names of manufacturing companies are given in the corresponding sections.

Uses. The most important areas of use of pigments are paints, varnishes, plastics, artists' colors, printing inks for paper and textiles, leather decoration, building materials (cement, renderings, concrete bricks and tiles—mostly based on iron oxide and chromium oxide pigments), leather imitates, floor coverings, rubber, paper, cosmetics, ceramic glazes, and enamels.

The paint industry uses high-quality pigments almost exclusively. An optimal, uniform particle size is important because it influences gloss, hiding power, tinting strength, and lightening power. Paint films must not be too thick, therefore pigments with good tinting strength and hiding power combined with optimum dispersing properties are needed.

White pigments are used for white coloring and covering, but also for reducing (lightening) colored and black pigments. They must have a minimal intrinsic color tone.

When choosing a pigment for a particular application, several points normally have to be considered. The coloring properties (e.g., color, tinting strength or lightening power, hiding power, see Section 1.3) are important in determining application efficiency and hence economics. The following properties are also important:

1) *General chemical and physical properties:* chemical composition, moisture and salt content, content of water-soluble and acid-soluble matter, particle size, density, and hardness (see Section 1.2)
2) *Stability properties:* resistance toward light, weather, heat, and chemicals, anticorrosive properties, retention of gloss (see Section 1.4)
3) *Behavior in binders:* interaction with the binder properties, dispersibility, special properties in certain binders, compatibility, and solidifying effect (see Section 1.5)

Important pigment properties and the methods for determining them are described later.

Classification. Inorganic pigments can be classified from various points of view. The classification given in Table 2 (for standards see Table 1, "Coloring materials, terms") follows a system recommended by ISO and DIN; it is based on coloristic and chemical considerations. As in many classification schemes, there are areas of overlap between groups so that sharp boundaries are often impossible. In this article white pigments are described in Chapter 2, colored pigments in Chapter 3, black pigments (carbon black) in Chapter 4, and specialty pigments in Chapter 5.

Table 2. Classification of inorganic pigments

Term	Definition
White pigments	the optical effect is caused by nonselective light scattering (examples: titanium dioxide and zinc sulfide pigments, lithopone, zinc white)
Colored pigments	the optical effect is caused by selective light absorption and also to a large extent by selective light scattering (examples: iron oxide red and yellow, cadmium pigments, ultramarine pigments, chrome yellow, cobalt blue)
Black pigments	the optical effect is caused by nonselective light absorption (examples: carbon black pigment, iron oxide black)
Luster pigments	the optical effect is caused by regular reflection or interference
Metal effect pigments	regular reflection takes place on mainly flat and parallel metallic pigment particles (example: aluminum flakes)
Nacreous pigments	regular reflection takes place on highly refractive parallel pigment platelets (example: titanium dioxide on mica)
Interference pigments	the optical effect of colored luster pigments is caused wholly or mainly by the phenomenon of interference (example: iron oxide on mica)
Luminescent pigments	the optical effect is caused by the capacity to absorb radiation and to emit it as light of a longer wavelength
Fluorescent pigments	the light of longer wavelength is emitted after excitation without a delay (example: silver-doped zinc sulfide)
Phosphorescent pigments	the light of longer wavelength is emitted within several hours after excitation (example: copper-doped zinc sulfide)

1.2. General Chemical and Physical Properties

1.2.1. Fundamental Aspects [1.2]

Chemical Composition. With few exceptions, inorganic pigments are oxides, sulfides, oxide hydroxides, silicates, sulfates, or carbonates (see Tables 3 and 4), and normally consist of single-component particles (e.g., red iron oxide, α-Fe_2O_3) with well defined crystal structures. However, mixed and substrate pigments consist of nonuniform or multicomponent particles.

Mixed pigments are pigments that have been mixed or ground with pigments or extenders in the dry state (e.g., chrome green pigments are mixtures of chrome yellow and iron blue). If the components differ in particle size and shape, density, reactivity, or surface tension, they may segregate during use.

In the case of *substrate pigments*, at least one additional component (pigment or extender) is deposited onto a substrate (pigment or extender), preferably by a wet method. Weak, medium, or strong attractive forces develop between these pigment

Table 3. Classification of white and black pigments

Chemical class	White pigments	Black pigments
Oxides	titanium dioxide zinc white, zinc oxide	iron oxide black iron–manganese black spinel black
Sulfides	zinc sulfide lithopone	
Carbon and carbonates	white lead	carbon black

components during drying or calcining. These forces prevent segregation of the components during use.

Special substrate pigments include the aftertreated pigments and the core pigments. To produce *aftertreated pigments* the inorganic pigment particles are covered with a thin film of inorganic or organic substances to suppress undesirable properties (e.g., catalytic or photochemical reactivity) or to improve the dispersibility of the pigments and the hydrophilic or hydrophobic character of their surfaces. The particles can be coated by precipitation (e.g., aftertreated TiO_2 pigments, see Section 2.1.3.4), by adsorption of suitable substances from solutions (usually aqueous), or by steam hydrolysis.

To produce *core pigments*, a pigment substance is deposited on an extender by precipitation or by wet mixing of the components. In the case of anticorrosive pigments (see Section 5.2), whose protective effect is located on their surfaces, the use of core pigments can bring about a significant saving of expensive material. Extender particles are also treated by fixing water-insoluble organic dyes on their surfaces via lake formation.

Analysis. The industrial synthesis of inorganic pigments is strictly controlled by qualitative and quantitative chemical analysis in modern, well-equipped physicochemical test laboratories. Quantitative chemical and X-ray analysis is carried out on raw materials, intermediates, and substances used for aftertreatment, but most importantly on the final products, byproducts, and waste products (wastewater and exhaust gas). This serves not only to fulfil quality requirements but also the demands of environmental protection. Quality control, carried out in specially equipped laboratories and supported by computers, includes testing of physical and technical application properties [1.3], [1.4]. Information on quality requirements for inorganic pigments is widely available in international, ISO, European (EN), and national standards (e.g., AFNOR, ASTM, BSI, DIN) [1.5], [1.6]. Standard analytical methods and conditions of delivery for the most important inorganic pigments are given in Table 1. Further information is given in later sections.

Crystallography and Spectra. The following are the most common crystal classes:

1) *Cubic:* zinc blende lattice (e.g., precipitated CdS), spinel lattice (e.g., Fe_3O_4, $CoAl_2O_4$)
2) *Tetragonal:* rutile lattice (e.g., TiO_2, SnO_2)
3) *Rhombic:* goethite lattice (e.g., α-FeOOH)
4) *Hexagonal:* corundum lattice (e.g., α-Fe_2O_3, α-Cr_2O_3)
5) *Monoclinic:* monazite lattice (e.g., $PbCrO_4$)

Table 4. Classification of inorganic colored pigments

Chemical class	Green	Blue-green	Blue	Violet	Red	Orange	Yellow	Brown
Oxides and oxide–hydroxides								
Iron oxide pigments					iron oxide red	iron oxide orange	iron oxide yellow	iron oxide brown
Chromium oxide pigments	chromium oxide	chromium oxide hydrate green						
Mixed metal oxide pigments	cobalt green and blue					chromium rutile orange	nickel rutile yellow, chromium rutile yellow	zinc iron spinell, Mn–Fe-brown
Sulfide and sulfoselenide pigments					cadmium sulfoselenide		cadmium sulfide (Cd, Zn) S	
Chromate pigments	chrome green				molybdate red	chrome orange	chrome yellow, zinc yellow, alkaline earth chromates	
Ultramarine pigments	ultramarine green, blue, violet, and red							
Iron blue pigments			iron blue					
Others			manganese blue	cobalt, manganese violet			naples yellow, bismuth vanadate	

In ideal solid ionic compounds, the absorption spectrum is composed of the spectra of the individual ions, as is the case in ionic solutions. For metal ions with filled *s, p,* or *d* orbitals, the first excited energy level is so high that only ultraviolet light can be absorbed. Thus, when the ligands are oxygen or fluorine, white inorganic compounds result. The absorption spectra of the chalcogenides of transition elements with incompletely filled *d* and *f* orbitals are mainly determined by the charge-transfer spectrum of the chalcogenide ion which has a noble gas structure. For the transition metals, lanthanides, and actinides, the energy difference between the ground state and the first excited state is so small that wavelength-dependent excitations take place on absorption of visible light, leading to colored compounds [1.7].

X-ray investigation of inorganic pigments yields information on the structure, fine structure, state of stress, and lattice defects of the smallest coherent regions that are capable of existence (i.e., crystallites) and on their size. This information cannot be obtained in any other way. Crystallite size need not be identical with particle size as measured by the electron microscope, and can, for example, be closely related to the magnetic properties of the pigment.

Particle Size. The important physical data for inorganic pigments comprise not only optical constants, but also geometric data: mean particle size, particle size distribution, and particle shape [1.8]. The standards used for the terms that are used in this section are listed in Table 1 ("Particle size analysis").

The concept of *particles and particle shape* corresponds to that used in the recommended and internationally accepted classification of pigment particles given in [1.9] (see Fig. 1 and Table 5).

The term *particle size* must be used with care, as is borne out by the large number of different "particle diameters" and other possible terms used to denote size (see Table 6). In granulometry, so-called *shape factors* are often used to convert equivalent diameters to "true" diameters. However, the determination and use of shape factors is problematic.

Table 5. Definitions of particles and associated terms (see also Fig. 1)

Term	Definition
Particle	individual unit of a pigment that can have any shape or structure
Primary or individual particles	particles recognizable as such by appropriate physical methods (e.g., by optical or electron microscopy)
Aggregate	assembly of primary particles that have grown together and are aligned side by side; the total surface area is less than the sum of the surface areas of the primary particles
Agglomerate	assembly of primary particles (e.g., joined together at the corners and edges), and/or aggregates whose total surface area does not differ appreciably from the sum of the individual surface areas
Flocculate	agglomerate present in a suspension (e.g., in pigment–binder systems), which can be disintegrated by low shear forces

Figure 1. Primary particles, agglomerates, and aggregates

Table 6. Particle size, particle size distribution, and characteristic quantities

Term	Definition
Particle size	geometrical value characterizing the spatial state of a particle
Particle diameter D_{eff}	diameter of a spherical particle or characteristic dimension of a regularly shaped particle
Equivalent diameter D	diameter of a particle that is considered as a sphere
Particle surface area S_T	surface area of a particle: a distinction is made between the internal and external surface areas
Particle volume V_T	volume of a particle: a distinction is made between effective volume (excluding cavities) and apparent volume (including cavities)
Particle mass m_T	mass of a particle
Particle density ϱ_T	density of a particle
Particle size distribution	statistical representation of the particle size of a particulate material
Distribution density	gives the relative amount of a particulate material in relation to a given particle size diameter. Density distribution functions must always be normalized
Cumulative distribution	normalized sum of particles that have a diameter less than a given particle size parameter
Fractions and class	a fraction is a group of particles that lies between two set values of the chosen particle size parameter that limits the class
Mean value and other similar parameters	the mean values of particle size parameters can be expressed in many ways, some values are used frequently in practice
Distribution spread	parameter for characterizing the nonuniformity of the particle size

In practice, empirically determined *particle size distributions* are represented by:

1) Tabulated results
2) Graphical representation in the form of a histogram (bar chart) or as a continuous curve
3) Approximation in the form of analytical functions

For standards see Table 1 "Particle size, representation".

Special distribution functions are specified in some standards (e.g., power distribution, logarithmic normal distribution, and RRSB distribution). Methods of determination for pigments are rated in Section 1.2.2.

The important parameters relating to particle size distribution are the *mean particle size* and the *spread of the distribution*. The way of expressing the mean particle size depends on the test method used or on which mean value best reflects the pigment property of interest. Depending on the spread of parameter, the various mean values for a given particulate material can differ considerably. The mean particle sizes of inorganic pigments lie in the range 0.01 – 10 μm, and are usually between 0.1 and 1 μm.

The *specific surface area* also represents a mean of the pigment particle size distribution. It can be used to calculate the mean diameter of the surface distribution. Care must be taken that the effect of the "internal surface area" is taken into account. If the product has an internal surface area which cannot be neglected in comparison to the external surface area, then the measured specific surface area no longer gives a true measure of the mean diameter. This applies, for instance, to aftertreated pigments because the treatment material is often very porous.

For anisometric particles (e.g., needle- or platelet-shaped particles) mathematical statistics may likewise be applied [1.10]. The two-dimensional logarithmic normal distribution of the length L and breadth B of the particles also allows the representation and calculation of the characteristic parameters and mean values. The eccentricity of the calculated standard deviation ellipse (Fig. 2) is a measure of the correlation between the length and breadth of the particle. By using more than two

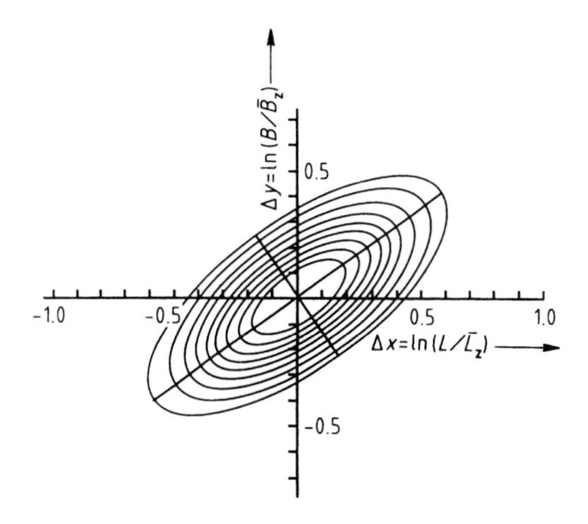

Figure 2. Standard deviation ellipses of a logarithmic normal distribution (yellow iron oxide pigment)
$\bar{L}_z$, $\bar{B}_z$ = median of L, B

particle fineness parameters, this principle can be further extended in a similar manner.

The effect of particle size on optical properties of pigments is described in Section 1.3.

1.2.2. Methods of Determination

General Methods. *Sampling* with a suitable device (beaker, sampling scoop) is applicable for solid materials, especially pigments, fillers, and resins in powder, granular, or lump form. For standards, see Table 1 ("Sampling").

Standard climates are constant artificial climatic conditions with a defined temperature and humidity and limited ranges of air pressure and flow velocity. They can be set up in "clima" cabinets, closets, or rooms. The standard climate 23/50 (23 °C and 50% relative humidity) is recommended. For standards, see Table 1 ("Climates").

Evaluation Methods. In testing paint materials, paints, and other coating materials, the properties or variations in properties often cannot be described quantitatively but must be assessed subjectively. A uniform system of evaluation has been established in the form of a numerical scale to facilitate the assessment of results and mutual understanding. This system should only be used if a result cannot be obtained as a directly measured value. For standards, see Table 1 ("Test evaluation").

Matter Volatile and Loss on Ignition. The content of *matter volatile* in a pigment is determined by drying a sample in an oven at 105 ± 2 °C. This normally gives a measure of the moisture content. For standards, see Table 1 ("Matter volatile").

Loss on ignition is determined by various means depending on the pigment. The principle is the same in all the methods: a weighed sample is heated to a predetermined temperature, cooled in a desiccator, and reweighed. Special standard procedures are specified for iron oxide pigments and carbon black. For standards, see Table 1 ("Iron oxide pigments: Specifications" and "Carbon black pigments: Specifications").

Aqueous Extracts. The content of *matter soluble in water* in pigments is determined by hot or cold extraction of the pigments under prescribed testing conditions. For standards, see Table 1 ("Matter soluble in water"). The decision whether to use hot or cold extraction depends on the properties of the pigment and should be agreed between the interested parties unless otherwise prescribed.

The *pH value* of a pigment denotes the pH of an aqueous suspension of the pigment prepared in a prescribed manner. For standards, see Table 1 ("pH value"). Apparatus: pH meter.

The *electrical conductivity* (or resistivity) of an aqueous pigment extract is determined from the electrical conductance (or resistance). For standards, see Table 1 ("Resistivity"). Apparatus: centrifuge with glass containers, resistance measurement bridge, electrolytic cell.

The *acidity* or *alkalinity* is measured as the quantity of cubic centimeters of 0.1 N alkali (NaOH or KOH) or acid solution that are required to neutralize an aqueous extract of 100 g pigment under prescribed conditions. Unless otherwise agreed, the pigment is extracted with hot water. If cold water is used, this must be specifically stated. For standards, see Table 1 ("Acidity/alkalinity"). Test reagents: 0.5 N HCl or H_2SO_4, 0.5 N NaOH or KOH, indicators (by agreement).

Water-Soluble Sulfates, Chlorides, and Nitrates. The procedures are applicable to pigments and extenders. The choice of hot or cold extraction depends on the properties of the pigment and should be agreed. The anions are determined by the usual analytical methods. For standards, see Table 1 ("Sulfates", "Chlorides", and "Nitrates"). Apparatus: Nessler tubes or spectrophotometer, equipment for pH measurement.

Water-Soluble Chromium(VI). Chromium(VI) is determined photometrically in an aqueous extract obtained by hot or cold extraction using the color produced with diphenylcarbazide, or directly as chromate. For standards, see Table 1 ("Matter soluble in water"). Apparatus: equipment for shaking conical flasks, photometer with glass cells, pH meter.

Matter Soluble in HCl. The pigment is extracted with 0.1 mol/L hydrochloric acid under prescribed conditions to determine the content of As, Ba, Cd, Co, Cr, Cr(VI), Cu, Hg, Mn, Ni, Pb, Sb, Se, and Zn. For standards, see Table 1 ("Matter soluble in HCl"). Apparatus: shaker for conical flasks.

Particle Size Distribution. Methods for determining the particle size of pigments should provide not only the mean particle size but also the complete particle size distribution. Thus, it is preferable to use methods that give not only the parameters of a particular distribution function, but also allow direct measurement of the true particle size distribution. Closely specified standard methods of determination exist for particle sizes >1 μm. During recent years several methods have been developed for the particle size region of particular relevance to pigments (<1 μm). The following methods are used:

1) Counting particles on an electron micrograph
2) Sedimentation analysis
3) Optical methods
4) Other methods that are based on physical volume effects (e.g., the Coulter counter, based on electrical conductivity)

Particle counting of electron micrographs is performed on a computer screen with an automatic counting apparatus. In order to minimize errors, at least 2500 (but preferably 10000) particles are counted.

For pigments, counting is the most suitable method for several reasons. The counting operation is carried out in the binder medium of interest, whereas with sedimentation analysis or the Coulter counter the medium cannot be freely chosen. Furthermore, counting can be carried out under the dispersion conditions used for examination (in contrast to methods in which a specified binder medium must be used that can lead to a different state of dispersion). Problems associated with concentration are much less frequent than in other methods, which sometimes require extremely dilute suspensions (especially optical analysis and the Coulter counter). Another problem with optical analysis is that calibration is often difficult because the absolute scattering and absorption functions of real particles are not well known, and because isodisperse particulate matter is usually not available to draw up a calibration curve. The most important advantages of the counting technique based on electron micrographs lie in the direct observation of the particle shapes and in the ability to distinguish between primary particles, agglomerates, and aggregates.

Sedimentation analysis uses either the earth's gravitational field or a centrifugal field. Owing to the small size of pigment particles (<1 μm), the time taken for sedimentation in the gravitational field is considerable. This has led to developments in centrifugal sedimentation techniques, both as regards apparatus (centrifuges, disk-shaped rotors) and methods (two-layer technique). For standards, see Table 1 ("Particle size analysis: Sedimentation method"). The rate of fall of a single particle in a liquid in a gravitational or centrifugal field gives a measure of the particle size. Methods of measurement differ in the approach used to determine the different particle size fractions:

1) *Suspension Methods.* Analysis starts with a suspension in which a representative sample of the dispersed solid is homogeneously distributed over the whole of the sedimentation depth.
2) *Two-Layer Method.* A suspension is spread in a thin layer on the surface of a clear, solids-free liquid. The particles then fall through the liquid in order of decreasing sedimentation velocity and reach the measuring plane in succession.

In *optical methods*, extremely dilute aqueous or air-fluidized pigment suspensions are almost exclusively used. Instruments for the 0.1 – 10 μm particle size region mostly use Frauenhofer diffraction with additional correction for Mie scattering on small particles. Optical methods allow quick and reproducible measurements. Unfortunately, information is not usually available concerning the optical constants of the sample; this causes drawbacks as regards absolute particle size data.

Sieve Analysis. The sieving residue can be determined by two methods:

1) *Wet Sieving by Hand.* In the utilization of pigments, it is important to know the content of pigment particles that are appreciably larger than the mean particle size. This material can consist of coarse impurities, pigment aggregates (agglomerates), or large primary particles. The dried pigment is washed with water through a sieve of the appropriate mesh size, and the retained material is determined gravimetrically after drying. For standards, see Table 1 ("Residue on sieve: By water").
2) *Wet Sieving by a Mechanical Flushing Procedure.* The sieve residue is the portion of coarse particles that cannot be washed through a specified test sieve with water. The result depends on the mesh size of the sieve. For standards, see Table 1 ("Residue on sieve: Mechanical method"). Apparatus: Mocker's apparatus.

The *specific surface* is usually understood to mean the area per unit mass of the solid material, but it is sometimes useful to relate the surface area to the volume of the solid (see Section 1.2.1). The specific surface area can only be determined indirectly owing to the small size of the pigment particles:

1) *Gas Adsorption by the Brunauer, Emmett, and Teller (BET) Method.* The specific surface area of porous or finely divided solids is measured. The method is limited to solids that do not react with the gas used (e.g., while the gas is adsorbed), and nonmicroporous materials. For standards, see Table 1 ("Specific Surface, BET Method" and "N_2 Adsorption").

2) *Carman's Gas Permeability Method.* A gas or a wetting liquid is made to flow through the porous material in a tube by applying vacuum or pressure. The pressure drop or flow rate is measured. For pigments, a modified procedure is used in which mainly nonlaminar flow takes place [1.11]. For standards, see Table 1 ("Specific surface: Permeability techniques").

Pigment Density. *Density* is determined by pycnometry at a standard temperature of 25 °C. For standards, see Table 1 ("Density"). Apparatus: pycnometer, vacuum pump, or centrifuge.

The *apparent density* of a powdered or granulated material after tamping is the mass (g) of 1 cm^3 of the material after tamping in a tamping volumeter under prescribed conditions. The *tamped volume* is the volume (cm^3) of 1 g of the material. Tamped volume and apparent density after tamping depend mainly on the true density, shape, and size of the particles. A knowledge of these parameters allows decisions to be made regarding dimensions of packing materials and product uniformity. For standards, see Table 1 ("Tamped volume"). Apparatus: tamping volumeter.

Hardness and Abrasiveness. The abrasiveness of a pigment is not identical to its intrinsic hardness, i.e., the hardness of its primary particles. In practice the Mohs hardness is not therefore a useful indication of the abrasiveness of a pigment. Abrasion rather depends on pigment particle size and shape and is usually caused by the sharp edges of the particles. A standard test procedure for determining abrasiveness does not exist. A method based on the abrasion of steel balls by the pigment is described in [1.12].

1.3. Color Properties

1.3.1. Fundamental Aspects [1.13]–[1.16]

When a photon enters a pigmented film, one of three events may occur:

1) It may be absorbed by a pigment particle
2) It may be scattered by a pigment particle
3) It may simply pass through the film (the binder being assumed to be nonabsorbent)

The important physical–optical properties of pigments are therefore their light-absorption and light-scattering properties. If absorption is very small compared with scattering, the pigment is a white pigment. If absorption is much higher than scattering over the entire visible region, the pigment is a black pigment. In a colored pigment, absorption (and usually scattering) are selective (i.e., dependent on wavelength).

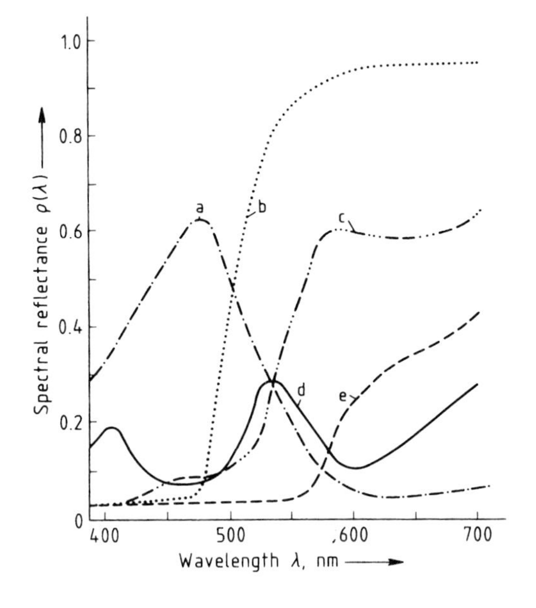

Figure 3. Spectral reflectance curves of some inorganic pigments in paints
a) Manganese blue;
b) CdS;
c) α-FeOOH; d) α-Cr_2O_3; e) α-Fe_2O_3

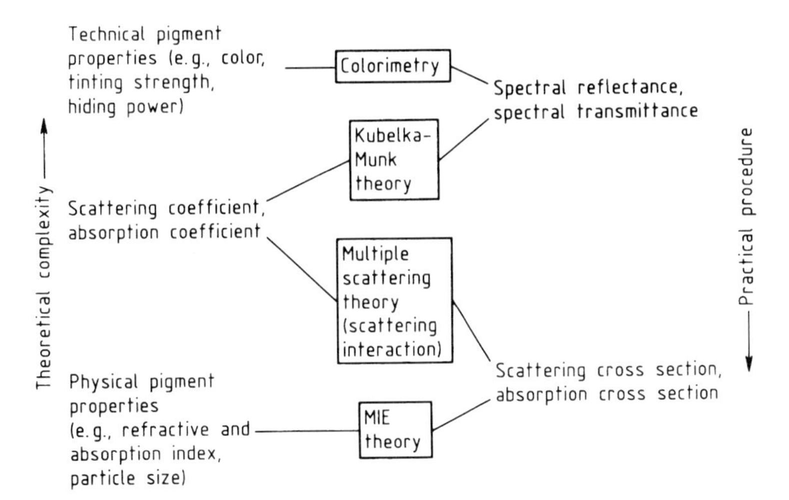

Figure 4. The relationships between the optical properties of pigments and their theoretical basis

Pigments and coatings may be unambiguously characterized by their spectral reflectance curves $\varrho(\lambda)$ or spectral reflectance factor curves $R(\lambda)$ (Fig. 3). The reflectance spectrum $\varrho(\lambda)$ or $R(\lambda)$ and hence the color properties can be almost completely derived from physical quantities [1.15] (Fig. 4):

1) *Colorimetry* relates the perceived color quality to the color stimulus, which in turn is based on the reflectance spectrum $\varrho(\lambda)$.
2) The *Kubelka–Munk theory* relates $\varrho(\lambda)$ to scattering, absorption, and film thickness (scattering coefficient S, absorption coefficient K, film thickness h).

3) The *theory of multiple scattering* (scattering interaction) relates the scattering coefficient S to the pigment volume concentration σ and to the scattering diameter Q_S of the individual particle. The absorption coefficient K is directly proportional to the absorption diameter Q_A and the concentration σ.
4) In *Mie's theory*, the scattering diameter Q_S and the absorption diameter Q_A are related to the particle size D, the wavelength λ, and the optical constants of the material (refractive index n and absorption index κ).

Colorimetry [1.17]–[1.19]. The principles of colorimetry are based on the fact that all color stimuli can be simulated by additively mixing only three selected color stimuli (trichromatic principle). A color stimulus can, however, also be produced by mixing the spectral colors. Thus, it has a spectral distribution, which in the case of nonluminous, perceived colors is called the spectral reflectance $\varrho(\lambda)$. After defining three reference stimuli, the trichromatic principle allows a three-dimensional color space to be built up in which the color coordinates (tristimulus values) can be interpreted as components of a vector (CIE system; for standards, see Table 1, "Colorimetry"; CIE = Commission Internationale de l'Éclairage). For uncolored illumination the three CIE tristimulus values depend on the spectral reflectance as follows:

$$X = \int \bar{x}(\lambda)\,\varrho(\lambda)\,\mathrm{d}\lambda \qquad Y = \int \bar{y}(\lambda)\,\varrho(\lambda)\,\mathrm{d}\lambda \qquad Z = \int \bar{z}(\lambda)\,\varrho(\lambda)\,\mathrm{d}\lambda$$

where $\bar{x}(\lambda)$, $\bar{y}(\lambda)$, and $\bar{z}(\lambda)$ are the CIE tristimulus values of the spectral colors and are called the CIE spectral tristimulus values (color matching function). The CIE chromaticity coordinates (x, y, and z) are given by

$$x = \frac{X}{X+Y+Z}; \quad y = \frac{Y}{X+Y+Z}; \quad z = 1 - x - y$$

They are represented as coordinates in a color plane. The chromaticity coordinates x and y are used to specify the saturation and hue of any color in the CIE chromaticity diagram. See Figure 4a for illumination D65. The CIE spectral tristimulus value $\bar{y}(\lambda)$ corresponds to the lightness sensitivity curve of the human eye. Therefore, a third color variable is specified in addition to x and y, namely the CIE tristimulus value Y, which is a measure of lightness.

This system allows exact measurement of color with worldwide agreement. For pigment testing, however, this is not sufficient because small color differences usually have to be determined and evaluated (e.g., between test and reference pigment). Using the CIE system, it is certainly possible to say which spectral distributions are visually identical, but this is not suitable for determining color differences. To establish color differences an "absolute color space" must be used. Here, colors are arranged three-dimensionally such that the distance between two colors in any direction in space corresponds to the perceived difference. Such a type of color space can be based on the color qualities lightness, hue, and saturation. Several such systems exist. The most widespread color system is probably the Munsell system, which is available in the form of an atlas.

For the quantitative determination of color differences, the transformation relationships between the CIE system (which has to be used for color measurement) and

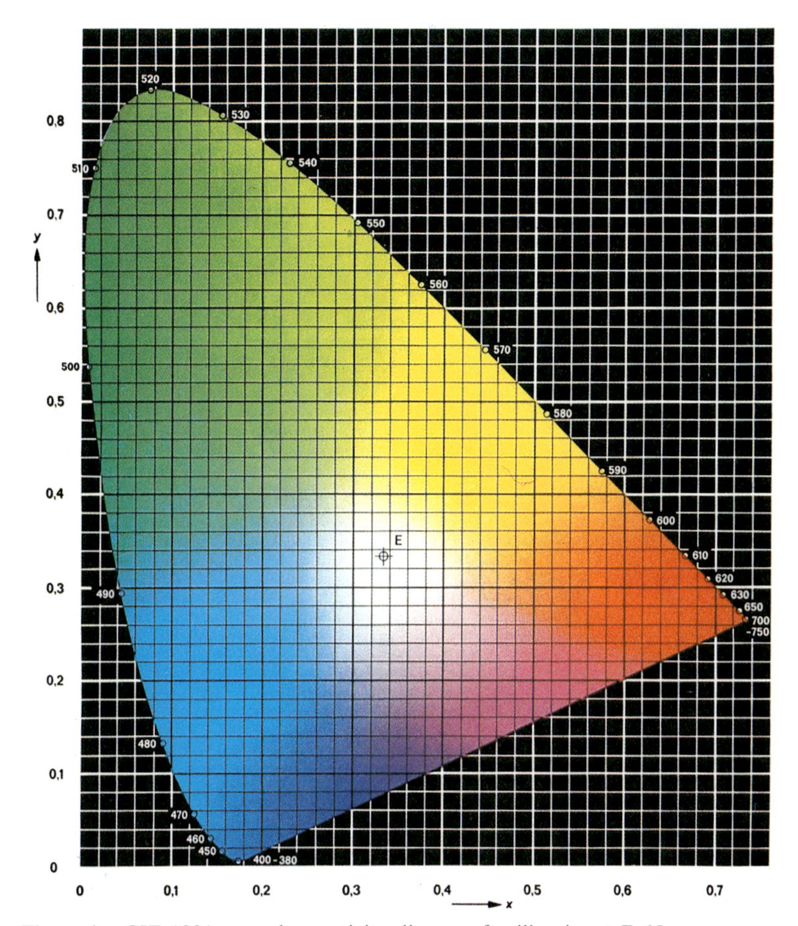

Figure 4a. CIE 1931 xy – chromaticity diagram for illuminant D65

the physiologically equidistant color system must be established. Color differences can then be calculated in the latter system. A large number of color difference systems have been developed, mainly as a need of industrial color testing.

The Adams–Nickerson (AN) system, well-known for many decades and derived from the Munsell system, was recommended for pigment testing by DIN (German Standards Institute) and later worldwide by the CIE ("CIELAB"; for standards, see Table 1, "Color differences"). The three coordinates are denoted by a^* (red-green axis), b^* (the yellow-blue axis), and L^* (the lightness axis). To calculate the CIELAB coordinates, X, Y, and Z are first converted into the functions X^*, Y^*, and Z^* by using a relationship that approximately takes account of the physiologically equidistant lightness steps:

$$X^* = \sqrt[3]{X/X_n}; \quad Y^* = \sqrt[3]{Y/Y_n}; \quad Z^* = \sqrt[3]{Z/Z_n}$$

where X_n, Y_n, and Z_n are the CIE tristimulus values of the illuminant, especially a standard illuminant. For radicands < 0.008856, these equations become:

$$X^* = (7.787\ X/X_n) + 0.138$$
$$Y^* = (7.787\ Y/Y_n) + 0.138$$
$$Z^* = (7.787\ Z/Z_n) + 0.138$$

Values of a^*, b^*, and L^* are obtained from the values of X^*, Y^*, and Z^*:

$$a^* = 500\ (X^* - Y^*)$$
$$b^* = 200\ (Y^* - Z^*)$$
$$L^* = 116\ Y^* - 16$$

The components of the color difference are obtained as differences between the test sample (T) and the reference pigment (R):

$$\Delta a^* = a_T^* - a_R^*; \quad \Delta b^* = b_T^* - b_R^*; \quad \Delta L^* = L_T^* - L_R^*$$

The color difference is finally calculated as the geometrical distance between the two positions in the CIELAB color space:

$$\Delta E_{ab}^* = \sqrt{\Delta a^{*2} + \Delta b^{*2} + \Delta L^{*2}}$$

An important advantage of the CIELAB system is that the resulting color difference can be split up into component contributions, namely lightness, saturation, and hue, corresponding to the arrangement of the color space:

Lightness difference

$$\Delta L^* = L_T^* - L_R^*$$

Chroma difference (saturation difference)

$$\Delta C_{ab}^* = \sqrt{a_T^{*2} + b_T^{*2}} - \sqrt{a_R^{*2} + b_R^{*2}}$$

Hue difference

$$\Delta H_{ab}^* = \sqrt{\Delta E_{ab}^{*2} - \Delta L^{*2} - \Delta C_{ab}^{*2}}$$

Kubelka–Munk Theory. The Kubelka–Munk theory [1.20], [1.21] is based on the fact that the optical properties of a film which absorbs and scatters light may be described by two constants: the absorption coefficient K and the scattering coefficient S. In a simplification, the flux of the diffuse incident light is represented by a single beam L^+, and the flux of the light scattered in the opposite direction by a beam L^-. Each beam is attenuated by absorption and scattering losses, but is reinforced by the scattering losses of the respectively opposite beam. The absorption and scattering losses are determined quantitatively by the two coefficients K and S. A simple system of two linked differential equations can be written. These can be integrated for the valid boundary conditions at the incident light side, and at the opposite side. Solutions for the transmittance τ and the reflectance ϱ are obtained from these integrals as a function of the absorption coefficient K, the scattering coefficient S, the film thickness h, and in special cases of the reflectance ϱ_0 of a given substrate.

The most important and widely used quantity derived from the Kubelka–Munk theory is the reflectance of an opaque (infinitely thick) film that is described by a very simple equation:

$$K/S = (1 - \varrho_\infty)^2/(2\varrho_\infty)$$

From this expression (Kubelka–Munk function) it follows that, within the range of validity of the theory, ϱ_∞ depends only on the ratio of the absorption coefficient to the scattering coefficient, and not on their individual values. The equation has been most useful where reflectance measurements are used to obtain information about absorption and scattering (e.g., in textile dyeing, thin layer chromatography, and IR spectroscopy).

This theory is especially useful for computer color matching of pigmented systems [1.22]–[1.24]: absorption and scattering coefficients are combined additively using the specific coefficients of the components multiplied by their concentrations.

Multiple Scattering. The absorption coefficient K obeys Beer's law even at high pigment volume concentrations σ, and is therefore proportional to σ (Fig. 5). The

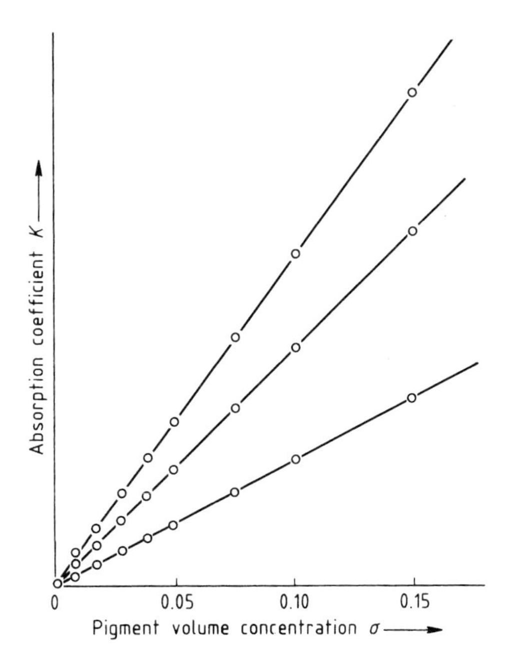

Figure 5. Absorption coefficient K for diffuse illumination as a function of the pigment volume concentration for three red iron oxide pigments

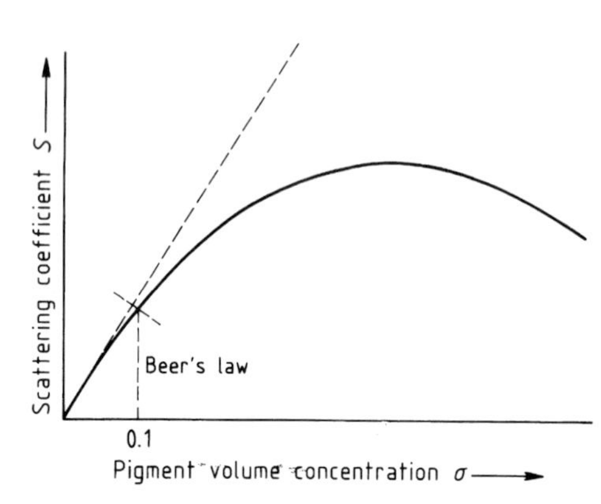

Figure 6. Scattering coefficients as a function of pigment volume concentration

relationship between the scattering coefficient S and the concentration gives rise to problems, however. The distance between the pigment particles decreases with increasing concentration; consequently there is interaction and hindrance between the light scattered by individual particles, and their scattering power usually falls. The scattering coefficient S is therefore linearly related to concentration only at low concentrations (the Beer's law region), at higher concentrations it remains below the linear value (Fig. 6). The concentration dependence of the scattering coefficient can be quantitatively represented by using empirical formulae [1.25], e.g., there is a linear relationship between S/σ and $\sigma^{\frac{2}{3}}$.

Mie's Theory. MIE applied the Maxwell equations to a model in which a plane wave front meets an optically isotropic sphere with refractive index n and absorption index κ [1.26]. Integration gives the values of the absorption cross section Q_A and the scattering cross section Q_S; these dimensionless numbers relate the proportion of absorption and scattering to the geometric diameter of the particle. The theory has provided useful insights into the effect of particle size on the color properties of pigments.

Scattering is considered first. Here, the crucial parameter α in the Q_A and Q_S formulae is a relative measure of particle size because it is proportional to the particle diameter D and is inversely proportional to the wavelength λ. At a constant wavelength λ and for various relative refractive indices n (i.e., relative to the binder, $n = n_p/n_B$ where n_p and n_B are the refractive indices of the particle and binder, respectively), it gives the relationship between scattering and particle size (Fig. 7) [1.27]. If, on the other hand, the particle size D is kept constant, α denotes the relationship between the scattering and the wavelength ($1/\lambda$ replaces D on the abscissa of Fig. 7). The well-marked maxima are typical, and their existence signifies (1) that an optimum particle size must exist with respect to lightening power [1.28], and (2) that for a given particle size, there must be a particular wavelength for maximum light scattering [1.29].

The first relationship can be used to predict the optimum particle size of white pigments (Table 7). The second relationship explains, for example, how white pigments in gray color mixtures can produce colored undertones as a result of selective light scattering (see Section 1.3.2).

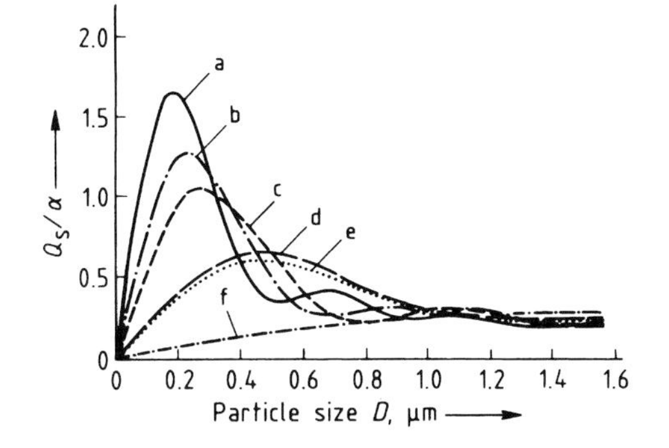

Figure 7. Scattering of white pigments as a function of particle size ($\lambda = 550$ nm)
a) Rutile; b) Anatase; c) Zinc sulfide; d) Zinc oxide; e) White lead; f) Barium sulfate

Table 7. Refractive indices and optimal particle sizes of some white pigments ($\lambda = 550$ nm) calculated according to the van de Hulst formula [1.27]

Pigment	Formula	Mean refractive index		Optimal particle size (relative to binder) D_{opt}, µm
		Vacuum	Binder	
Rutile	TiO_2	2.80	1.89	0.19
Anatase	TiO_2	2.55	1.72	0.24
Zinc blende	α-ZnS	2.37	1.60	0.29
Baddeleyite	ZrO_2	2.17	1.47	0.37
Zincite	ZnO	2.01	1.36	0.48
Basic lead carbonate		2.01	1.36	0.48
Basic lead sulfate		1.93	1.30	0.57

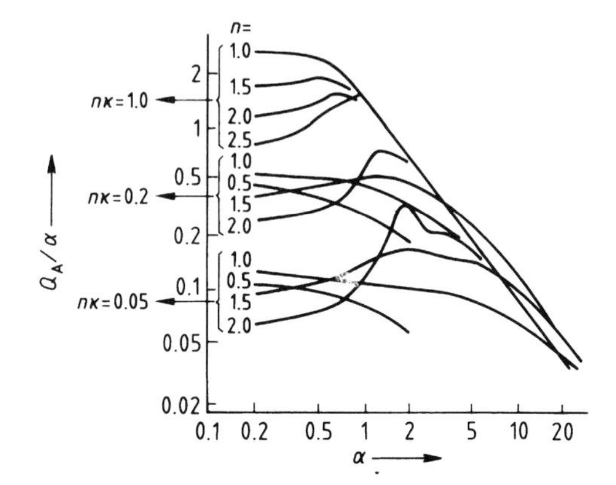

Figure 8. Absorption as a function of particle size

The consequences of Mie's theory for *absorption* (i.e., for tinting strength) are now considered. Calculations from Mie's theory, using the relative refractive index n and the absorption index κ, are given in Figure 8 [1.30]. The parameter α on the abscissa can once more be taken as a relative measure of the particle size. The following conclusions may be drawn:

1) For very small particles, the absorption is independent of the particle size, and hence any further reduction in particle size does not produce additional absorption.
2) With increasing absorption index κ, the absorption of very small particles increases.
3) Absorption values for large particles are approximately equal for all relative refractive indices n and absorption indices κ, and decrease hyperbolically.
4) The top curve in Figure 8 applies to pigments with a high absorption index κ and low refractive index n (e.g., carbon black) and shows that the optimal particle size lies below a given limit.
5) The lowest curve applies to pigments with a small absorption index κ and high relative refractive index n, as is usually the case with inorganic pigments (e.g., red iron oxide). Here, there is a distinct maximum [1.11], [1.16].

The above relationships (Fig. 8) show that the optical pigment properties depend on the particle size D and the complex refractive index $n^* = n(1 - i\kappa)$, which incorporates the real refractive index n and the absorption index κ. As a result, the reflectance spectrum and hence the color properties of a pigment can be calculated if its complex refractive index, concentration, and particle size distribution are known [1.31]. Unfortunately, reliable values for the necessary optical constants (refractive index n and absorption index κ) are lacking. These two parameters generally depend on the wavelength and, owing to the optical anisotropy of most pigments, on the illuminant and viewing direction. In pigments with a high absorption, the refractive index shows "anomalous dispersion". Refractive indices of inorganic pigment materials are given at selected wavelengths in Tables 7 and 8. However, these values were measured on large crystals, and not on pigment particles. Unfortunately, no direct methods exist for determining n and κ in colored inorganic pigments.

Table 8. Refractive indices of important colored inorganic pigments [1.31 a]

Mineral	Formula	Wavelength λ, nm	Refractive index		
			n_ω	n_ε	
			n_α	n_β	n_γ
Bismuth vanadate	$BiVO_4$	670	2.45 [a]		
Cobalt blue	$CoAl_2O_4$	(blue)	1.74		
Eskolaite	α-Cr_2O_3	671	2.5 [a]		
Greenockite	CdS	589	2.506	2.529	
Goethite	α-FeOOH	589	2.275	2.409	2.415
Hematite	α-Fe_2O_3	686	2.988	2.759	
Carbon [b]	C	578	1.97		
Crocoite	$PbCrO_4$	671	2.31	2.37	2.66
Magnetite	Fe_3O_4	589	2.42		
Red lead	Pb_3O_4	671	2.42		
Ultramarine [c]			1.50		

[a] Mean refractive index. [b] For carbon arc lamps. [c] Formula is $[AlSiO_4]_6(SO_4)(S,Cl)_2(Na,Ca)_8$.

1.3.2. Color Measurement

General. Two methods are used for color measurement: the tristimulus method using a three-filter colorimeter and the spectral method using a reflectance spectrophotometer. Manufacturers have developed three-filter reflectometers and recording spectrophotometers. Instruments are continuously being improved due to technological advances. Photometer illuminating/viewing geometries are standardized as methods A and B, and are designed to suit the individual application (see Section 1.4.2). For standards, see Table 1 ("Color Differences, Conditions/Evaluation...").

Method A. The geometry d/8 (diffuse illumination, viewing from an angle of 8°) or 8/d (illumination from an angle of 8°, diffuse viewing)—including specular reflection—enables total surface reflection and reflection from the interior of a sample to be measured. An amount representing the surface reflection has to be subtracted from the measured value. Thus, measured color variations can be ascribed to differences or changes of the colorants in the interior of the sample.

Method B. Here, color differences in samples are evaluated in almost the same way as in visual evaluation by exclusion of gloss effects. Suitable geometries are d/8, d/0, 8/d, and 0/d with a gloss trap, and 45/0 and 0/45.

After the color of the sample and reference pigments has been measured, color differences are usually calculated by transformation of the X, Y, and Z values into the CIELAB system to calculate color difference. Color measurement results of black and white pigments can be expressed more simply because they only amount to a determination of the relative color undertone. For this, the environment of the reference pigment is divided into eight sectors, these being filled with color names from "red" to "violet". The octant in which the CIELAB color position of the sample is located is found by calculation.

When colored or black systems are reduced with white pigments, an undertone is observed, which is a particle-size effect of the white pigment (see Section 1.3.1). These undertones can be conveniently expressed as CIELAB color differences. The effects can, however, also be measured by using the difference $R_z - R_x$ between the values obtained with the blue and red tristimulus filters. The undertone measured in this way depends on the lightness, and has a maximum at $Y = 41.4$. The lightness of a gray paste should therefore have this value to ensure that undertone differences between white pigments are comparable [1.32], [1.33].

Measured color differences are only true (i.e., significant) when they are not falsified by measuring errors. A *significance* test standard (see Table 1, "Color differences, significance") has been developed to check this [1.33]. The numerical value of a color difference must be higher than a critical value, which is statistically calculated using the standard deviation.

Problems concerning the *acceptance* (tolerance) of color differences (e.g., in production quality control or in computer color matching) should also be solved by mathematical statistics [1.16].

The *gloss* of pigmented coatings is not a true pigment property. The pigment can, however, influence the luster quality, mainly via its dispersing properties (see Section 1.5.2). The degree of gloss of a coating can range from high gloss (specular reflection) to an ideally matt surface (complete scattering). *Gloss haze* is due to a disturbance of specular reflection: the reflected objects appear as seen through a veil, this is caused by halation effects. Gloss retention is discussed in Section 1.4.1, a method of gloss measurement is described in Section 1.4.2. Special problems arise in the measurement of black [1.34] and fluorescent pigments [1.35].

Methods of Determination. *Lightness.* The white pigment powder is compressed in a suitable powder press to give an even, matt surface. The CIE tristimulus value Y is measured with color measuring equipment. For standards, see Table 1 ("Lightness"). Apparatus: spectrophotometer or tristimulus colorimeter, powder press, white standard.

Full Shade. Full-shade systems are media that contain only a single pigment. The color of the full-shade system in an optically infinitely thick (opaque) coating is referred to as full shade. The *mass tone* denotes the color obtained when the pigmented medium is applied as a layer that does not hide the substrate completely (e.g., on a white substrate). Evaluation can be carried out visually or by color measurement. For standards, see Table 1 ("Color in full-shade systems"). Apparatus: spectrophotometer or tristimulus colorimeter.

Color Difference. The computer printout in Figure 9 shows the CIELAB color positions of a reference pigment and a test sample: the color difference ΔE^*_{ab}, and the derived differences (Section 1.3.1), i.e., lightness difference ΔL^*, chroma difference ΔC^*_{ab}, and hue difference ΔH^*_{ab}. The color difference of colored pigments in reduction can be similarly determined, including the "color difference after color reduction" which arises when tinting strength matching is carried out (see Section 1.3.3). An improved color-difference formula (CIE 94) considers that chroma and hue differences can themselves be dependent on chroma (see [1.36], to be published as part of ISO 7724).

Undertone of Near-White Samples and Gray Undertones. The undertone of an almost uncolored sample is the small amount of color by which the color of a sample differs from ideal white or achromatic material. It is described by hue and chroma. The distance and direction of the CIELAB color position of the test sample (a^*_T, b^*_T) from the achromatic position (0, 0) are used to characterize the hue. The *relative undertone* is expressed by the distance and direction between the CIELAB color position of the test sample (a^*_T, b^*_T) and that of the reference pigment (a^*_R, b^*_R). In both cases, the distance is expressed by a figure and the direction by a color name. For standards, see Table 1 ("Hue relative of near white specimens"). Apparatus: spec-

Full shade (CIELAB, C/2°)	R_x	R_y	R_z	L^*	a^*	b^*	C^*_{ab}	h_{ab}
Reference	37.27	29.07	7.95	60.8	8.3	46.5	47.2	79.9°
Test	36.83	28.23	7.93	60.1	10.3	45.3	46.5	77.2°
Color difference: test minus reference		ΔE^*_{ab}	ΔL^*	Δa^*	Δb^*	ΔC^*_{ab}	ΔH^*_{ab}	Δh_{ab}
		2.4	−0.7	2.0	−1.2	−0.7	−2.2	−2.7

Figure 9. CIELAB color differences between two yellow oxide pigments

	Reflectometer coordinates			CIELAB color differences		Undertone	
	R_x	R_y	R_z	ΔE^*_{ab}	ΔL^*	Color	Distance, Δs
Reference	94.38	93.62	91.09				
Test	94.34	93.95	92.05	0.55	0.13	blue-green	0.54

Figure 10. Undertone of two nearly white samples (TiO_2 pigments)

trophotometer or tristimulus colorimeter for determining the CIE tristimulus values X, Y, and Z with standard illuminants D 65 or C. If a computer is available, the color name can be printed out instead of the number of the octant (for example, see Fig. 10).

1.3.3. Tinting Strength, Lightening Power, and Scattering Power

The *tinting strength* is a measure of the ability of a colorant to confer color to a light-scattering material by virtue of its absorption properties. The *lightening power* can be considered as the tinting strength of a white pigment, and is a measure of its ability to increase the reflectance of an absorbing (black or colored) medium by virtue of its scattering power. Tinting strength is expressed as the mass ratio in which the reference pigment (mass, m_R) can be replaced by the test pigment (mass, m_T) to give the same color quality in a white system. Analogously, lightening power is the mass ratio in which the reference pigment can be replaced by the test pigment to give the same lightness in a colored system. Thus, the same equation defines tinting strength and lightening power:

$$P = (m_R/m_T)_{\Delta Q = \mathrm{const}}$$

where ΔQ = const. expresses the tinting strength matching. Both parameters are yield properties; if, for example, tinting strength is doubled, only half the weight of pigment is required. Optical properties can therefore provide information about the economic performance of a colored pigment ("value for money"). Testing of tinting strength and lightening power can be rationalized by means of the Kubelka–Munk theory (see Section 1.3.1).

Tinting Strength. Within certain limits, relative tinting strength can be interpreted as the ratio of the absorption coefficients of equal masses of test and reference pigments. This procedure avoids visual matching of the test and reference pigments employed in a previously used method (tinting strength matching). For standards, see Table 1 ("Tinting strength, relative"). Materials and apparatus: white paste, spectrophotometer or tristimulus photometer.

In the determination of the tinting strength, specification of the matching between the test and the reference pigments is extremely important [1.37]. Since the criterion used for matching greatly influences the value of the tinting strength, it is not permissible to speak of tinting strength purely and simply. There are as many tinting strength values as there are matching criteria. In the German Standard, the following criteria are permitted:

1) The lowest of the three CIE tristimulus values X, Y and Z
2) The CIE tristimulus value Y
3) The depth of shade (see Section 1.4.2.1)

When determining the tinting strength of inorganic pigments, the tristimulus value Y (lightness) is usually used [1.37].

Another standard method uses depth of shade as the matching criterion, but it is employed almost exclusively for testing organic pigments. This method can be applied by means of the "principle of spectral evaluation" (see Section 1.3.4) [1.38], which uses the wavelength-dependent Kubelka–Munk coefficients $S(\lambda)$, $K(\lambda)$ to calculate the match [1.39]. The tinting strength for all other matching criteria can be determined by applying the principle of spectral evaluation. (For tinting strength in cement, see Section 1.5.2.3; for change in tinting strength, see Section 1.5.2.2).

Lightening Power. If the gray paste method is used to determine the lightening power (for standards, see Table 1, "Lightening power of white pigments"), the concentration at which the white pigment should be assessed has to be agreed between the intended parties in accordance with the type of application [1.40]. This process is especially recommended for the routine testing of pigments because the lightening power, the relative scattering power (see below), and the undertone can be determined with the same gray milling paste of the test pigment.

The method can be rationalized by means of the Kubelka–Munk theory. For standards, see Table 1 ("Tinting strength, relative: Photometric").

Relative Scattering Power. For standards, see Table 1 ("Scattering power, relative"). The relative scattering power S is the ratio of the scattering power S_T of the test white pigment to the scattering power of a white reference pigment S_R [1.41]. It can be determined in two ways:

1) *Black-Ground Method.* The relative scattering power is determined from the tristimulus values Y of the pigmented medium applied in various film thicknesses to black substrates. Compared with the gray paste method, the black-ground method has the advantage that it is not restricted to any particular test medium. Apparatus: spectrophotometer or tristimulus colorimeter.
2) *Gray Paste Method.* The relative scattering power is determined from the tristimulus values Y of gray pastes. The method has the advantage of being less time consuming than the black-ground method. The results of the two methods are not, however, generally in agreement. Materials and apparatus: black paste, spectrophotometer or tristimulus colorimeter.

1.3.4. Hiding Power and Transparency

The definition of hiding power is based on a black and white contrasting support upon which the film of coating is applied. The thickness h of the applied film is determined at the point at which the contrasting surface just disappears, as judged by eye. The film thickness (mm) which fulfils this condition is called the *hiding thickness*. Its reciprocal, the *hiding power* ($mm^{-1} = m^2/L$) is, like tinting strength, an indicator of yield because it gives the area (m^2) that can be covered up with 1 L of applied paint.

This traditional visual testing method has been improved by the use of a photometer and a colorimetric criterion to evaluate the hiding film thickness, e.g., $\Delta E^*_{ab} = 1$ for equally thick films applied to black and white backgrounds. The method can, however, be further rationalized. In this method only a single coating has to be prepared. The total reflectance spectrum is measured and after measuring the film thickness, the spectral scattering and absorption coefficients $S(\lambda)$ and $K(\lambda)$ are calculated at selected wavelengths (e.g., at intervals of 20 nm) by the Kubelka–Munk method. Once these values are known, the expected reflectances over black and white at the chosen wavelengths of the spectrum can be calculated for a given film thickness. Using the CIE and CIELAB systems, the CIELAB position of a color which would have this spectrum is calculated. This provides an iteration method for determining the film thickness at which the hiding power criterion $\Delta E^*_{ab} = 1$ is satisfied [1.38], [1.42], [1.43]. This principle of spectral evaluation [1.38] is the key to handling test methods based on the visual matching of two color samples (see also "Transparency" and "Tinting Strength", Section 1.3.3).

The *transparency* of a pigmented system denotes its ability to scatter light as little as possible. The color change of a transparent pigmented system when applied to a black substrate has to be very small; the lower the color change, the higher is the transparency [1.44]. Measurement of transparency is important for assessing transparent varnishes and printing inks.

Hiding Power. The hiding criterion is an agreed color difference between two contrasting areas of a coated contrast substrate. The hiding power can be determined and expressed as D_v (m^2/L) or D_m (m^2/kg), which give the area (m^2) of the contrasting substrate that can be coated with 1 L or 1 kg, respectively, of the pigmented medium when the hiding criterion is satisfied. Layers of various thicknesses or mass-to-area ratios of the test material are applied to the contrast background. The color difference ΔE^*_{ab} between the two contrasting areas is determined by color measurement on each of the layers. The color difference is plotted against the reciprocal of the film thickness (mm^{-1}) or the reciprocal of the mass to area ratio (m^2/kg). The hiding power D_v (m^2/L) or D_m (m^2/kg) at the hiding criterion (e.g., $\Delta E^*_{ab} = 1$) is then obtained from the diagram (Fig. 11). For standards, see Table 1 ("Hiding power"). Apparatus: spectrophotometer or tristimulus colorimeter, film thickness measuring equipment.

A modern well-equipped color measurement laboratory can use the "principle of spectral evaluation" described above [1.38] to simulate this procedure with a computer. The thickness of the hiding film can then be calculated in advance from the reflectance curves of a single film on a black/white substrate at a known film thickness.

Hiding Power of Achromatic Coatings. Equally thick coatings of the pigmented coating material are applied to a black and white background, and their reflectances measured with a photometer after drying. The hiding power is then calculated from the reflectances on black and white substrates and the film thickness. The hiding criterion is $\Delta L^* = 1$, in analogy with that for colored pigments [1.14]. The use of

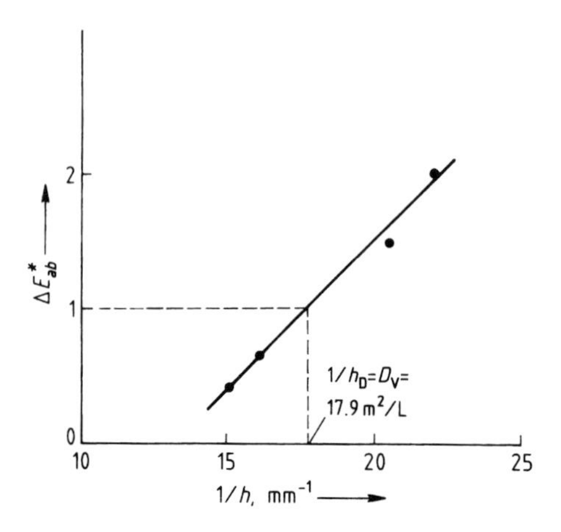

Figure 11. Determination of the hiding power of a yellow iron oxide pigment

hiding power as a basis for assessing the economy and effectiveness of pigments is described in [1.45]. Apparatus: spectrophotometer or tristimulus colorimeter, black and white glass plates.

Transparency. Transparency is expressed quantitatively as the transparency number. This is defined as the reciprocal of the increase in color difference ΔE^*_{ab} on a black substrate obtained on increasing the film thickness h of the pigmented medium. The transparency number has the unit mm (= L/m^2). It indicates the number of liters of pigmented medium needed to coat 1 m^2 of a black substrate in order to obtain a color difference of $\Delta E^*_{ab} = 1$ relative to this substrate. In a simplified method the transparency number can be determined by evaluating one or two points on the straight part of the $\Delta E^*_{ab}(h)$ curve. A computer method is more exact, furthermore calculations can be made using the spectral "principle of spectral evaluation" (see above). For standards, see Table 1 ("Transparency").

1.4. Stability Towards Light, Weather, Heat, and Chemicals

1.4.1. Fundamental Aspects [1.46]

Many pigmented systems show typical color or structural changes when subjected to intense radiation or weathering [1.47]. The best-known of these are yellowing [1.48], chalking, and loss of gloss. These processes involve photochemical reactions in which the pigment can act as a catalyst or in which the pigment itself undergoes chemical changes.

Inorganic pigments are chemically very stable and are classed as one of the most stable coloring matters. This is especially true for oxide pigments which often have a highly protective effect on the substrate [1.49]. Apart from the purely mechanical stabilization imparted by the pigment, this protective effect can give a coating an increased economic advantage. On the other hand, sulfide pigments can be oxidized by the atmosphere to form sulfates which can be washed away by rain [1.50].

The photochemical processes that take place when TiO_2-containing coatings undergo chalking have been elucidated [1.51]. The shorter wavelength radiation of sunlight acts on rainwater and atmospheric oxygen to form extremely reactive radicals ($^{\bullet}OH$, $HO_2^{\bullet}$) that cause deterioration of the coating matrix by oxidative attack. Titanium dioxide pigments may be stabilized by reducing the number of radical-producing hydroxy groups on the surface of the TiO_2 particles (e.g., by doping with zinc oxide). Alternatively, coatings of oxide hydrates are produced by aftertreatment of the pigment surface to give "reaction walls" on which the radicals are destroyed [1.52].

Continuous breakdown of the binder leads to the loss of gloss. Surface gloss gives an attractive appearance and is usually desired, it is obtained by ensuring that the pigment particles are well dispersed. The stability of gloss to weathering is therefore of great importance. If breakdown of the binder is so extensive that the pigment may be loosened, *chalking* takes place. Chalking is defined as loosening of pigment or extender particles following destruction of the binder at the surface. For standards, see Table 1 ("Coating materials").

Resistance to light and weather generally depends on the chemical composition, structure, defects, particle shape and size, and concentration of the pigment [1.53]. However, these properties also depend on the medium in which the pigment is used. Testing is carried out by open-air weathering, accelerated weathering, and chemical test methods.

Accelerated weathering is carried out in special weathering devices (e.g., Weatherometer, Xenotest) which simulate exposure to sunlight and periods of rain. The spectral composition of the light used (Fig. 12) should match the "global radiation"

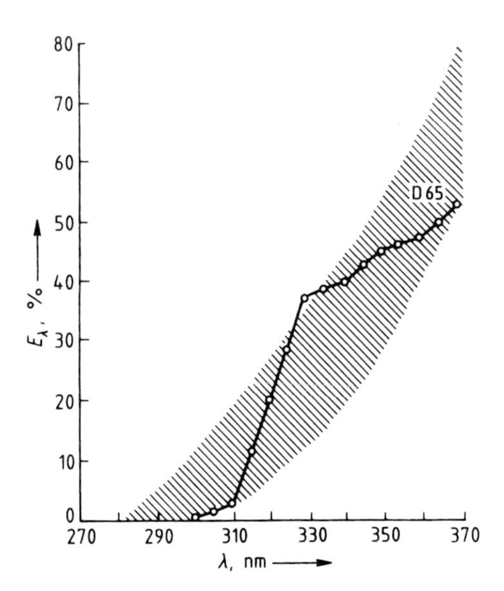

Figure 12. Spectral energy distribution of daylight phase D65 with permitted range of deviation (shading)

in temperate latitudes (global radiation denotes the sum of the direct solar radiation and indirect radiation from the sky). However, accelerated tests only allow limited conclusions to be drawn about light and weather resistances because it is not possible to accurately simulate all climatic conditions, their combinations and sequences, or sudden changes. Even so, natural weathering tests are not without problems: weather cannot be standardized, and a "uniform" weather therefore does not exist [1.54], [1.55]. Weathering results, even from successive years, can thus vary considerably [1.56]. Weather also varies with the geographical coordinates. For standards, see Table 1 ("Climates: Standardized"); the earth is classified into climatic regions and weathering adjustment scales exist [1.57].

None of the chemical methods (e.g., mandelic acid or methanol test) gives a perfect indication of weather resistance. Most of them only measure partial qualities of the photochemical reactivity [1.58].

The oxide pigments have the highest *heat stability*, followed by the sulfide pigments which can even be used in enamels and glass melts (e.g., cadmium pigments). The oxide hydroxides and carbonates are less heat stable. Thermal stability is of great importance for modern coating techniques (e.g., for stoving finishes and other products requiring high temperatures). The heat resistance of a pigment primarily depends on the binder and the duration of heating. It is conveniently determined by observing hue changes in the pigmented coating system after a defined heat treatment, e.g., with white coatings yellowing occurs. For standards, see Table 1 ("PVC").

The *chemical resistance* of inorganic pigments, especially oxide pigments, is generally very high [1.59]. Other fastness properties such as overspray fastness, fastness to blooming, and fastness to plasticizers are of more relevance to organic pigments. For corrosion inhibiting (anticorrosive) pigments, see Section 5.2.

1.4.2. Test Methods

1.4.2.1. Light Stability

Accelerated Tests. Light stability is a material property. It is defined as the resistance of coatings towards changes caused by the action of global radiation (daylight phase: standard illuminant D 65), possibly in the presence of moisture. In coatings, light and moisture can lead to chalking and changes in gloss and color. For standards, see Table 1 ("Light stability"). Apparatus: testing cabinets, equipment for wetting the test plates, equipment for producing air flow.

Evaluation of Color Changes (see Section 1.3.2). Color changes are differences in lightness, chroma, and hue such as exist between identical samples with different histories. Evaluation methods are particularly suitable for the determination and evaluation of color changes that occur following physical and chemical stresses on pigmented materials. The CIE tristimulus values X, Y, and Z of the samples or of different test locations on a sample are determined with a colorimeter. The CIELAB color difference is calculated for each sample or test location from the tristimulus values before and after the stress (see Section 1.3.2). Apparatus: spectrophotometer, tristimulus colorimeter.

Standard Depth of Shade. The depth of shade is a measure of the intensity of a color sensation. It increases with increasing chroma and generally decreases with increasing lightness. The colorfastness of colorations (e.g., textile dyeing) is dependent on depth of shade; the fastness properties of pigments or pigmented systems must therefore be tested at a predetermined depth of shade. This standard method therefore ensures that samples with standard depths of shade are used to determine the fastness properties of pigments or pigmented systems [1.60], [1.61]. For standards, see Table 1 ("Standard depth of shade"). Apparatus: spectrophotometer or tristimulus filter photometer with gloss trap. (For light stability in cement, see Section 1.5.2.3.)

1.4.2.2. Weather Resistance

The empirical tests described below determine the chalking and weathering differences between two pigments, but do not always give the true differences. Other test methods have therefore been developed, e.g., determination of mass losses on weathering (gravimetric test) [1.51].

Accelerated Tests. Weather resistance in an accelerated test is defined as the resistance of plastics towards changes caused by simulated open-air weathering (simulation of global radiation by means of filtered xenon arc radiation and periodic rain). After the weathering (measured by the product of intensity and duration), defined properties of the test sample are compared with those of an identical unweathered sample. Properties should be considered which are of practical importance, such as color or surface properties. For standards, see Table 1 ("Weathering in apparatus"). Apparatus: test chamber, rain and air humidification equipment, air flow equipment, radiation measuring equipment.

Degree of Chalking. The degree of chalking of a coating is measured by the quantity of loose pigment particles. It may be determined by three methods:

1) *Kempf Method.* For standards, see Table 1 ("Chalking degree: Kempf method"). Water-treated photographic paper is applied to the coating (with the humid gelatin side facing the coating surface) and pressed by means of a special stamp device. Loose pigment particles that are not attached to the binder, are picked up by the gelatin coating. The degree of chalking is judged visually (for scale of measurement see Table 1, "Test evaluation"). See also Section 1.2.2. Apparatus: Kempf stamp equipment.
2) *Adhesive Tape Method.* For standards, see Table 1 ("Chalking degree; Adhesive tape method"). Loose pigment particles are picked up by transparent adhesive tape. The degree of chalking is determined from a reference scale (see Table 1, "Test Evaluation" and Section 1.2.2) or by using a comparative scale that is agreed upon by the interested parties.
3) *Photographic Methods.* Chalking prints on paper are compared with photographic standards.

Assessment of Gloss. Gloss is not a purely physical quantity, it is also dependent on physiological and psychological factors. The usefulness of measured reflectance values lies in the possibility of detecting surface changes of the test sample and observing them over a long period. These changes can be produced by weathering, abrasion, and similar causes of surface wear, or by the method used to produce the surface. For standards, see Table 1 ("Reflectometer"). Apparatus: reflectometer with incident light angles of 20°, 60°, or 85°.

Effects of Humid Climates. The specimen is exposed to warm air saturated with water vapor, with or without intermittent cooling to room temperature. The test is very suitable for assessing the corrosion resistance of metals, protective coatings, and composite materials in buildings. For standards, see Table 1 ("Climates: Containing evaporated water"). Apparatus: Kesternich condensation equipment.

1.4.2.3. Heat Stability [1.62]

Measured values of the heat stability of pigments depend on the type and duration of the heat applied, and also on the binder used. Low thermal stability in white coatings leads to yellowing, even when the unpigmented binder is resistant to yellowing. With colored coatings, changes in the hue of the pigmented binder can be measured colorimetrically (see Section 1.3.2). Standards only exist for special systems (e.g., hard and soft PVC; and cement, see Section 1.5.2.3). Thermal resistances of pigments are customarily quoted at two temperatures: (1) the maximum temperature at which the hue remains stable, and (2) the temperature at which a defined hue change is observed. For standards, see Table 1 ("Heat stability" and "Coloration of building materials").

1.4.2.4. Fastness to Chemicals [1.62]

General chemical resistance can be classified as short-, medium-, or long-time resistances. Changes in the coating are assessed visually. For standards, see Table 1 ("Chemical resistance"). Apparatus: film thickness measuring equipment.

Resistance to Water in Sulfur-Dioxide-Containing Atmospheres. For standards, see Table 1 ("Corrosion testing; SO_2" and "SO_2 resistance"). General conditions are standardized for exposing samples to a varying climate where water condenses in the presence of sulfur dioxide, so that tests in different laboratories give comparable results. The test allows rapid detection of defects in corrosion-inhibiting systems.

Furthermore, a method for determining the colorfastness of pigments in binders in the presence of sulfur dioxide is also prescribed. The test is carried out simultaneously on ten identical samples and consists of three cycles. Apparatus: condensation equipment.

Salt Spray Fog Test with Sodium Chloride Solutions. The sample is sprayed continuously with a 5% aqueous sodium chloride solution. There are three variations of the test: the salt spray fog test, the acetic acid salt spray test, and the copper chloride–acetic acid salt spray test.

For standards, see Table 1 ("Corrosion testing: NaCl"). Materials and apparatus: spray chamber, test solutions (5% NaCl solutions).

Resistance to Spittle and Sweat. This test indicates whether a pigment on a child's colored toy is likely to be transferred to the mouth, mucous membranes, or skin during use. Strips of filter paper are wetted with $NaHCO_3$ and NaCl solutions and pressed against the test samples. The discoloration of the paper is judged visually.

1.5. Behavior of Pigments in Binders

1.5.1. Fundamental Aspects [1.63], [1.64]

A pigment–binder dispersion is a suspension before it is dried; after drying it is a solid sol. In pigment–binder systems, the concepts and laws of colloid chemistry therefore apply. The dispersing of pigments and extenders in binders is an extremely complex process of a series of steps that can be interlinked [1.63]. Dispersing involves the following steps:

1) *Wetting*. Removal of the air from the surface of the pigment particles and formation of a solvate layer [1.65], [1.66].
2) *Disintegration*. Breaking up of pigment agglomerates with external energy (e.g., by dispersing equipment).
3) *Stabilization*. Maintenance of the disperse state by creating repulsive forces between the particles (e.g., by coating them with solvate layers). These forces must be greater than the van der Waals attraction forces [1.67] that cause flocculation [1.68], [1.69].

The degree of dispersing usually has a large influence on the properties of the coating system [1.70], [1.71]. Some of these properties depend strongly on the degree of dispersing, and are therefore used to measure dispersibility. For standards, see Table 1 ("Ease of dispersion"). Examples are viscosity (pigment–binder dispersions always show non-Newtonian behavior properties before drying) and color properties, especially hue, tinting strength, gloss, and gloss haze [1.72] (see Sections 1.3 and 1.4).

In many systems, tinting strength depends so highly on the degree of dispersing that the rate of tinting strength development can be taken as a direct measure of dispersibility [1.73]–[1.76] (see Section 1.5.2.2). The numerical values of the half-life times (characterizing the dispersibility) and of the increases in strength (characterizing the strength potential) depend on the binder and the dispersing equipment used.

In systems where different pigments are combined (e.g., white and colored pigments), segregation effects can occur which change the optical appearance (flooding

[1.77]). Flooding can be counteracted by preflocculation of the pigments with suitable gelling agents, and in some cases by addition of extenders.

In solvent-containing systems with several pigments, but also in full-shade systems, color changes (rub-out effect) can take place due to flocculation, especially if the systems are subjected to mechanical stress during application and drying. The flocculation tendency can be determined by means of the "rub-out test".

1.5.2. Test Methods

1.5.2.1. Pigment–Binder Interaction

Oil Absorption. The oil absorption gives the mass or volume of linseed oil required to form a coherent putty-like mass with 100 g pigment under specified conditions. The mixture should just not smear on a glass plate. For standards, see Table 1 ("Oil absorption"). Materials and apparatus: rough glass plate, spatula with steel blade, raw linseed oil.

Binder Absorption, Smear Point, and Yield Point [1.78]. Smear point and yield point are used to determine the binder needed to formulate a suitable millbase for grinding by dissolvers, roll mills, ball mills, attritors, sand mills, and pearl mills. The amount of binder is given in volume or mass units. Apparatus: rough glass plate, spatula with steel blade.

Viscosity. Viscosity is a useful parameter for coatings applied by brushing. It is determined by measuring the torque applied to a rotating cylinder or disk in or on the surface of a suspension. This method is applicable to Newtonian and non-Newtonian systems. In non-Newtonian suspensions, the apparent viscosity is obtained by dividing the shear stress by the shear rate. The viscosities provide information about the force required in the initial brushing phase of a coating. For standards, see Table 1 ("Viscosity"). Apparatus: cone/plate or rotating cylinder viscometer.

Fineness of Grind. The grind gauge (grindometer) consists of a steel block with a groove. The depth of the groove at one end is approximately twice the diameter of the largest pigment particle, and decreases continuously to zero at the other end. The sample is placed at the deep end of the groove and drawn to the other end with a scraper. The depth at which a large number of particles become visible on the surface of the pigment–binder system as pinholes or scratches is read off from a scale graduated in micrometers. For standards, see Table 1 ("Fineness of grind"). Apparatus: Hegman's grindometer gauge (block and scraper).

1.5.2.2. Dispersing Behavior in Paint Systems

Low-Viscosity Media. The time is measured that is required to produce a homogeneous suspension of particles in the dispersion medium using an oscillatory shaking machine equipped with several containers. Not only can small quantities of the millbase with the same composition be tested (as with other types of apparatus), but various millbases can also be tested under the same conditions. A low-viscosity alkyd resin system of the stoving or oxidatively drying type can be used as a test medium. For standards, see Table 1 ("Ease of dispersion: Oscillatory shaking machine").

High-Viscosity Media (Pastes). The dispersion properties are determined with an automatic muller. The advantage is that small amounts of material can be simply tested under reproducible conditions. For standards, see Table 1 ("Ease of dispersion: Automatic muller").

Development of Fineness of Grind. The quantity measured is the dispersing effect needed to achieve a given fineness of grind. Samples of the product are taken at various stages of the dispersion process and the fineness of grind is determined with a grindometer gauge. The method is used to compare diverse millbases material types of dispersing equipment, or dispersing methods used with the same pigment. For standards, see Table 1 ("Fineness of grind").

Change in Tinting Strength, Half-Life Times, and Increase in Strength. For standards see Table 1 ("Change in strength"). Dispersibilities of pigments may be compared by means of their dispersing resistance, dispersing equipment may be characterized by means of dispersing effects. Measurement can be based on determination of half-life times in relation to the final tinting strength. The increase in strength, however, only gives the difference between the initial and final tinting strength [1.76]. Half-life times are functions of the dispersing rate. They are only valid for a given combination of dispersing equipment, dispersing process, and medium. Samples are taken from the millbase at specified stages in the dispersing process and mixed with a white paint based on the same binder material, or with a compatible white paste. The tinting strength is then determined (see Section 1.3.3) and a graph is drawn; development of tinting strength is plotted against the dispersion stages (Fig. 13) [1.79].

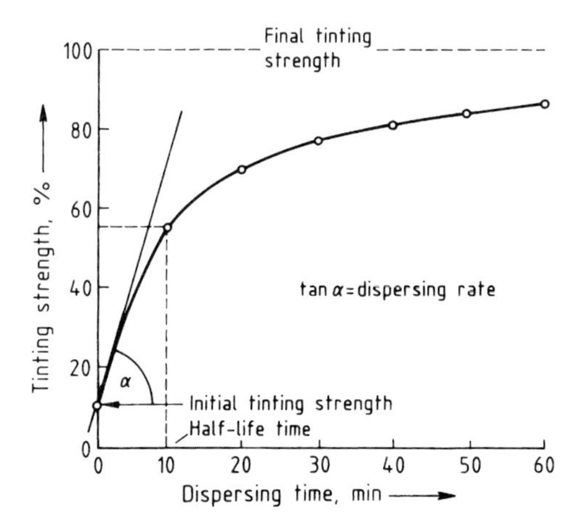

Figure 13. Change in tinting strength as a function of dispersing time

Pigment Volume Concentration *(PVC)* and Critical Pigment Volume Concentration *(CPVC)*. The pigment volume concentration (σ) is the fractional volume of pigment in the total solids volume of the dry paint film:

$$\sigma = \frac{V_p}{V_p + V_b}$$

where V_p is the pigment volume and V_b the binder volume. The *PVC* [1.80] is determined by separating the pigment fraction from a weighed paint sample by, for example, extracting the (liquid) binder with suitable organic solvents or solvent mixtures (sometimes using a centrifuge) or using combustion methods (sometimes including fuming with sulfuric acid) and analysis of the residue [1.81].

Under *CPVC* conditions, the pigment particles are at a maximum packing density, and the interstices are completely filled with binder. With smaller amounts of binder, the interstices are incompletely filled. The *CPVC* thus represents a pigment concentration boundary at which abrupt changes in the properties of the film occur.

The methods of determining the *CPVC* [1.82] are divided into two groups:

1) Methods based on the observation of a sudden change in properties in a series of *PVCs* (e.g., ion permeability, permeability to water vapor, color, gloss stability after coating with a silk-luster paint)
2) Methods based on the experimental production of the dense pigment packing typical of *CPVC*, e.g., by means of oil absorption, filtration, vacuum filtration, or by measuring the volume of the supercritical film

Measurement of Deposit. Common methods used to measure sediments are manual testing, sedimentation balance, radioactive methods, dipping of bodies, and probe tools.

The special measuring equipment known as the Bayer system [1.83], can test for deposits with great accuracy and good reproducibility. The force acting on a needle probe is measured which penetrates into the coating material from above.

1.5.2.3. Miscellaneous Pigment–Binder Systems

Plastics. Specifications for thermoplastics are given in Table 1 ("Thermoplastics"). Specifications for poly(vinyl chloride) are also listed ("PVC"). These specifications include methods for producing basic mixtures for testing pigments in PVC and for specimen preparation. A method is described for determining bleeding, i.e., the migration of coloring matter into a material in contact with the sampling equipment. Special procedures are also included for the determination of heat stability and increase in strength caused by cold rolling.

Building Materials. For standards, see Table 1 ("Coloration of building materials"). The following tests are prescribed for assessing the suitability of a pigment for coloring cement and lime-bonded building materials:

Relative tinting strength
Colorfastness in cement
Colorfastness in lime
Lightfastness
Heat stability
Influence of the pigment on hardness of the building material
Influence of the pigment on setting properties

Paper and Board. For standards, see Table 1 ("Opacity" and "Transparency"). Methods of measurement are specified for the reflectance of paper and board (non-fluorescent), and the opacity or transparency of paper (to measure the transmitted light).

2. White Pigments

White pigments include TiO_2, zinc white (ZnO), zinc sulfide, lithopone (a mixed pigment produced from zinc sulfide and barium sulfate), and white lead (basic lead carbonate). The optical properties of white pigments are a result of their low light absorption and their strong, mainly nonselective, scattering of light.

2.1. Titanium Dioxide

Titanium dioxide [*13463-67-7*], TiO_2, M_r 79.90, occurs in nature in the modifications rutile, anatase, and brookite. Rutile and anatase are produced industrially in large quantities and are used as pigments and catalysts, and in the production of ceramic and electronic materials.

Titanium dioxide is of outstanding importance as a white pigment because of its scattering properties (which are superior to those of all other white pigments), its chemical stability, and lack of toxicity. Titanium dioxide is the most important inorganic pigment in terms of quantity, 3.2×10^6 t were produced in 1995. World production of titanium dioxide pigment is shown in Table 9 [2.1], [2.2], [2.3].

2.1.1. Properties [2.4], [2.5]

Physical Properties. Of the three modifications of TiO_2, rutile is the most thermodynamically stable. Nevertheless, the lattice energies of the other phases are similar and hence are stable over long periods. Above 700 °C, the monotropic conversion of anatase to rutile takes place rapidly. Brookite is difficult to produce, and therefore has no value in the TiO_2 pigment industry.

In all three TiO_2 modifications one titanium atom in the lattice is surrounded octahedrally by six oxygen atoms, and each oxygen atom is surrounded by three titanium atoms in a trigonal arrangement. The three modifications correspond to different ways of linking the octahedra at their corners and edges. Crystal lattice constants and densities are given in Table 10.

Rutile and anatase crystallize in the tetragonal system, brookite in the rhombic system. The melting point of TiO_2 is ca. 1800 °C. Above 1000 °C, the oxygen partial

Table 9. World production of TiO_2 pigment

Year	Sulfate process		Chloride process		Total
	10^3 t/a	%	10^3 t/a	%	10^3 t/a
1965	1254	90.3	135	9.7	1389
1970	1499	77.4	437	22.6	1936
1977	1873	72.3	716	27.7	2589
1988	1781	60.2	1178	39.8	2959
1995	1481	46.0	1739	54.0	3220
2000*	1540	40	2310	60.0	3850

* Estimated.

Table 10. Crystallographic data for TiO_2 modifications

Phase	CAS registry no.	Crystal system	Lattice constants, nm			Density, g/cm³
			a	*b*	*c*	
Rutile	[*131-80-2*]	tetragonal	0.4594		0.2958	4.21
Anatase	[*1317-70-0*]	tetragonal	0.3785		0.9514	4.06
Brookite	[*12188-41-9*]	rhombic	0.9184	0.5447	0.5145	4.13

pressure increases continuously as oxygen is liberated and lower oxides of titanium are formed. This is accompanied by changes in color and electrical conductivity. Above 400 °C, a significant yellow color develops, caused by thermal expansion of the lattice; this is reversible. Rutile has the highest density and the most compact atomic structure, and is thus the hardest modification (Mohs hardness 6.5–7.0). Anatase is considerably softer (Mohs hardness 5.5).

Titanium dioxide is a light-sensitive semiconductor, and absorbs electromagnetic radiation in the near UV region. The energy difference between the valence and the conductivity bands in the solid state is 3.05 eV for rutile and 3.29 eV for anatase, corresponding to an absorption band at <415 nm for rutile and <385 nm for anatase.

Absorption of light energy causes an electron to be excited from the valence band to the conductivity band. This electron and the electron hole are mobile, and can move on the surface of the solid where they take part in redox reactions.

Chemical Properties. Titanium dioxide is amphoteric with very weak acidic and basic character. Accordingly, alkali-metal titanates and free titanic acids are unstable in water, forming amorphous titanium oxide hydroxides on hydrolysis.

Titanium dioxide is chemically very stable, and is not attacked by most organic and inorganic reagents. It dissolves in concentrated sulfuric acid and in hydrofluoric acid, and is attacked and dissolved by alkaline and acidic molten materials.

At high temperature, TiO_2 reacts with reducing agents such as carbon monoxide, hydrogen, and ammonia to form titanium oxides of lower valency; metallic titanium is not formed. Titanium dioxide reacts with chlorine in the presence of carbon above 500 °C to form titanium tetrachloride.

Surface Properties of TiO_2 Pigments. The specific surface area of TiO_2 pigments can vary between 0.5 and 300 m^2/g depending on its use. The surface of TiO_2 is saturated by coordinatively bonded water, which then forms hydroxyl ions. Depending on the type of bonding of the hydroxyl groups to the titanium, these groups possess acidic or basic character [2.6], [2.7]. The surface of TiO_2 is thus always polar. The surface covering of hydroxyl groups has a decisive influence on pigment properties such as dispersibility and weather resistance.

The presence of the hydroxyl groups makes photochemically-induced reactions possible, e.g., the decomposition of water into hydrogen and oxygen and the reduction of nitrogen to ammonia and hydrazine (see also Section 2.1.5) [2.8].

2.1.2. Raw Materials [2.9], [2.10]

The raw materials for TiO_2 production include natural products such as ilmenite, leucoxene, and rutile, and some very important synthetic materials such as titanium slag and synthetic rutile. Production values for the most important titanium-containing raw materials are listed in Table 11. Total production of titanferous raw materials, excluding material produced and consumed in the former Soviet Union and China, grew in 1994 to 3.69 million tonnes of contained TiO_2 [2.11]. Australia was the largest producing country, followed by South Africa, Canada and Norway.

2.1.2.1. Natural Raw Materials

Titanium is the ninth most abundant element in the earth's crust, and always occurs in combination with oxygen. The more important titanium minerals are shown in Table 12. Of the natural titanium minerals, only ilmenite, leucoxene, and rutile are of economic importance. Leucoxene is a weathering product of ilmenite.

The largest titanium reserves in the world are in the form of anatase and titanomagnetite, but these cannot be worked economically at the present time. About 95% of the world's production of ilmenite and rutile is used to produce TiO_2 pigments, the remainder for the manufacture of titanium metal and in welding electrodes.

Ilmenite and Leucoxene. *Ilmenite* is found worldwide in primary massive ore deposits or as secondary alluvial deposits (sands) that contain heavy minerals. In the massive ores, the ilmenite is frequently associated with intermediary intrusions (Tellnes in Norway and Lake Allard in Canada). The concentrates obtained from these massive ores often have high iron contents in the form of segregated hematite or magnetite in the ilmenite. These reduce the TiO_2 content of the concentrates (see Table 13). Direct use of these ilmenites has decreased owing to their high iron

Table 11. Production of titanium-containing raw materials (1994)

Product	Country	Production (t/a)
Ilmenite*	Australia	1 770 000
	Canada	1 850 000
	India	280 000
	Norway	700 000
	Nouth Africa	1 500 000
	USA	320 000
	Former Soviet Union	250 000
	Other	450 000
	Total ilmenite	7 120 000
Rutile	Australia	216 000
	Sierra Leone	144 000
	South Africa	90 000
	Former Soviet Union	30 000
	Other	70 000
	Total rutile	550 000
Synthetic rutile	Australia	417 000
	India	43 000
	Japan	10 000
	Malaysia	15 000
	USA	140 000
	Total synthetic rutile	625 000
Titania slag	Canada	764 000
	Norway	170 000
	South Africa	773 000
	Total titania slag	1 707 000

* includes ilmenite used in the production of titania slag and synthetic rutile

Table 12. Titanium minerals

Mineral	Formula	TiO_2 content (wt%)
Rutile	TiO_2	92–98
Anatase	TiO_2	90–95
Brookite	TiO_2	90–100
Ilmenite	$FeTiO_3$	35–60
Leucoxene	$Fe_2O_3TiO_2$	60–90
Perovskite	$CaTiO_3$	40–60
Sphene (titanite)	$CaTiSiO_5$	30–42
Titanomagnetite	$Fe(Ti)Fe_2O_4$	2–20

content. A digestion process is employed to produce iron sulfate heptahydrate. In cases where iron sulfate is not required as a product, metallurgical recovery of iron from the iron-rich ilmenites and production of a titanium-rich slag are being increasingly used.

Table 13. Composition of ilmenite deposits (wt %)

Component	Tellnes (Norway)	Richard's Bay (Republic of South Africa)	Capel (Western Australia)	Quilon (India)
TiO_2	43.8	46.5	54.8	60.3
Fe_2O_3	14.0	11.4	16.0	24.8
FeO	34.4	34.2	23.8	9.7
Al_2O_3	0.6	1.3	1.0	1.0
SiO_2	2.2	1.6	0.8	1.4
MnO	0.3	n.d.	1.5	0.4
Cr_2O_3	n.d.	0.1	0.1	0.1
V_2O_5	0.3	0.3	0.2	0.2
MgO	3.7	0.9	0.15	0.9

The enrichment of ilmenite in beach sand in existing or fossil coastlines is important for TiO_2 production. The action of surf, currents, and/or wind results in concentration of the ilmenite and other heavy minerals such as rutile, zircon, monazite, and other silicates in the dunes or beaches. This concentration process frequently leads to layering of the minerals. Attack by seawater and air over geological periods of time leads to corrosion of the ilmenite. Iron is removed from the ilmenite lattice, resulting in enrichment of the TiO_2 in the remaining material. The lattice is stable with TiO_2 contents up to ca. 65 %, but further removal of iron leads to the formation of a submicroscopic mixture of minerals which may include anatase, rutile, and amorphous phases. Mixtures with TiO_2 contents as high as 90 % are referred to as *leucoxene*. Leucoxene is present in corroded ilmenite and in some deposits is recovered and treated separately. However, the quantities produced are small in comparison to those of ilmenite.

The concentrates obtained from ilmenite sand, being depleted in iron, are generally richer in TiO_2 than those from the massive deposits. Other elements in these concentrates include magnesium, manganese, and vanadium (present in the ilmenite) and aluminum, calcium, chromium, and silicon which originate from mineral intrusions.

Two-thirds of the known ilmenite reserves that could be economically worked are in China, Norway (both massive deposits), and the former Soviet Union (sands and massive deposits). On the basis of current production capacities, these countries could cover all requirements for ca. 150 years. However, the countries with the largest outputs are Australia (sands), Canada (massive ore), and the Republic of South Africa (sands). Other producers are the United States (sands, Florida), India (sands, Quilon) the former Soviet Union (sands, massive ore), Sri Lanka (sands), and Brazil (rutilo e ilmenita do Brasil). In 1994 the production of ilmenite was about 1.2×10^6 t of contained TiO_2.

Rutile is formed primarily by the crystallization of magma with high titanium and low iron contents, or by the metamorphosis of titanium-bearing sediments or magmatites. The rutile concentrations in primary rocks are not workable. Therefore, only sands in which rutile is accompanied by zircon and/or ilmenite and other heavy minerals can be regarded as reserves. The world reserves of rutile are estimated to

Table 14. Composition of rutile deposits [2.12]

Rutile component	Content (wt %)		
	Eastern Australia	Sierra Leone	Republic of South Africa
TiO_2	96.00	95.70	95.40
Fe_2O_3	0.70	0.90	0.70
Cr_2O_3	0.27	0.23	0.10
MnO	0.02	n.d.	n.d.
Nb_2O_5	0.45	0.21	0.32
V_2O_5	0.50	1.00	0.65
ZrO_2	0.50	0.67	0.46
Al_2O_3	0.15	0.20	0.65
CaO	0.02	n.d.	0.05
P_2O_5	0.02	0.04	0.02
SiO_2	1.00	0.70	1.75

be 28×10^6 t, including the massive Piampaludo ore reserves in Italy, whose workability is in dispute.

As in the case of ilmenite, the largest producers are in Australia, the Republic of South Africa and Sierra Leone. There is not enough natural rutile to meet demand, and it is therefore gradually being replaced by the synthetic variety. In 1994 the world wide production of rutile was about 0.5×10^6 t of contained TiO_2. Compositions of typical rutile concentrates are given in Table 14.

Anatase, like rutile, is a modification of TiO_2. The largest reserves of this mineral are found in carboniferous intrusions in Brazil. Ore preparation techniques allow production of concentrates containing 80% TiO_2, with possible further concentration to 90% TiO_2 by treatment with hydrochloric acid [2.13]. The TiO_2 amount of these mineral deposits is estimated up to 100×10^6 t.

Ore Preparation. Most of the world's titanium ore production starts from heavy mineral sands. Figure 14 shows a schematic of the production process. The ilmenite is usually associated with rutile and zircon, so that ilmenite production is linked to the recovery of these minerals. Geological and hydrological conditions permitting, the raw sand (usually containing 3–10% heavy minerals) is obtained by wet dredging (a). After a sieve test (b), the raw sand is subjected to gravity concentration in several stages with Reichert cones (d) and/or spirals (e) to give a product containing 90–98% heavy minerals. This equipment separates the heavy from the light minerals (densities: 4.2–4.8 g/cm^3 and < 3 g/cm^3 respectively) [2.14].

The magnetic minerals (ilmenite) are then separated from the nonmagnetic (rutile, zircon, and silicates) by dry or wet magnetic separation (f). If the ores are from unweathered deposits, the magnetite must first be removed. An electrostatic separation stage (h) allows separation of harmful nonconducting mineral impurities such as granite, silicates, and phosphates from the ilmenite, which is a good conductor.

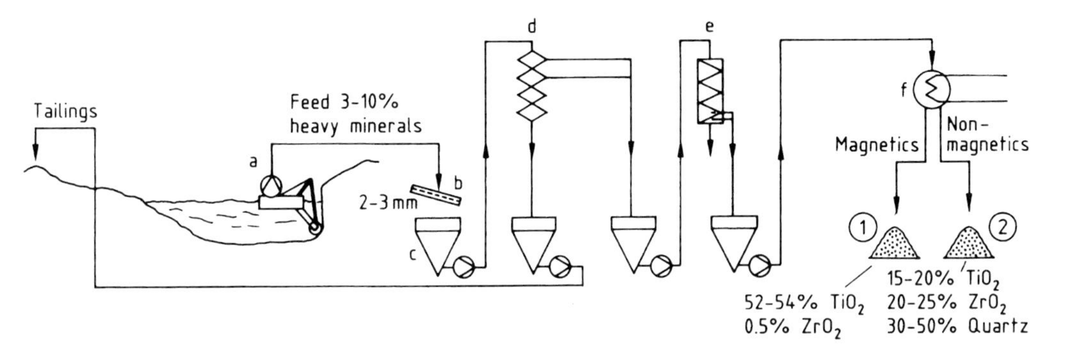

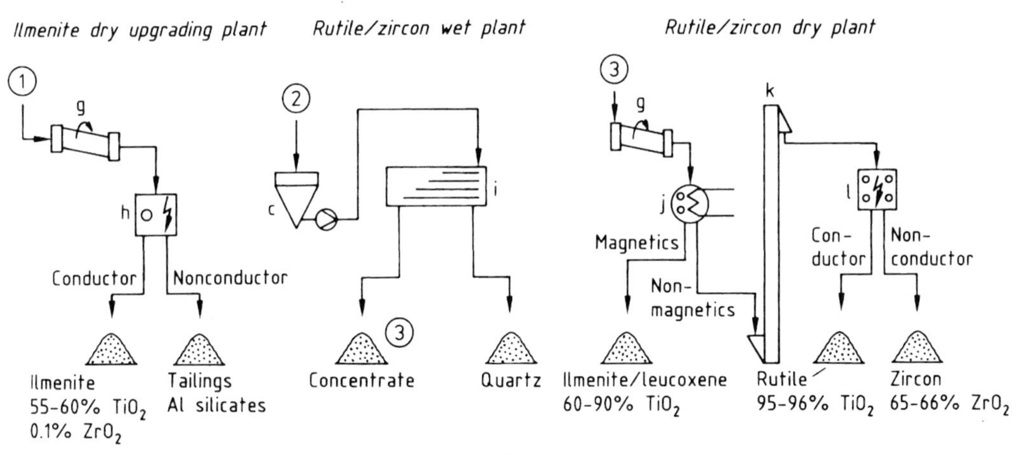

Figure 14. The processing of heavy mineral sands
a) Dredger; b) Sieve; c) Bunker; d) Reichert cones; e) Spirals; f) Magnetic separator; g) Dryer; h) Electrostatic separator; i) Shaking table; j) Dry magnetic separator; k) Vertical belt conveyer; l) Electrostatic separator

The nonmagnetic fraction (leucoxene, rutile, and zircon) then undergoes further hydromechanical processing (i) (shaking table, spirals) to remove the remaining low-density minerals (mostly quartz). Recovery of the weakly magnetic weathered ilmenites and leucoxenes is by high-intensity magnetic separation (j) in a final dry stage. The conducting rutile is then separated from the nonconducting zircon electrostatically in several stages (l). Residual quartz is removed by an air blast.

2.1.2.2. Synthetic Raw Materials

Increasing demand for raw materials with high TiO_2 content has led to the development of synthetic TiO_2 raw materials. In all production processes, iron is removed from ilmenites or titanomagnetites.

Titanium Slag. The metallurgical process for removing iron from ilmenite is based on slag formation in which the iron is reduced by anthracite or coke to metal at 1200–1600 °C in an electric arc furnace, and then separated. Titanium-free pig iron is produced together with slags containing 70–85 % TiO_2 (depending on the ore used) that can be digested with sulfuric acid because they are high in Ti^{3+} and low in carbon. Raw materials of this type are produced in Canada by the Quebec Iron and Titanium Corporation (QIT), in the Republic of South Africa by Richard's Bay Minerals (RBM), and to a smaller extent by Tinfos Titan and Iron K.S. (Tyssedal, Norway). Total slag production grew in 1994 to 1.4×10^6 t of contained TiO_2.

Synthetic Rutile. In contrast to ilmenite, only a small number of rutile deposits can be mined economically, and the price of natural rutile is therefore high. Consequently, many different processes have been developed to remove the iron from ilmenite concentrates without changing the grain size of the mineral because this is highly suitable for the subsequent fluidized-bed chlorination process. All industrial processes involve reduction of Fe^{3+} with carbon or hydrogen, sometimes after preliminary activation of the ilmenite by oxidation. Depending on the reducing conditions, either Fe^{2+} is formed in an activated ilmenite lattice, or metallic iron is produced.

The activated Fe^{2+}-containing ilmenite can be treated with hydrochloric or dilute sulfuric acid (preferably under pressure), and a "synthetic rutile" with a TiO_2 content of 85–96% is obtained [2.15]. The solutions containing iron(II) salts are concentrated and then thermally decomposed to form iron oxide and the free acid, which can be used again in the digestion process [2.16].

Metallic iron can be removed in various ways. The following processes are described in the patent literature:

1) Size reduction followed by physical processes such as magnetic separation or flotation
2) Dissolution in iron(III) chloride solutions [2.17], the resulting iron(II) salt is oxidized with air to give iron oxide hydroxides and iron(III) salts
3) Dissolution in acid
4) Oxidation with air in the presence of electrolytes. Various iron oxide or iron oxide hydroxide phases are formed depending on the electrolyte used: Possible electrolytes include iron(II) chloride solutions [2.18], ammonium chloride [2.19], or ammonium carbonate–carbonic acid [2.20]
5) Oxidation with the iron(III) sulfate from ilmenite digestion (see Section 2.1.3.1) [2.21], followed by crystallization of the iron(II) sulfate
6) Chlorination to form iron(III) chloride [2.22]
7) Reaction with carbon monoxide to form iron carbonyls [2.23] which can be decomposed to give high-purity iron

Another possible method of increasing the TiO_2 content of ilmenite is by partial chlorination of the iron in the presence of carbon. This is operated on a large scale by several companies [2.24], [2.25]. The most important companies which are producing synthetic rutile are located in Australia (Renison Goldfields Consolidated, Tiwest, Westralian Sands), United States (Kerr-McGee Synthetic Rutile), India (Kevala Minerals and Metals Ltd., DCW Ltd., Bene-Chlor Chemicals Ltd.), and Malaysia (Hitox). In 1994 production of synthetic rutile was about 0.6×10^6 t of contained TiO_2.

2.1.3. Production

Titanium dioxide pigments are produced by two different processes. The older *sulfate process* depends on the breakdown of the titanium-containing raw material ilmenite or titanium slag with concentrated sulfuric acid at 150–220 °C. Relatively pure TiO_2 dihydrate is precipitated by hydrolysis of the sulfate solution, which contains colored heavy metal sulfates, sometimes in high concentration. The impurities are largely removed in further purification stages. The hydrate is then calcined, ground, and further treated.

In the *chloride process,* the titanium-containing raw materials ilmenite, leucoxene, natural and synthetic rutile, titanium slag, and anatase are chlorinated at 700–1200 °C. Titanium tetrachloride is separated from other chlorides by distillation. Vanadium tetrachloride (VCl_4) and vanadium oxychloride ($VOCl_3$) must, however, first be reduced to solid chlorides. The $TiCl_4$ is burnt at temperatures of 900–1400 °C to form TiO_2. This extremely pure pigment undergoes further treatment depending on the type of application.

2.1.3.1. Sulfate Method

The sulfate method is summarized in Figure 15.

Grinding. The titanium-bearing raw materials are dried to a moisture content of < 0.1 %. Drying is mainly intended to prevent heating and premature reaction on mixing with sulfuric acid. The raw materials are ground in ball mills to give a mean particle size of ca. 40 μm. The combination of grinding and drying shown in Figure 15 (a) is recommended. The small amount of metallic iron present in titanium slag is removed magnetically (c), which almost completely eliminates hydrogen evolution during subsequent digestion.

Digestion. *Batch digestion* is usually employed. The ground raw materials (ilmenite, titanium slag, or mixtures of the two) are mixed with 80–98 % H_2SO_4. The ratio of H_2SO_4 to raw material is chosen so that the weight ratio of free H_2SO_4 to TiO_2 in the suspension produced by the hydrolysis is between 1.8 and 2.2 (the so-called "acid number"). The reaction in the digestion vessel (f) is started by adding water, dilute sulfuric acid, oleum, or sometimes steam. The temperature initially increases to 50–70 °C due to the heat of hydration of the acid. The exothermic sulfate formation then increases the temperature to 170–220 °C. If dilute acid or sparingly soluble raw materials are used, external heating is required.

After the maximum temperature has been reached, the reaction mixture must be left to mature for 1–12 h, depending on the raw material, so that the titanium-containing components become as soluble as possible. Digestion can be accelerated by blowing air through the mass while the temperature is increasing, and also during the maturing period.

Several *continuous digestion processes* have been proposed [2.26]. A proven method is to continuously feed a mixture of ilmenite and water together with the acid

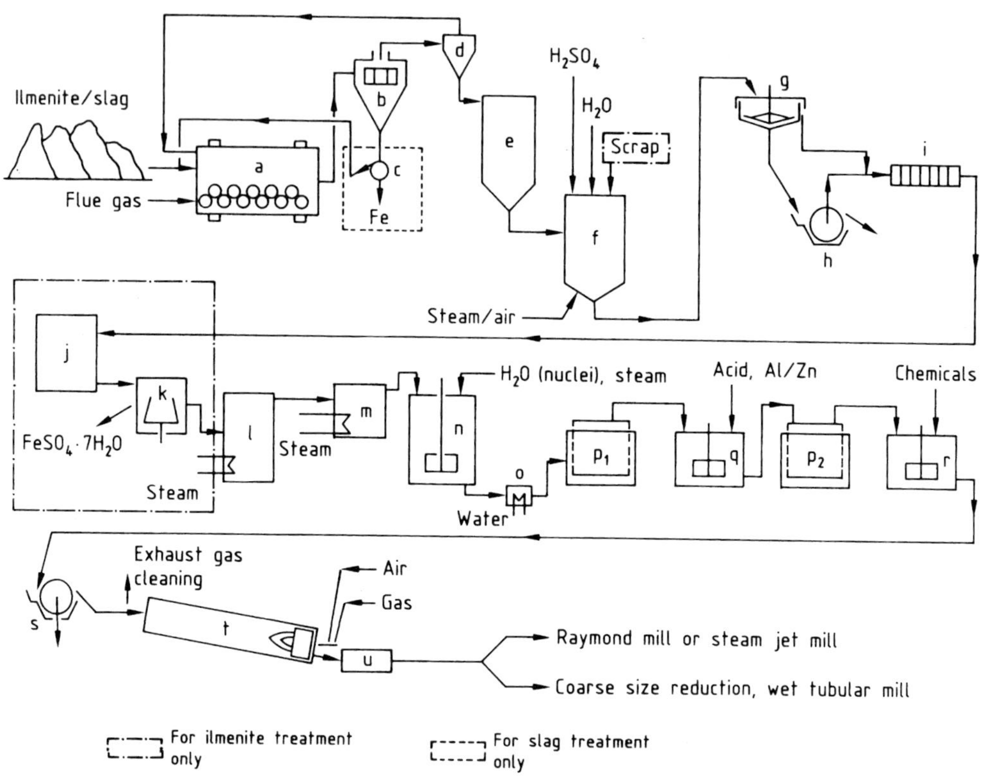

Figure 15. Production of TiO_2 by the sulfate process
a) Ball mill/dryer; b) Screen; c) Magnetic separator; d) Cyclone; e) Silo; f) Digestion vessel; g) Thickener; h) Rotary filter; i) Filter press; j) Crystallizer; k) Centrifuge; l) Vacuum evaporator; m) Preheater; n) Stirred tank for hydrolysis; o) Cooler; p) Moore filters; q) Stirred tank for bleaching; r) Stirred tank for doping; s) Rotary filter for dewatering; t) Rotary kiln; u) Cooler

into a double-paddle screw conveyor. After a relatively short dwell time (<1 h), a crumbly cake is produced [2.10]. This process utilizes a more limited range of raw materials than the batch process because they need to be very reactive.

Dissolution and Reduction. The cake obtained by digestion is dissolved in cold water or in dilute acid recycled from the process. A low temperature must be maintained ($< 85\,°C$) to avoid premature hydrolysis, especially with the product from ilmenite. Air is blown in to agitate the mixture during dissolution. With the ilmenite product, the TiO_2 concentration of the solution is 8–12 wt%, and with the slag product between 13 and 18 wt%.

The trivalent iron is hydrolyzed together with the titanium compounds, and adheres to the titanium oxide hydrate. Therefore all the Fe^{3+} is reduced to Fe^{2+} by scrap iron during dissolution of the ilmenite product, or immediately afterwards. Reoxidation of the iron during subsequent processing is prevented with Ti^{3+} which is obtained by reducing a small part of the Ti^{4+}. Alternatively, reduction of Ti^{4+} to Ti^{3+} can be carried out in part of the solution under optimized conditions; this concentrated Ti^{3+} solution is then added in a controlled manner to the reaction solution [2.27]. In solutions obtained from titanium slag, the Ti^{3+} content of the solution must be decreased by oxidation with atmospheric oxygen so that no loss of yield occurs during hydrolysis.

With both ilmenite and titanium slag, mixed digestion can be carried out in which the Ti^{3+} content of the slag reduces all the Fe^{3+} to Fe^{2+}. The dissolved products obtained from the separate digestion of ilmenite and titanium slag can also be mixed [2.28], [2.29].

Clarification. All undissolved solid material must be removed as completely as possible from the solution. The most economical method is to employ preliminary settling in a thickener (g), followed by filtration of the sediment with a rotary vacuum filter (h). The filtrate and the supernatant from the thickener are passed through filter presses (i) to remove fines. Owing to the poor filtering properties of the solution, the rotary filter must be operated as a precoat filter. Preliminary separation in the thickener must be assisted by adding chemicals to promote sedimentation. Attempts to carry out the entire clarification process in a single stage using automated filter presses have been reported [2.30].

Crystallization. The solutions from slag digestion contain 5–6 wt% $FeSO_4$, and those from ilmenite digestion 16–20 wt% $FeSO_4$ after reduction of the Fe^{3+}. The solution is cooled under vacuum to crystallize out $FeSO_4 \cdot 7H_2O$ (j) and reduce the quantity of $FeSO_4$ discharged with the waste acid. The concentration of the TiO_2 in the solution is thereby increased by ca. 25%. The salt is separated by filtration or centrifugation (k).

The iron sulfate is used in water purification, and as a raw material for the production of iron oxide pigments. Alternatively, it can be dehydrated and thermally decomposed to give iron(III) oxide and sulfur dioxide.

Hydrolysis. Titanium oxide hydrate is precipitated by hydrolysis at 94–110 °C. Other sulfuric-acid-soluble components of the raw material are precipitated simultaneously, mainly niobium as its oxide hydrate.

Hydrolysis is carried out in brick-lined, stirred tanks (n) into which steam is passed. The hydrolysate does not have any pigment properties, but these are strongly influenced by the particle size and degree of flocculation of the hydrolysate (mean particle size of hydrolysate is ca. 5 nm, and of TiO_2 pigments 200–300 nm).

The properties of the hydrolysate depend on several factors:

1) The hydrolysis of concentrated solutions of titanium sulfate (170–230 g TiO_2/L) proceeds very sluggishly and incompletely (even if boiled) unless suitable nuclei are added or formed to accelerate hydrolysis. The nuclei are usually produced by two methods. In the Mecklenburg method, colloidal titanium oxide hydrate is precipitated with sodium hydroxide at 100 °C; 1% of this hydrate is sufficient. In the Blumenfeld method a small part of the sulfate solution is hydrolyzed in boiling water and then added to the bulk solution [2.31]. The particle size of the hydrolysate depends on the number of nuclei.
2) The particle size and degree of flocculation of the hydrolysate depend on the intensity of agitation during the nuclei formation by the Blumenfeld method and also during the initial stage of the hydrolysis.

3) The titanium sulfate concentration has a great influence on the flocculation of the hydrolysate. It is adjusted, if necessary by vacuum evaporation, to give a TiO_2 content of 170–230 g/L during hydrolysis. Lower concentrations result in a coarser particle size.
4) The acid number should be between 1.8 and 2.2. It has a considerable effect on the TiO_2 yield and on the particle size of the hydrolysate. For a normal hydrolysis period (3–6 h) the TiO_2 yield is 93–96%.
5) The properties of the hydrolysate are affected by the concentrations of other salts present, especially $FeSO_4$. High concentrations lead to finely divided hydrolysates.
6) The temperature regime mainly affects the volume–time yield and hence the purity of the hydrolysate.

Purification of the Hydrolysate. After hydrolysis, the liquid phase of the titanium oxide hydrate suspension contains 20–28% H_2SO_4 and various amounts of dissolved sulfates, depending on the raw material. The hydrate is filtered off from the solution (p_1) (weak acid), and washed with water or dilute acid. Even with acid washing, too many heavy metal ions are adsorbed on the hydrate for it to be directly usable in the production of white pigment. Most of the impurities can be removed by reduction (bleaching), whereby the filter cake is slurried with dilute acid (3–10%) at 50–90 °C and mixed with zinc or aluminum powder (q). Bleaching can also be carried out with powerful nonmetallic reducing agents (e.g., $HOCH_2-SO_2Na$). After a second filtration and washing process (p_2), the hydrate only has low concentrations (ppm) of colored impurities but still contains chemisorbed 5–10% H_2SO_4. This cannot be removed by washing and is driven off by heating to a high temperature.

Doping of the Hydrate. When producing titanium dioxide of maximum purity, the hydrate is heated (calcined) without any further additions. This gives a fairly coarse grade of TiO_2 with a rutile content that depends on the heating temperature. However, to produce specific pigment grades, the hydrate must be treated with alkali-metal compounds and phosphoric acid as mineralizers (< 1%) prior to calcination (r). Anatase pigments contain more phosphoric acid than rutile pigments. To produce rutile pigments, rutile nuclei (< 10%) must be added; ZnO, Al_2O_3, and/or Sb_2O_3 (< 3%) are sometimes also added to stabilize the crystal structure.

Nuclei are produced by converting the purified titanium oxide hydrate to sodium titanate, which is washed free of sulfate and then treated with hydrochloric acid to produce the rutile nuclei. Rutile nuclei can also be prepared by precipitation from titanium tetrachloride solutions with sodium hydroxide solution.

Calcination. The doped hydrate is filtered with rotary vacuum filters (s) to remove water until a TiO_2 content of ca. 30–40% is reached. Pressure rotary filters or automatic filter presses can also be used to obtain a TiO_2 content of ca. 50%. Some of the water-soluble dopants are lost in the filtrate and can be replaced by adding them to the filter cake before it is charged into the kiln. Calcination is performed in rotary kilns (t) directly heated with gas or oil in countercurrent flow. Approximately two-thirds of the residence time (7–20 h in total) is needed to dry the material. Above ca. 500 °C, sulfur trioxide is driven off which partially decomposes to sulfur dioxide and oxygen at higher temperatures. The product reaches a maximum temperature of 800–1100 °C depending on pigment type, throughput, and temperature profile of the kiln. Rutile content, particle size, size distribution, and aggregate formation are extremely dependent on the operating regime of the kiln. After leaving the kiln, the clinker can be indirectly cooled or directly air-cooled in drum coolers (u).

The exhaust gas must have a temperature of $> 300\,°C$ at the exit of the kiln to prevent condensation of sulfuric acid in the ducting. Energy can be saved by recirculating some of the gas to the combustion chamber of the kiln and mixing it with the fuel gases as a partial replacement for air. Alternatively, it can be used for concentrating the dilute acid (see Section 2.1.3.5). The gas then goes to the waste-gas purification system.

Grinding. The agglomerates and aggregates in the clinker can be reduced to pigment fineness by wet or dry grinding. Coarse size reduction should be carried out in hammer mills prior to wet grinding in tube mills (with addition of wetting agents). The coarse fraction can be removed from the suspension by centrifugation, and recycled to the mills. Hammer mills, cross-beater mills, and particularly pendular and steam-jet mills are suitable for dry grinding. Special grinding additives can be used that act as wetting agents during subsequent pigment treatment or improve the dispersibility of untreated pigments.

2.1.3.2. The Chloride Process

The chloride process is summarized in Figure 16.

Chlorination. The titanium in the raw material is converted to titanium tetrachloride in a reducing atmosphere. Calcined petroleum coke is used as the reducing agent because it has an extremely low ash content and, due to its low volatiles content, very little HCl is formed. The titanium dioxide reacts exothermically as follows:

$$TiO_2 + 2\,Cl_2 + C \rightarrow TiCl_4 + CO_2$$

As the temperature rises, an endothermic reaction also occurs to an increasing extent in which carbon monoxide is formed from the carbon dioxide and carbon. Therefore, oxygen must be blown in with the chlorine to maintain the reaction temperature between 800 and 1200 °C. The coke consumption per tonne of TiO_2 is 250–300 kg. If CO_2-containing chlorine from the combustion of $TiCl_4$ is used, the coke consumption increases to 350–450 kg.

The older *fixed-bed chlorination method* is hardly used today. In this process, the ground titanium-containing raw material is mixed with petroleum coke and a binder, and formed into briquettes. Chlorination is carried out at 700–900 °C in brick-lined reactors.

Fluidized-bed chlorination was started in 1950. The titanium raw material (with a particle size similar to that of sand) and petroleum coke (with a mean particle size ca. five times that of the TiO_2) are reacted with chlorine and oxygen in a brick-lined fluidized-bed reactor (Fig. 16, c) at 800–1200 °C. The raw materials must be as dry as possible to avoid HCl formation. Since the only losses are those due to dust entrainment the chlorine is 98–100% reacted, and the titanium in the raw material

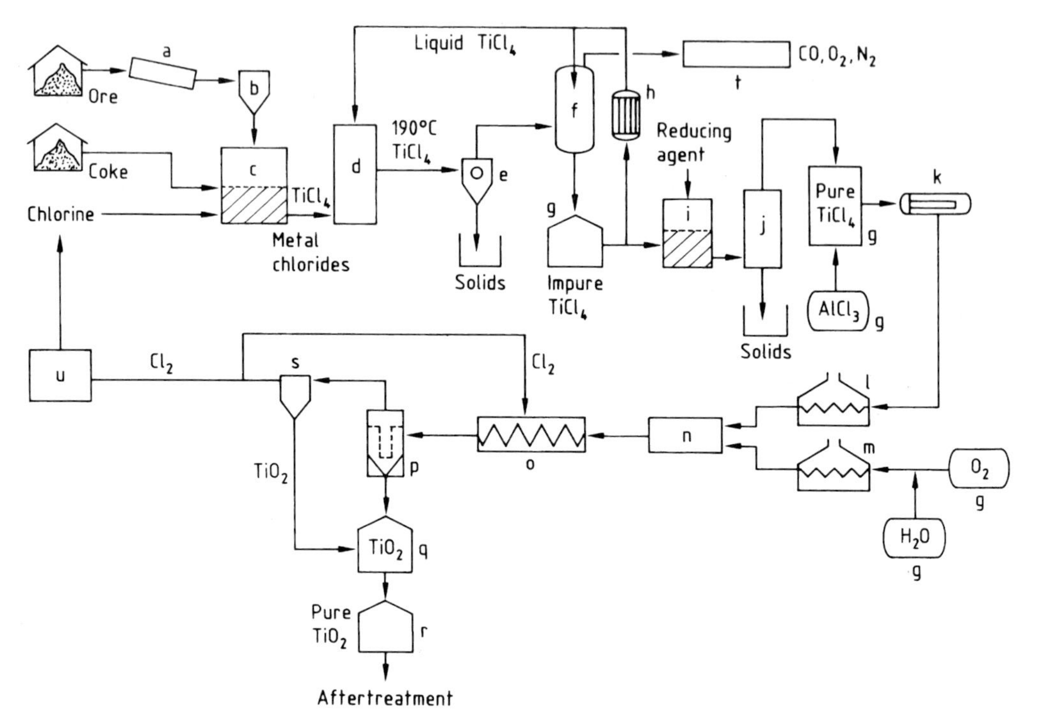

Figure 16. Flow diagram of TiO_2 production by the chloride process
a) Mill; b) Silo; c) Fluidized-bed reactor; d) Cooling tower; e) Separation of metal chlorides; f) $TiCl_4$ condensation; g) Tank; h) Cooler; i) Vanadium reduction; j) Distillation; k) Evaporator; l) $TiCl_4$ superheater; m) O_2 superheater; n) Burner; o) Cooling coil; p) Filter; q) TiO_2 purification; r) Silo; s) Gas purification; t) Waste-gas cleaning; u) Cl_2 liquefaction

is 95–100% reacted, depending on the reactor design and the gas velocity. Magnesium chloride and calcium chloride can accumulate in the fluidized-bed reactor due to their low volatility. Zirconium silicate also accumulates because it is chlorinated only very slowly at the temperatures used. All the other constituents of the raw materials are volatilized as chlorides in the reaction gases.

The ceramic cladding of the fluidized-bed reactor is rather rapidly destroyed by abrasion and corrosion. If chlorination is interrupted, there is a further danger that the raw materials may sinter and eventually cannot be fluidized.

Gas Cooling. The reaction gases are cooled with liquid $TiCl_4$ either indirectly or directly (d). Crystallization of the chlorides of the other components causes problems because they tend to build up on the cooling surfaces, especially the large quantities of iron(II) and iron(III) chlorides formed on chlorination of ilmenite [2.32]. In this first stage, the reaction gases are cooled only down to a temperature ($< 300\,°C$) at which the accompanying chlorides can be satisfactorily separated from the $TiCl_4$ by condensation or sublimation (e).

The gas then consists mainly of $TiCl_4$, and is cooled below 0 °C, causing most of the $TiCl_4$ to condense (f). The small amounts of $TiCl_4$ and Cl_2 remaining in the exhaust gas (CO_2, CO, and N_2) are removed by scrubbing with alkali (t).

Purification of $TiCl_4$. The chlorides that are solid at room temperature and the entrained dust can be separated from the $TiCl_4$ by simply evaporating (distilling) this off (j). Dissolved chlorine can be removed by heating or reduction with metal powders (Fe, Cu, or Sn).

Removal of vanadium tetrachloride (VCl_4) and vanadium oxychloride ($VOCl_3$) from the $TiCl_4$ by distillation is very difficult owing to the closeness of their boiling points. They are therefore reduced to form solid, low-value vanadium chlorides (i). An enormous number of reducing agents have been recommended; important examples are copper, titanium trichloride, hydrogen sulfide, hydrocarbons, soaps, fatty acids, and amines. After subsequent evaporation (j) the titanium chloride should contain < 5 ppm vanadium. If organic reducing agents are used, the residues may cause problems by baking onto the surfaces of the heat exchanger.

Phosgene and $SiCl_4$ can be removed by fractional distillation.

Combustion of $TiCl_4$ and Recovery of TiO_2. Titanium tetrachloride is combusted with oxygen at 900–1400 °C to form TiO_2 pigment and chlorine (n). The purified $TiCl_4$ is vaporized (k) and the vapor is indirectly heated to ca. 500–1000 °C (l). The reaction

$$TiCl_4 + O_2 \rightarrow 2\ Cl_2 + TiO_2$$

is weakly exothermic, and requires a high reaction temperature, so that the oxygen must also be heated to $>$ 1000 °C (m). This can be achieved with an electric plasma flame, by reacting part of the oxygen with carbon monoxide, or by indirect heating. Hot $TiCl_4$ and oxygen (110–150% of the stoichiometric amount) are fed separately into a reaction chamber where they must be mixed as rapidly and completely as possible to give a high reaction rate. For this reason, and also because the TiO_2 has a strong tendency to cake onto the walls [2.33]–[2.35], many different reactor designs have been proposed and used. The same considerations apply to the cooling unit (o) where the pigment is very rapidly cooled to below 600 °C. Cooling zones of various geometries are used. If caking occurs, the material can be removed by introducing abrasive particles [2.36], [2.37].

The mixture of gases (Cl_2, O_2, and CO_2) and pigment can be further cooled during dry separation of the pigment either indirectly or directly by solid particles, e.g., sand. The pigment-containing gas is then filtered (p). The gas stream is recycled to the cooling zone (o) of the combustion furnace and to the chlorination process as oxygen-containing chlorine via the liquefaction unit (u). The chlorine adsorbed on the pigment can be removed by heating or by flushing with nitrogen or air.

The wet separation process, in which the pigment-containing gas mixture (Cl_2, O_2, and CO_2) is quenched in water, has not become established.

2.1.3.3. Pigment Quality

The quality of the TiO_2 pigment is influenced by various factors. Reaction temperature, excess oxygen, and flow conditions in the reactor affect particle size and size distribution. Therefore, optimum conditions must be established for every reactor design. Caking of the TiO_2 on the walls of the reactor leads to impairment of quality.

The presence of water and a Cs-compound during combustion of the $TiCl_4$ gives rise to nuclei which promote the formation of finely divided pigment particles with high scattering power [2.38]. It can be added directly to the oxygen or can be produced by the combustion of hydrogen-containing materials.

The presence of $AlCl_3$ promotes the formation of rutile and a more finely divided pigment. It is added in amounts of up to 5 mol%. Many methods have been proposed for rapidly generating and directly introducing the $AlCl_3$ vapor into the $TiCl_4$ vapor. Addition of PCl_3 and $SiCl_4$ suppresses rutile formation, so that anatase pigment is obtained [2.39]. However, pigments of this type have not appeared on the market.

Pigments produced by the chloride process (chloride pigments) have better lightness and a more neutral hue than pigments produced by the sulfate process (sulfate pigments). Pigments used in demanding applications are almost always subjected to inorganic aftertreatment.

2.1.3.4. Aftertreatment

Aftertreatment of the pigment particles improves the weather resistance and lightfastness of the pigmented organic matrix, and dispersibility in this matrix. The treatment consists of coating the individual pigment particles with colorless inorganic compounds of low solubility by precipitating them onto the surface. However, this reduces the optical performance of the pigment approximately in proportion to the decrease in the TiO_2 content. The surface coatings prevent direct contact between the binder matrix and the reactive surface of the TiO_2. The effectiveness of these coatings largely depends on their composition and method of application, which may give too porous or too dense a coating. The treatment process also affects the dispersibility of the pigment, and therefore a compromise often has to be made. High weather resistance and good dispersibility of the pigment in the binder or matrix are usually desired. These effects are controlled by using different coating densities and porosities. Other organic substances can be added during the final milling of the dried pigment.

Several types of treatment are used:

1) Deposition from the gas phase by hydrolysis or decomposition of volatile substances such as chlorides or organometallic compounds. Precipitation onto the pigment surface is brought about by adding water vapor. This method is especially applicable to chloride pigments, which are formed under dry conditions.
2) Addition of oxides, hydroxides, or substances that can be adsorbed onto the surface during pigment grinding. This can produce partial coating of the pigment surface.
3) Precipitation of the coating from aqueous solutions onto the suspended TiO_2 particles. Batch processes in stirred tanks are preferred; various compounds are deposited one after the other under optimum conditions. There is a very extensive patent literature on this subject. Continuous precipitation is sometimes used in mixing lines or cascades of stirred tanks. Coatings of widely differing compounds are produced in a variety of sequences. The most common are oxides, oxide hydrates, silicates, and/or phosphates of titanium, zirconium, silicon, and aluminum. For special applications, boron, tin, zinc, cerium, manganese, antimony, or vanadium compounds can be used [2.40], [2.41].

Three groups of pigments have very good lightfastness or weather resistance:

1) Pigments with dense surface coatings for paints or plastics formed by:

   a) Homogeneous precipitation of SiO_2 with precise control of temperature, pH, and precipitation rate [2.42]: ca. 88% TiO_2
   b) Two complete aftertreatments, calcination is performed at 500–800 °C after the first or second aftertreatment [2.43]: ca. 91% TiO_2
   c) Aftertreatment with Zr, Ti, Al, and Si compounds, sometimes followed by calcination at 700–800 °C [2.44]: ca. 95% TiO_2

2) Pigments with porous coatings for use in emulsion paints obtained by simple treatment with Ti, Al, and Si compounds, giving a silica content of 10% and a TiO_2 content of 80–85%
3) Lightfast pigments with dense surface coatings for the paper industry that have a stabilized lattice and a surface coating based on silicates or phosphates of titanium, zirconium, and aluminum: ca. 90% TiO_2

Coprecipitation of special cations such as antimony or cerium can improve lightfastness further [2.45]. After treatment in aqueous media, the pigments are washed on a rotary vacuum filter or a filter press until they are free of salt, and then dried using, e.g., belt, spray, or fluidized-bed dryers.

Before micronizing the pigment in air-jet or steam-jet mills, and sometimes also before drying, the pigment surface is improved by adding substances to improve dispersibility and facilitate further processing. The choice of compounds used, which are mostly organic, depends on the intended use of the pigment. The final surface can be made either hydrophobic (e.g., using silicones, organophosphates, and alkyl phthalates) or hydrophilic (e.g., using alcohols, esters, ethers and their polymers, amines, organic acids). Combinations of hydrophobic and hydrophilic substances have proved especially useful for obtaining surface properties that give better dispersibility and longer shelf life [2.46].

2.1.3.5. Problems with Aqueous and Gaseous Waste

Aqueous Waste. In the *sulfate process*, 2.4–3.5 t concentrated H_2SO_4 are used per tonne of TiO_2 produced, depending on the raw material. During processing, some of this sulfuric acid is converted to sulfate, primarily iron(II) sulfate, the rest is obtained as free sulfuric acid (weak acid). Filtration of the hydrolysate suspension can be carried out to give 70–95% of the SO_4^{2-} in a weak acid fraction containing ca. 20–25 wt% free sulfuric acid, the remaining sulfate (5–30%) is highly diluted with wash-water.

In the past it has been common practice to discharge the waste acid directly into the open sea or coastal waters. For a long time the weak acid problem has been the subject of public discussion and criticism. As a result the European Community has decided to stop the discharge of weak acid into open waters until 1993.

The European titanium dioxide producers have developed different effluent treatment processes to meet the environmental requirements [2.47]. The most important processes are the precipitation of gypsum ($CaSO_4$) from the weak acid [2.48] and the concentration and recovery of the free and bound acid.

In the gypsum process the acid effluent is treated in a first stage with fine divided calcium carbonate ($CaCO_3$) to precipitate white gypsum. After filtering off, washing and drying the white gypsum is used for the manufacturing of plaster boards. In a second stage the remaining metal sulfates in the filtrate are precipitated by treating with calcium hydroxide as metal hydroxides and further gypsum. This mixture, the red gypsum, can be used e.g. for landfill.

In the recycling process both the free and the bound sulfuric acid (as metal sulfates) can be recovered from the weak acid in the calcination furnace (Fig. 17, k) and in metal sulfate calcination (Fig. 18). The process consists of two stages:

1) Concentration and recovery of the free acid by evaporation
2) Thermal decomposition of the metal sulfates and production of sulfuric acid from the resulting sulfur dioxide

As a result of energy requirements only acid containing >20% H_2SO_4 can be economically recovered by evaporation. The weak acid is concentrated from ca. 20–25% to ca. 28% with minimum heat (i.e., energy) consumption, e.g., by using waste heat from sulfuric acid produced by the contact process [2.49], or from the waste gases from the calcination kilns used in TiO_2 production [2.50] (Fig. 18).

Following preliminary evaporation, further concentration is carried out in multi-effect vacuum evaporators. Since the water vapor pressure decreases strongly as the H_2SO_4 concentration increases, in general only two-stage evaporation can effectively exploit the water vapor as a heating medium. Evaporation produces a suspension of metal sulfates in 60–70% sulfuric acid (stage 1 in Fig. 17). The suspension is cooled to 40–60 °C in a series of stirred tanks (stage 2, d) [2.51], giving a product with good filtering properties and an acid of suitable quality for recycling to the digestion process. Filtration (stage 3, e) is usually carried out with pressure filters [2.52] because they give a filter cake with an extremely low residual liquid content.

The concentration of the acid recycled to the digestion process depends on the quality of the titanium-containing raw material. For raw materials with a high titanium content, the 65–70% sulfuric acid separated from the metal sulfates must be further concentrated to give 80–87% acid (stage 5).

Concentration can be carried out in steam-heated vacuum evaporators, or by using the heat from the TiO_2 calcination kilns [2.53]. Cooling the acid obtained after this concentration process yields a suspension of metal sulfates that can be directly used for digestion of the raw material. The metal sulfates recovered from the sulfuric acid in stage 3 are moist because they contain 65–70% sulfuric acid; they therefore have no direct use. They can be converted to a disposable material by reaction with calcium compounds [2.54]. Thermal decomposition of the metal sulfates to form the metal oxides, sulfur dioxide, water, and oxygen is energy intensive, but is advantageous from the ecological point of view. The energy requirement is ca. 4×10^9 J per tonne of filter cake. Thermal decomposition is carried out at 850–1100 °C in a fluidized-bed furnace (stage 6). The energy is supplied by coal, pyrites, or sulfur. The sulfur dioxide produced by the thermal decomposition is purified by the usual methods, dried, and converted into sulfuric acid or oleum. This pure acid or oleum is mixed with the recovered sulfuric acid and used in the digestion process.

The metal oxides produced by thermal decomposition contain all the elements initially present in the raw material apart from the titanium which has been convert-

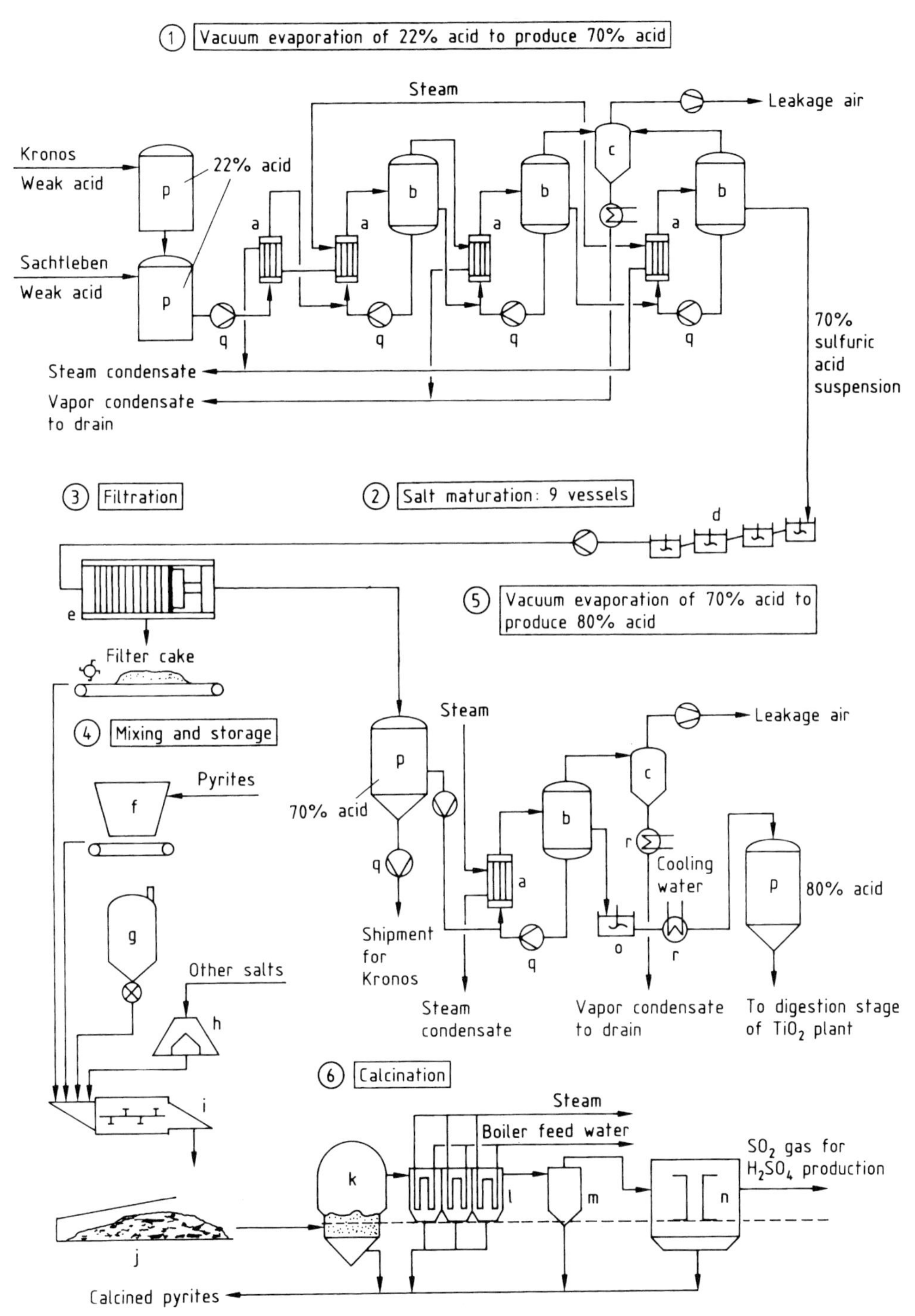

Figure 17. Weak acid recovery plant used by Sachtleben Chemie (based on know-how of Bayer AG)
a) Heat exchanger; b) Evaporator; c) Injection condenser; d) Stirred salt maturing vessels; e) Filter press; f) Bunker for pyrites; g) Coal silo; h) Bunker; i) Mixing screw unit; j) Covered store for mixed filter cake; k) Calcination furnace; l) Waste-heat boiler; m) Cyclone; n) Electrostatic precipitator; o) Stirred tank; p) Storage tank; q) Pump; r) Cooler

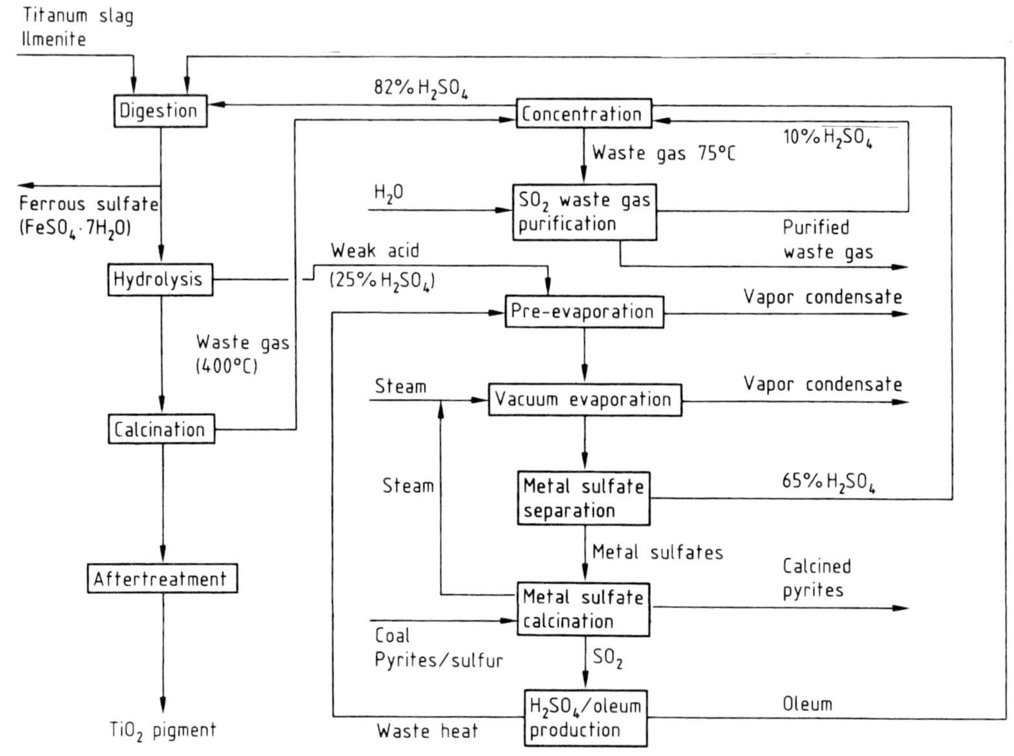

Figure 18. Waste heat recovery and sulfuric acid recycling during weak acid treatment (Bayer AG)

ed into pigment. The mixture of metal oxides, mainly iron oxide, can be used as an iron compound in the construction industry.

The continually increasing demand for environmentally friendly industrial processes has also led to the development of techniques for recycling of the remaining 5–30% sulfate contained in the acidic wash water [2.55]. In modern processes, up to 99% of sulfuric acid can be recovered and reused in production. In the *chloride process*, wastewater problems arise if the raw material contains $< 90\%$ TiO_2. The metal chloride by products are sometimes disposed of in solution by the "deep well" method (e.g., at Du Pont). The metal chloride solutions are pumped via deep boreholes into porous geological strata. Special geological formations are necessary to avoid contamination of the groundwater by impurities.

Increasing restrictions also apply to the chloride process, so that efforts are continually being made to use the iron chloride byproduct, e.g., in water treatment and as a flocculation agent [2.56]. Another process for treating metal chorides with cement and alkaline compounds to produce rock-like aggregates for road building is described in [2.57].

Waste Gas Problems. The gases produced in the calcination kiln are cooled in a heat exchanger, and entrained pigment is removed, washed, and recycled to the process. The SO_2 and SO_3 formed during calcination are then scrubbed from the gases to form dilute sulfuric acid which is recycled.

2.1.4. Economic Aspects

The burning of $TiCl_4$ with oxygen or the calcination of TiO_2 hydrolysates produces either anatase or rutile pigments, depending on the doping and lattice stabilization. They are marketed directly or after being coated with oxides or hydroxides of various elements. Different treatments are necessary depending on the field of application, and all major pigment producers have a large number of pigment grades. Product groups are listed in Table 15. Pigments of all grades are available with or without organic treatment. Over 400 different TiO_2 pigment grades are currently on the market. Table 16 gives the capacities and processes of the most important pigment producers.

A considerable increase of capacity is planned by debottlenecking and building new plants. Pigment plant capacity is forcast to grow to 5.2×10^6 t in 2005 [2.58].

Powdered TiO_2 pigments are usually supplied in 25 kg sacks (50 lbs, USA) or in large bags containing 0.5–1 t pigment. Aqueous suspensions with solids contents of 68–75% are also available and have great advantages as regards the distribution and metering of the pigment in aqueous systems. The dust formation that occurs with dry pigment is also avoided. With the development of products with improved flow properties and modern pneumatic delivery technology, supply in silo wagons is becoming increasingly important.

2.1.5. Pigment Properties

The pigment properties are extremely important when TiO_2 is used as a white pigment; they include lightening power, hiding power, lightness, hue, gloss formation, gloss haze, dispersibility, lightfastness, and weather resistance. These properties are a function of chemical purity, lattice stabilization, particle size and size distribution, and the coating produced by aftertreatment. They also depend on the medium and cannot generally be accurately described in scientific terms. Some of the important properties of TiO_2 pigments are described below.

Scattering Power. The refractive indices of rutile and anatase are very high (2.70 and 2.55, respectively). Even after incorporation in a wide range of binders, they lie

Table 15. Classification of TiO_2 pigments according to composition (DIN 55912, sheet 1, issue 07.85, ISO 591–1977)

Pigment	Class	TiO_2 (min.), wt%	Water-soluble salts, wt%	Volatiles (max.), wt%
Anatase	A 1	98	0.6	0.5
(Type A)	A 2	92	0.5	0.8
Rutile	R 1	97	0.6	0.5
(Type R)	R 2	90	0.5	to be agreed
	R 3	80	0.7	to be agreed

Table 16. World TiO_2 pigment producers (1995)

Country	Company	Location	Capacity, 10^3 t/a		
			Chloride	Sulfate	Total
United States	Du Pont	New Johnsonville	305		305
		De Lisle	245		245
		Edge Moore	127		127
		Antioch	38		38
	SCM Corporation	Baltimore	50	66	116
		Ashtabula	165		165
	Kemira Oy	Savannah	91	54	145
	Kerr McGee Corporation	Hamilton	130		130
	Tioxide	Lake Charles	50		50
	NL Chemical Incorporated	Lake Charles	50		50
Canada	NL Chemicals Incorporated	Varennes	50	36	86
Brazil	Titanio do Brasil/Bayer	Salvador		55	55
Mexico	Du Pont	Tampico	80		80
Total, America					*1592*
Germany	Kronos Titan	Leverkusen	80	35	115
	NL Chemicals	Nordenham		60	60
	Bayer	Krefeld–Uerdingen		105	105
	Sachtleben Chemie, Metallgesellschaft	Duisburg–Homberg		80	80
United Kingdom	Tioxide UK	Grimsby		100	100
	Tioxide UK	Greatham	80		80
	SCM Chemicals	Stallingborough	109	10	119
France	Thann et Mulhouse (Rhône-Poulenc)	Le Havre		95	95
		Thann		30	30
	Tioxide France	Calais		100	100
Finland	Kemira Oy	Pori		90	90
Italy	Tioxide Italia	Scarlino		80	80
Belgium	Kronos Titan	Langenbrügge	50		50
	Bayer	Antwerpen		33	33
Spain	Tioxide Espan.	Huelva		80	80
Netherlands	Kemira	Botlek	45		45
Norway	Kronos Titan	Fredrikstad		32	32
Total, Western Europe					*1294*

Russia	Lakokraska	Chelyabinsk		10	10
Ukraine	Lakokraska	Armyansk		80	80
	Agrokhim	Sumy		40	40
Poland	Zachem	Police		36	36
Yugoslavia	Cinkarna	Celje		25	25
Czechoslavakia	Precheza	Prerov		25	25
Total, Eastern Europe					*216*
Republic of South Africa	SA Tioxide	Umbogintwini		38	38
Total Africa					*38*
Australia	SCM Chemical	Kemerton	79		79
	Tioxide Australia	Burnie		35	35
	Kerr Mc Gee Corporation	Kwinana	64		64
Total, Australia					*178*
Japan	Ishihara Sangyo Kaisha	Yokkaichi	55	100	155
	Tayca	Saidaiji		60	60
	Sakai Chemical	Onahama		43	43
	Furukawa Mining	Osaka		23	23
	Fuji Titanium	Kobe		16	16
	Titan Kogyo	Ube		17	17
	Tohoku Chemical	Akita		30	30
Total, Japan					*344*
India	Kerala Minerals & Metals	Kerala	22		22
	Travancore Titanium	Trivandrum		13	13
South Korea	Hankook Titanium	Incheon		36	36
Taiwan	China Metal & Chemicals	Chin Shin		10	10
	Du Pont	Kuan Yin	60		60
Malaysia	Tioxide	Terengganu		50	50
S. Arabia	Christal	Yanbu	55		55
Total, Far East (excluding Japan)					*246*
Total world capacity			*2080*	*1828*	*3908*

in the range between 1.33 (water) and 1.73 (polyester fibers). The scattering power depends on the particle size, and for TiO_2 is at its maximum at a particle size of 0.2 μm (Mie's theory) [1.26]. The scattering power also depends on the wavelength; TiO_2 pigment particles with a size ≤ 0.2 μm scatter light of shorter wavelengths more strongly and therefore show a slight blue tinge, while larger particles have a yellow tone.

Hue. The whiteness (lightness and hue) of TiO_2 pigments depends primarily on the crystalline modification, the purity, and the particle size of the TiO_2 (see above). As the absorption band (385 nm) of anatase pigments is shifted into the UV region, compared with rutile pigments they have less yellow undertone. Any transition elements present in the crystal structure have an adverse effect on the whiteness, so manufacturing conditions are of the greatest importance. Thus, pigments produced by the chloride process (which includes distillative purification of $TiCl_4$ before the combustion stage) have a higher color purity and very high lightness values.

Dispersion. Good disintegration and dispersion of the TiO_2 pigments in the medium are necessary to obtain high gloss and low gloss haze. These requirements are satisfied by intensive grinding and by coating the pigment surface with organic compounds. The compounds used for this surface treatment depend on the field of application (see Section 2.1.3.4).

Lightfastness and Weather Resistance. Weathering of paints and coatings containing TiO_2 leads to pigment chalking [2.59]. If weathering occurs in the absence of oxygen, or in binders with low permeability to oxygen (e.g., in melamine–formaldehyde resins), no chalking is observed, but graying takes place, which decreases on exposure to air. Graying is greatly reduced in the absence of water. Both effects are more severe with anatase pigments. Empirical stabilization processes have been developed by pigment producers, e.g., doping with zinc or aluminum prior to calcination.

According to modern theories, impairment of the lightfastness and weather resistance of TiO_2 pigments proceeds according to the following cycle [2.60]:

1) Molecules of water are bound to the TiO_2 surface, forming hydroxyl groups on the surface.
2) Absorption of light of short wavelength (anatase < 385 nm, rutile < 415 nm) occurs, producing an electron and an electron defect or "hole" (exciton) in the crystal lattice which migrate to the surface of the pigment.
3) At the surface of the pigment, an OH^- ion is oxidized to an $OH^{\cdot}$ radical by an electron "hole". The $OH^{\cdot}$ radical is then desorbed and can oxidatively break down the binder. A Ti^{3+} ion is simultaneously produced by reduction of Ti^{4+} with the remaining electron of the exciton.
4) The Ti^{3+} ion can be oxidized by adsorbed oxygen with formation of an O_2^- ion. The latter reacts with H^+ and is converted into an $HO_2^{\cdot}$ radical.
5) The cycle ends with the binding of water to the regenerated TiO_2 surface.

The chalking process can be regarded as the reaction of water and oxygen to form $OH^{\cdot}$ and $HO_2^{\cdot}$ radicals under the influence of shortwave radiation and the catalytic activity of the TiO_2 surface:

$$H_2O + O_2 \xrightarrow[TiO_2]{h\nu} OH^{\cdot} + HO_2^{\cdot}$$

The enthalpy requirement for this reaction (312 kJ/mol) is provided by radiation of wavelength 385 nm. The cycle (1)–(5) is broken by excluding air or water. If oxygen is excluded or a binder is chosen in which the diffusion of oxygen is rate determining, a concentration of Ti^{3+} ions builds up. Graying then takes place, but this decreases with gradual exposure to oxygen. If water is excluded, rehydration and formation of surface hydroxyl groups do not take place; breakdown of the binder therefore ceases. Despite this photochemical breakdown of the binder, treated rutile pigments are used to stabilize many binders. This is because nonpigmented coatings are degraded by exposure to light and weathering; the added TiO_2 pigments prevent light from penetrating the deeper layers of the coating film and thus inhibit breakdown of the binder. High-quality TiO_2 pigments must satisfy stringent requirements with respect to weather resistance. They must withstand the severe climatic conditions of the Florida test, resisting a two-year exposure without appreciable chalking or deterioration of gloss.

2.1.6. Analysis

The crystal structure of the pigments is determined by X-ray analysis which is sensitive enough to determine 0.3–0.5% anatase in the presence of 99.7–99.5% rutile. For standards, see Table 1 (Titanium dioxide pigments; "Methods of analysis" and "Specification").

A qualitative test for TiO_2 is a blue-violet coloration of beads of microcosmic salt ($NaNH_4HPO_4 \cdot 4H_2O$), or a yellow-orange coloration produced when hydrogen peroxide is added to a test solution in hot, concentrated sulfuric acid containing ammonium sulfate. For quantitative determination, the pigment is dissolved or digested in sulfuric acid and the solution is reduced to Ti^{3+} with cadmium, zinc, or aluminum. The Ti^{3+} ions are then usually titrated with a standard solution of iron(III) ammonium sulfate solution, with potassium rhodanide as an indicator, or using potentiometric end point determination.

Impurities can be determined by wet analysis, X-ray fluorescence, or spectrographic analysis (e.g., atomic absorption).

Typical analysis figures for an untreated rutile pigment are TiO_2 99.4%, K_2O 0.24%, P_2O_5 0.21%, Fe_2O_3 40 ppm, Sb_2O_3 24 ppm, Al_2O_3 20 ppm, Mg 5 ppm, Zn 3 ppm, Cr 2 ppm, Mn, Cu, Hg, Cd, Co, Ni, Se, Sn, Ag <1 ppm.

2.1.7. Uses of Pigmentary TiO_2

Titanium dioxide is used universally, having almost completely replaced other white pigments. Consumption figures for 1996 are given in Table 17 [2.61]. The greatest annual increase in use has been for coloring plastics (5,5%), followed by the coloring of paper (3.0%). Geographically, the increase in consumption of TiO_2 has been the greatest in Asia (see Table 18).

Table 17. Consumption of TiO_2 pigments in 1996 [2.61]

Use	World total	
	[10^3 t]	[%]
Coatings	1988	59
Paper	424	13
Plastics	686	20
Other	286	8
Total	3384	100

Table 18. Predicted percentage annual growth rates for use of TiO_2 (1993–2000) [2.3]

End use	United States	Europe, Middle East, Africa	Asia and Pacific	World total
Coatings	3,4	2,0	5,0	2,5
Paper	2,0	4,0	4,0	3,0
Plastics	4,5	5,0	10,0	5,5
Total	3,0	2,5	6,5	3,3

Paints and coatings account for the largest volume of TiO_2 production. The presence of the pigment enables the protective potential of the coating material to be fully exploited. As a result of continuing developments in TiO_2 pigments, coatings only a few micrometers thick fully cover the substrate. Commercially available pigments permit paint manufacture with simple dispersion equipment, such as disk dissolvers. Organic treatment (see Section 2.1.3.4) prior to steam jet micronization yields pigments with improved gloss properties and reduced gloss haze for use in stoving enamels. Sedimentation does not occur when these products are stored, and they possess good lightfastness and weather resistance.

Printing Inks. Modern printing processes operate at coating thicknesses of < 10 μm, and therefore require the finest possible TiO_2 pigments. These very low film thicknesses are only possible with TiO_2 pigments that have a lightening (reducing) power seven times that of lithopone. Because of its neutral hue, TiO_2 is especially suitable for lightening (reducing) colored pigments.

Plastics. Titanium dioxide is used to color durable and non durable goods like toys, appliances, automobiles, furniture and packaging films. Furthermore, TiO_2 pigments absorb UV radiation with a wavelength < 415 nm and thus protect the pigmented goods from these harmful rays.

Fibers. Titanium dioxide pigments give a matt appearance to synthetic fibers, eliminating the greasy appearance caused by their translucent properties. Anatase pigments are used for this because their abrasive effect on the spinning operation is about one quarter that of the rutile pigments. The poor lightfastness of anatase pigments in polyamide fibers can be improved by treatment with manganese or vanadium phosphate.

Paper. In Europe, fillers such as kaolin, chalk, or talc are preferred as brightening agents and opacifiers in paper manufacture. Titanium dioxide pigments are suitable for very white paper that has to be opaque even when very thin (air mail or thin printing paper). The TiO_2 can be incorporated into the body of the paper or applied as a coating to give a superior quality ("art" paper).

Laminated papers are usually colored with extremely lightfast rutile pigments before being impregnated with melamine–urea resin for use as decorative layers or films.

Other areas of application for TiO_2 pigments include the enamel and ceramic industries, the manufacture of white cement, and the coloring of rubber and linoleum.

Titanium dioxide pigments are also used as UV absorbers in sunscreen products, soaps, cosmetic powders, creams, toothpaste, cigar wrappers, and in the cosmetics industry. Their most important properties are their lack of toxicity, compatibility with skin and mucous membranes, and good dispersibility in organic and inorganic solutions and binders.

Electrically-conducting TiO_2 pigments have been produced by an aftertreatment to give a coating of mixed oxides of indium and tin, or antimony and tin [2.62]. These pigments are applied to fibers used in photosensitive papers for electrophotography, and for the production of antistatic plastics.

2.1.8. Uses of Nonpigmentary TiO_2

A number of industrial products require TiO_2 starting materials with well-defined properties for a specific application. Some of the most important of these grades of titanium dioxide are those with a high specific surface area, a small particle size, and very high reactivity. Stringent requirements often exist regarding purity and property consistency. The most important applications for nonpigmentary TiO_2 are vitreous enamels, glass and glass ceramics, electroceramics, catalysts and catalyst supports, welding fluxes, colored pigments, electrical conductors, chemical intermediates like potassium fluorotitanate, structural ceramics, UV-absorbers and refractory coatings [2.63]. The annual growth rate in these markets is expected to be a few per cent.

Electroceramics. Titanates like barium, strontium, calcium and lead titanate prepared from finely divided, high-purity TiO_2 hydrolysates are used in capacitors, PTC-resistors and piezoelectric materials. The specifications of the TiO_2 starting materials with respect to purity, reactivity, and sintering properties are expected to become more stringent. The market is estimated to be several thousand of tonnes a year as TiO_2. A strong annual growth is expected.

Catalysts. Titania is an active catalyst for different reactions, inorganic and organic, thermal and photochemical. It may be self-supported, or it may be supported on other material. For catalysis it is usually doped with other elements in order to enhance the desired effect.

The most important application for TiO_2 is the removal of nitrogen oxides from waste gases from power stations and industry. The nitrogen oxides in the waste gas react with ammonia in the presence of oxygen over the catalyst to produce nitrogen and water (SCR, Selective Catalytic Reduction) [2.64]. The world market for SCR catalysts is believed to be several thousand tonnes a year as TiO_2. In addition to TiO_2 the catalysts usually contain ca. 10 wt% tungsten oxide and 1 wt% V_2O_5, and are extruded into a honeycomb shape or supported in thin layers on metal sheets. The TiO_2 has high specifications with respect to purity, particle size, and porosity to ensure that the desired catalytic activity is obtained.

An increasing demand is expected for the removal of nitrogen oxides from waste gases from stationary and instationary diesel engines for e.g. emergency plants, ships and trucks. Another application is the combined removal of nitrogen oxides and dioxines from waste gases of refuse disposal plants [2.65].

Mixed Metal Oxide Pigments. The starting material is a TiO_2 hydrolysate, which is calcined with oxides of transition metals to form chromium rutile or nickel rutile pigments (see Section 3.1.3.1).

UV Absorption. Nanostructured TiO_2 particles (5–50 nm particle size) are used as sunscreen filter in the cosmetic industry for skin protection. Nanosized TiO_2 is an effective UV-absorber against UV-B (280–320 nm) and UV-A rays (320–400 nm). Because of its small particle size it appears transparent [2.66]. Intensive research work is in progress worldwide aimed at utilizing the photoactivity of TiO_2. Titanium dioxide catalyzes the decomposition of organic compounds in wastewater [2.67]; water is decomposed into hydrogen and oxygen in the presence of sunlight.

2.1.9. Toxicology

Titanium dioxide is highly stable and is regarded as completely nontoxic. Investigations on animals which have been fed TiO_2 over a long period give no indication of titanium uptake [2.68]. Absorption of finely divided TiO_2 pigments in the lungs does not have any carcinogenic effect [2.69].

2.2. Zinc Sulfide Pigments [2.70]–[2.72]

White pigments based on zinc sulfide were first developed and patented in 1850 in France. Although they are still of economic importance, they have continually lost market volume since the early 1950s when titanium dioxide was introduced. Only one modern production installation for zinc sulfide pigments still exists in the market-economy-oriented industrial nations (Sachtleben Chemie, Germany). There are other production plants in eastern Europe and in China.

The zinc-sulfide-containing white pigment with the largest sales volume is *lithopone* [*1345-05-7*], which is produced by coprecipitation and subsequent calcination of a mixture of zinc sulfide [*1314-98-3*], ZnS, M_r 97.43, and barium sulfate [*7727-43-7*], $BaSO_4$, M_r 233.40. For standards, see Table 1 (Lithopone pigments; "Specification"). Pure zinc sulfide is marketed as *Sachtolith*.

White zinc sulfide pigments maintain their market position in areas of use where not only their good light scattering ability but other properties such as low abrasion, low oil number, and low Mohs hardness are required.

They are often produced from many types of industrial waste. This recycling relieves pressure on the environment, as these materials would otherwise have to be disposed of.

2.2.1. Properties

Some physical and chemical properties of ZnS and $BaSO_4$ are given in Table 19.

A white pigment must not absorb light in the visible region (wavelength 400–800 nm), but should disperse incident radiation in this region as completely as possible. The spectral reflectance curves of zinc sulfide and barium sulfate (Fig. 19) fulfil these conditions to a large extent. The absorption maximum for ZnS at ca. 700 nm is a result of lattice stabilization with cobalt ions, whose function is explained in Section 2.2.2. The absorption edge in the UV-A region is responsible for the bluish-white tinge of zinc sulfide. Depending on the production process, zinc sulfide has a sphalerite or wurtzite lattice type.

The refractive index n of ZnS, which determines its scattering properties, is 2.37 and is much greater than that of plastics and binders ($n = 1.5$–1.6). Spheroidal ZnS particles have their maximum scattering power at a diameter of 294 nm. Barium sulfate does not directly contribute to the light scattering due to its relatively low refractive index ($n = 1.64$), but acts as an extender, and increases the scattering efficiency of the ZnS.

Table 19. Properties of the components of zinc sulfide pigments

Property	Zinc sulfide	Barium sulfate*
Physical properties		
Refractive index n	2.37	1.64
Density, g/cm^3	4.08	4.48
Mohs hardness	3	3.5
Solubility in water (18 °C), wt%	1.8×10^{-4}	2.5×10^{-4}
Chemical properties		
Resistance to acids/bases	soluble in strong acids	insoluble
Resistance to organic solvents	insoluble	insoluble

* Component of lithopone.

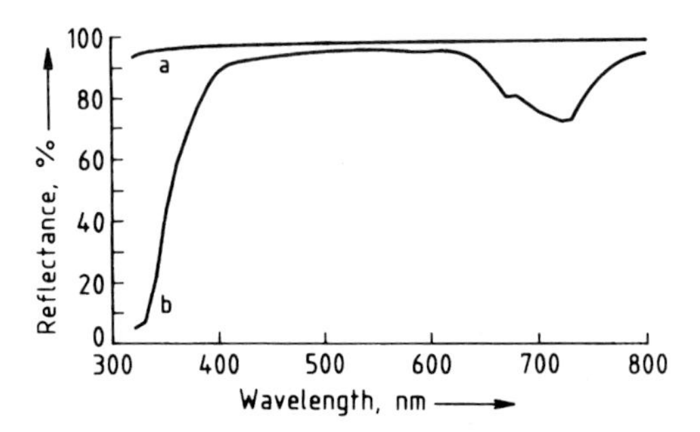

Figure 19. Spectral reflectance curves of barium sulfate and zinc sulfide
a) Barium sulfate; b) Zinc sulfide (Co-doped)

The barium sulfate in lithopone can be identified thermoanalytically by a reversible endothermic transformation at 1150 °C. Both Sachtolith and lithopone are thermally stable up to ca. 550 °C in the presence of air. Due to their low Mohs hardness, they are less abrasive than other white pigments. Barium sulfate is practically inert toward acids, bases, and organic solvents. Zinc sulfide is stable in aqueous media between pH 4 and 10, and is largely inert toward organic media. In the presence of water and oxygen, it can be oxidatively decomposed by the action of UV radiation.

2.2.2. Production

Raw Materials. The source of zinc can be zinc oxide from a smelter, zinc dross or sweepings, ammonium chloride slag from hot dip galvanizing, or liquid waste such as pickle liquors from galvanizing plants. Variations in the price of zinc have a large effect on the economics of zinc sulfide pigments.

The starting material for water-soluble barium compounds is fused barium sulfide produced by coke reduction of naturally occurring barite with a low silica and strontium content. Suitable barite is readily available from many deposits worldwide.

Lithopone. The reaction of equimolar quantities of $ZnSO_4$ and BaS produces a white, water-insoluble coprecipitate with the theoretical composition 29.4 wt% ZnS and 70.6 wt% $BaSO_4$:

$$ZnSO_4 + BaS \rightarrow ZnS + BaSO_4 \quad (1)$$

By using a different molar ratio, this composition can be changed; for example, precipitation according to Equation (2) gives a product containing 62.5 wt% ZnS and 37.5 wt% $BaSO_4$:

$$ZnSO_4 + 3\,ZnCl_2 + 4\,BaS \rightarrow 4\,ZnS + BaSO_4 + 3\,BaCl_2 \quad (2)$$

Figure 20 is a flow diagram of lithopone production. The *solutions of zinc salts* contain impurities (e.g., salts of iron, nickel, chromium, manganese, silver, cadmium) that depend on their origins. The main sources of zinc sulfate solutions are zinc electrolyses and the reprocessing of zinc scrap and zinc oxide. The first stage of purification consists of chlorination. Iron and manganese are precipitated as oxide-hydroxides, and cobalt, nickel, and cadmium as hydroxides. The solutions are then mixed with zinc dust at 80 °C. All the elements more noble than zinc (cadmium, indium, thallium, nickel, cobalt, lead, iron, copper, and silver) are almost completely precipitated, while zinc goes into solution. The metal slime is filtered off and taken to copper smelters for recovery of the noble metal components. A small quantity of a water-soluble cobalt salt is added to the purified zinc salt solution. The cobalt (0.02–0.5%) becomes incorporated into the ZnS lattice during subsequent calcination to stabilize the final product against light. Zinc sulfide that is not treated in this way becomes gray in sunlight.

The *barium sulfide solution* is produced by dissolving fused barium sulfide in water. The barium sulfide is obtained by reducing an intimate mixture of crushed barite (ca. 1 cm lumps) with petroleum coke according to Equation (3) in a directly heated rotary kiln at 1200–1300 °C:

$$BaSO_4 + 2\,C \rightarrow BaS + 2\,CO_2 \tag{3}$$

The warm solution (60 °C) containing ca. 200 g/L barium sulfide is filtered and immediately pumped to the precipitation stage. Further purification is not necessary. Unreacted gangue and heavy metals are collected as insoluble sulfides in the filter cake. The almost clear solution can be stored only for a short period. Longer storage leads to undesirable polysulfide formation.

The zinc salt and BaS solutions are mixed thoroughly under controlled conditions (vessel geometry, temperature, pH, salt concentration, and stirring speed, see (a) in Fig. 20). The precipitated "raw lithopone" does not possess pigment properties. It is filtered off (b_1) and dried (c); ca. 2 cm lumps of the material are calcined in a rotary kiln (d) directly heated with natural gas at 650–700 °C. Crystal growth is controlled by adding 1–2 wt% NaCl, 2 wt% Na_2SO_4 and traces of Mg^{2+} (ca. 2000 ppm), and K^+ (ca. 100–200 ppm). The temperature profile and residence time in the kiln are controlled to obtain ZnS with an optimum particle size of ca. 300 nm.

The hot product from the kiln is quenched in water (e), passed via classifiers and hydroseparators (f) into thickeners (g), filtered on rotary filters (b_2), and washed until salt-free. The dried product is ground in high-intensity mills (g) and may undergo organic treatment (with a polyalcohol) depending on the application.

Figure 21 shows a scanning electron micrograph of lithopone. The ZnS and $BaSO_4$ particles can be distinguished by means of their size. The average particle diameter of $BaSO_4$ is 1 µm.

Sachtolith. Production is similar to that of lithopone. For process engineering reasons, a sodium sulfide solution is used as the sulfide component, and is formed according to Equation (4):

$$BaS + Na_2SO_4 \rightarrow Na_2S + BaSO_4 \tag{4}$$

The $BaSO_4$, which is produced as a byproduct, is filtered, washed, dried, and ground. It is a high-quality extender (Blanc fixe) used in the paint industry.

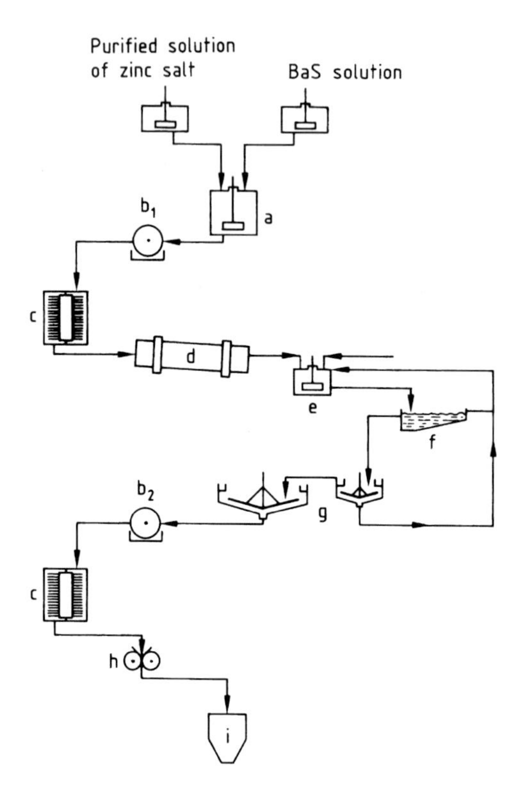

Figure 20. Flow diagram for lithopone production
a) Precipitation vessel; b) Rotary filter; c) Turbo dryer; d) Rotary kiln; e) Chilling vessel; f) Rake classifier; g) Thickener; h) Grinder; i) Silo

The Na_2S solution is mixed with a cobalt-treated zinc salt solution under precisely controlled conditions. The resulting zinc sulfide precipitate is calcined and processed to give the finished product.

Hydrothermal Process. Crystal growth of ZnS can be achieved by a hydrothermal process instead of by calcination. The raw lithopone is precipitated with a slight excess of sulfide at pH 8.5. The pH is then adjusted to 12–13 with sodium hydroxide solution, and 0.5% sodium carbonate is added. The suspension is then autoclaved for ca. 15–20 min at 250–300 °C. In contrast to the wurtzite structure of the calcined product, the hydrothermal product has a sphalerite structure with a ca. 10% greater scattering power. Although the product is of better quality, the hydrothermal process is less economic due to the high cost of the materials required for lining the autoclave (e.g., tantalum or a zirconium alloy).

Environmental Protection. During the reduction of barite and the calcination of Sachtolith and lithopone, sulfur dioxide is liberated. This is removed from the waste gas in a purification stage which is based on the reversible, temperature-dependent solubility of sulfur dioxide in polyglycol. The absorbed sulfur dioxide can be recovered as a liquid product or as a raw material for sulfuric acid. Any soluble barium in the residue from the dissolution of the fused BaS is removed by treatment with

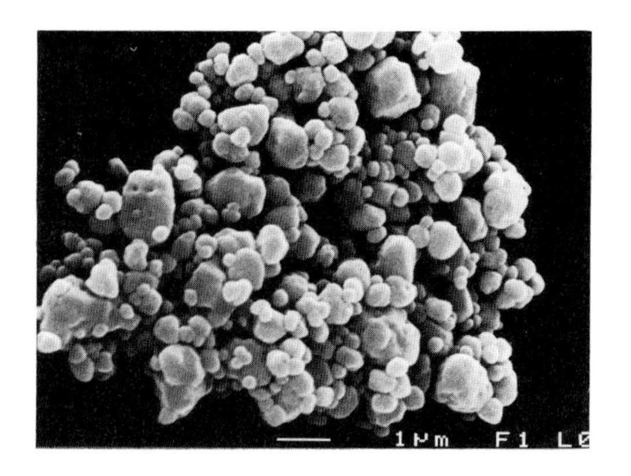

Figure 21. Scanning electron micrograph of lithopone. The larger particles are barium sulfate (mean size 1.0 μm) and the smaller particles are zinc sulfide (mean size 0.3 μm).

iron-containing waste hydrochloric acid. The gangue, in the form of a slime, is water-insoluble and can therefore be disposed of. The barium chloride solution is used as a raw material for the production of precipitated barium sulfate fillers.

2.2.3. Commercial Products

Commercial lithopone grades contain 30% ZnS (red seal) and 60% ZnS (silver seal). The ZnS content of Sachtolith is > 97%. Various chemical (e.g., polyalcohols, siloxanes, silanes) and mechanical treatments (e.g., jet milling) are used to obtain other products for special applications. The technical data for commercial red seal lithopone and Sachtolith are given in Table 20.

Table 20. Technical data for red seal lithopone and Sachtolith

Parameter	Standard	Sachtolith	Red seal lithopone
ZnS, wt%	DIN 55910	≥ 97	ca. 29
ZnO, wt%	DIN 55910	0.2	0.1
$BaSO_4$, wt%	DIN 55910	≤ 3	ca. 70
Brightness*		98	98–99
Lightening power	DIN 53192	400	120
Water-soluble salts, wt%	DIN 53197	< 0.2	< 0.2
Sieve residue**, wt%		0.02	< 0.1
Oil number	DIN 53192	12	8
pH	DIN 53200	6–8	7
Specific surface area, m^2/g		8	3

* $BaSO_4$ white standard = 100. ** Test sieve 45 μm, DIN 4188, ISO DIS 3310/1.

2.2.4. Uses

Lithopone is mainly used in *coating materials* with relatively high pigment concentrations (Table 21). Examples are primers, plastic masses, putties and fillers, artists' colors, and emulsion paints. An important property of lithopone is its low binder requirement, giving paint products with good flow and application properties. It is suitable for almost all binder media, and has good wetting and dispersing properties. With optimum feed composition, good dispersion can be achieved simply by the action of a dissolver. It can be economically advantageous to use lithopone in combination with TiO_2 pigments; the good hiding power of the TiO_2 pigments is combined with the economic and technical advantages of lithopone. Due to the strong shift of the absorption band towards the blue, lithopone is especially useful as a white pigment for UV-cured paint systems. Zinc compounds have a fungicidal and algicidal action, and inclusion of lithopone or Sachtolith in paint formulations for exterior use therefore prevents attack by algae or fungi.

The material advantages of lithopone are used in *plastics* (e.g., good lightfastness and clear bluish-white shade). The product also imparts very good extruding properties to the plastic resulting in high throughput rates and economic extruder operation. In fire-resistant systems, ca. 50 % of the flame retardant antimony trioxide can be replaced by nontoxic lithopone without any adverse effect.

Sachtolith is mainly used in plastics (Table 21). Functional properties such as lightening and hiding power are criteria for the use of Sachtolith. It has proved to be very useful for coloring many thermoplasts. During the dispersion process it does not cause abrasion of metallic production machinery or adversely effect the polymer, even at high operating temperatures or during multistage processing. Even ultrahigh molecular mass thermoplasts can be colored without problems. In glass-fiber-reinforced plastics, the soft texture of Sachtolith prevents mechanical fiber damage during extrusion. Sachtolith is also used as a dry lubricant during the fabrication of these materials.

The low abrasiveness of Sachtolith prolongs the operating life of stamping tools used in the manufacture of industrial rubber articles. The lightfastness and ageing resistance of many elastomers are improved by Sachtolith. It is also used as a dry lubricant for roller and plain bearings, and as a white pigment for greases and oils.

Table 21. Uses of Sachtolith and lithopone (as percentage of total consumption)

Use	Sachtolith	Lithopone	
		Western Europe	World
Paints	20	79	94
Plastics	64	17	2
Lubricants	6		
Others	10	4	4

2.2.5. Economic Aspects

Total world production of lithopone in 1990 was 220×10^3 t. This was subdivided as follows (10^3 t):

Germany	30 (+7 Sachtolith)
Yugoslavia	5
Czechoslovakia	15
Soviet Union	50
People's Republic of China	120

Only estimated figures are available for the Soviet Union and the People's Republic of China.

A decrease in output is to be expected because replacement by TiO_2 is not yet at an end, especially in coating materials. In the long term, only the high-quality grades can maintain their place in the market, i.e., those in which technical properties are required in addition to light scattering.

2.2.6. Toxicology

The use of zinc sulfide and barium sulfate in contact with foods is permitted by the FDA (United States) and in most European countries. Some restrictions apply in France, Italy, the United Kingdom, and Czechoslovakia.

Soluble zinc is toxic in large amounts, but the human body requires small quantities (10–15 mg/d) for metabolism. Zinc sulfide is completely harmless in the human due to its low solubility. The acid concentration in the stomach and the rate of dissolution following ingestion are not sufficient to produce physiologically significant quantities of soluble zinc. LD_{50} values in the rat exceed 20 g/kg. No cases of poisoning or chronic damage to health have been observed in the manufacture of zinc sulfide pigments despite exposure to dust that occurs mainly during grinding and packing.

2.3. Zinc Oxide (Zinc White) [2.73]–[2.77]

2.3.1. Introduction

Zinc oxide [*1314-13-2*], ZnO, M_r 81.38, was formerly used only as a white pigment, and was named zinc white (C.I. Pigment White 4), Chinese white, or flowers of zinc. The term zinc white now denotes zinc oxide produced by the combustion of zinc metal according to the indirect or French process.

Historical Aspects. Zinc oxide has long been known as a byproduct of copper smelting. The Romans called it "cadmia" and used it as such in the production of brass. They also purified it for use in ointments by reduction followed by oxidation. In the Middle Ages, the alchemists thought that cadmia could be converted into gold.

In the mid-18th century, the German chemist CRAMER discovered that cadmia could be obtained by the combustion of metallic zinc. COURTOIS began to produce zinc white in 1781 in France, but it was not until 1840 that industrial production was started by LECLAIRE (indirect or French process). The use of this white pigment spread rapidly. Zinc oxide replaced white lead because it had the advantage of being nontoxic, of not darkening in the presence of sulfurous gases, and of having better hiding power.

Around 1850, S. WETHERILL of the New Jersey Zinc Company perfected a roasting furnace in which a grate was charged with coal and then covered with a mixture of zinc ore and coal. The zinc was reduced by the partial combustion of the coal and reoxidized at the furnace exit (direct or American process). These furnaces were subsequently improved but are now no longer used. During the second half of the 19th century, the use of ZnO in rubber was introduced to reduce the time needed for vulcanization. The discovery of the first organic accelerators for vulcanization in 1906 added to the importance of ZnO, which acts as an activator in these materials.

A third industrial production process exists but this wet process is less widely used.

2.3.2. Properties

Physical Properties. Zinc oxide is a fine white powder that turns yellow when heated above 300 °C. It absorbs UV light at wavelengths below 366 nm. Traces of monovalent or trivalent elements introduced into the crystal lattice impart semiconducting properties. The elementary particles of ZnO obtained by the thermal method may be granular or nodular (0.1–5 μm) or acicular (needle-shaped). Some physical properties are given below:

Density	5.65–5.68 g/cm^3
Refractive index	1.95–2.1
mp	1975 °C
Heat capacity	
25 °C	40.26 J mol^{-1} K^{-1}
100 °C	44.37 J mol^{-1} K^{-1}
1000 °C	54.95 J mol^{-1} K^{-1}
Thermal conductivity	25.2 W m^{-1} K^{-1}
Crystal structure	hexagonal, wurtzite
Mohs hardness	4–4.5

Chemical Properties. Zinc oxide is amphoteric; it reacts with organic and inorganic acids, and also dissolves in alkalis and ammonia solution to form zincates. It combines readily with acidic gases (e.g., CO_2, SO_2, and H_2S). It reacts at high temperatures with other oxides to form compounds such as zinc ferrites.

2.3.3. Production

About 1–2 % of zinc oxide is produced by the wet process, 10–20 % by the direct process, and the remainder by the indirect process.

Raw Materials. In the early days, the raw materials were mainly zinc ores or concentrates for the direct process, or metal from zinc producers for the indirect process. Nowadays, zinc oxide manufacturers mainly use residues and secondary zinc. This fact, combined with the demand for chemical purity imposed by the users, means that processes have had to be modified and a number of purification techniques are used.

Direct or American Process. The direct process is noted for its simplicity, low cost, and excellent thermal efficiency. It consists of an initial high-temperature reduction (1000–1200 °C) of a zinc-containing material (as oxide), the reducing agent being coal. Reduction takes place according to Boudouard's equations:

$$ZnO + C \rightleftharpoons Zn + CO$$
$$ZnO + CO \rightleftharpoons Zn + CO_2$$
$$C + O_2 \rightleftharpoons CO_2$$
$$CO_2 + C \rightleftharpoons 2CO$$

The zinc vapor and the CO gas are then oxidized to zinc oxide and carbon dioxide above the reaction bed or at the furnace exit. Various zinc-containing materials are used, e.g., zinc concentrates, metallization residues, byproduct zinc hydroxide, and above all zinc dross from casting furnaces or galvanizing. The dross must first be treated to remove chloride and lead by heating at ca. 1000 °C in rotary kilns.

Only rotary kilns are now used for the direct process; the use of static furnaces has been discontinued. The zinc content of raw materials is between 60 and 75 %. There are two types of rotary kiln:

1) One type is a long (ca. 30 m), fairly narrow (2.5 m diameter) kiln, heated by gas or oil. The raw material (a mixture of zinc-containing material and coal) is charged continuously either countercurrent or cocurrent to the combustion gases. The residues, which still contain some zinc and unburnt coal, leave the furnace continuously at the end opposite to the feed end. The excess coal is sieved out and recycled. The combustion gases, containing zinc vapor, ZnO, and CO, pass into a chamber where oxidation is completed and large particles of impurities settle out. The gases are then cooled in a heat exchanger or by dilution with air. The zinc oxide is collected in bag filters.
2) The second type of rotary kiln is shorter (5 m) and has a larger diameter (ca. 3 m). Charging is continuous, but the dezincified residues are removed batchwise.

In both cases, operating conditions are controlled to obtain a high yield and to give the required particle shape and size. Provided no contamination is introduced, chemical purity is determined solely by the composition of the raw materials used.

Indirect or French Process. The zinc is boiled, and the resulting vapor is oxidized by combustion in air under defined conditions. The crystallographic and physical properties of the ZnO can be controlled by adjustment of the combustion conditions (e.g., flame turbulence and air excess). The chemical composition of the ZnO is solely a function of the composition of the zinc vapor.

Many types of furnace are available to produce vapor of the required purity from various raw materials and obtain a high yield of zinc. Pure zinc (super high grade, SHG; high grade, HG) or, to an increasing extent, metal residues (e.g., scrap zinc, die casting dross, or galvanizer's dross) are used as raw materials. Various liquid- or vapor-phase separation techniques are used for separating Cd, Pb, Fe, and Al from zinc metal before it is oxidized.

1) *Muffle Furnaces or Retorts of Graphite or Silicon Carbide*. The metal is fed into the furnace either batchwise as a solid or continuously as a liquid. The heat of vaporization is supplied by heating the outside of the retort with a burner. The nonvolatile residues (iron and lead in the case of dross from smelting) accumulate in the retort and must be removed at intervals. This is facilitated by tipping the retorts.
2) *Fractional Distillation*. The vapor, containing Cd, Pb, Fe, Al, and Cu, can be purified by fractional distillation in columns (New Jersey Zinc Co.) with silicon carbide plates. Oxidation takes place at the exit of the column.
3) *Furnaces with Two Separate Chambers*. The metallic raw material, which can be in large pieces, is fed into the first chamber where it melts. This is connected to the second, electrically heated chamber where distillation takes place in the absence of air. The first version of this type of furnace was constructed by LUNDEVALL [2.78].

   The nonmetallic residues are removed at the surface of the melting chamber. Impurities, such as Fe, Al, and some of the Pb, accumulate in the distillation chamber and are periodically removed in the liquid state. The last traces of lead are then removed by fractional distillation.
4) *Smelting Process in a Rotary Kiln*. Indirect zinc oxide is also made by smelting in a rotary kiln, starting from the same raw materials. Melting, distillation, and part of the oxidation all take place in the same zone, allowing utilization of a large part of the heat of combustion of the zinc. By controlling the temperature and partial pressures of carbon dioxide and oxygen, the impurity content (Pb) can be limited and the shape and size of the ZnO particles can be adjusted, though to a lesser extent than in the other processes.

Wet Process. Zinc oxide is also produced industrially from purified solutions of zinc sulfate or chloride by precipitating the basic carbonate, which is then washed, filtered, and finally calcined. This method produces a grade of zinc oxide with a high specific surface area.

Products of this type are also obtained from waste hydroxides which are purified by a chemical route and then calcined.

Aftertreatment. Thermal treatment at temperatures up to 1000 °C improves the pigment properties of the ZnO and is mainly applied to oxide produced by the direct method. Controlled atmospheric calcination also improves the photoconducting properties of the high-purity oxide used in photocopying.

The ZnO surface is made more organophilic by coating it with oil and propionic acid. The ZnO is often deaerated and sometimes pelletized or granulated to improve handling properties.

2.3.4. Quality Specifications

Many standard specifications have been laid down for the more important uses of ZnO (rubber, paints, and the pharmaceutical industry). For standards, see Table 1 ("Zinc oxide pigments; Methods of analysis" and "Specification"). Various methods of classification are used, often based on the production process and the chemical

Table 22. Classification of commercially available grades of zinc oxide

Parameter	A	B	C	D
	Indirect process	Indirect process	Direct process	Wet process
ZnO (min.), wt%	99.5	99	98.5	93
Pb (max.), wt%	0.004	0.25	0.25	0.001
Cd (max.), wt%	0.001	0.05	0.03	0.001
Cu (max.), wt%	0.0005	0.003	0.005	0.001
Mn (max.), wt%	0.0005	0.001	0.005	0.001
Water-soluble salts (max.), wt%	0.02	0.1	0.65	1
Loss on ignition (max.), wt%	0.3	0.3	0.3	4
Acidity, g H_2SO_4/100 g	0.01	0.1	0.3	0.2
Specific surface area, m^2/g	3–8	3–10	1–5	25 (min.)

composition. The most well-known are pharmacopeias, RAL 844 C3 (Reichs-Ausschuß für Lieferbedingungen), ASTM D79, BS 254, DAB 8 (Deutsches Arzneibuch), T 31 006 NF (French standard), and ISO R 275. Table 22 shows the classification of commercially available zinc oxide grades.

Classification based on color codes is common in Europe, but is of limited value. Manufacturers have their own standards. In general, the terms silver seal and white seal indicate category A, and red seal category B.

2.3.5. Uses

Zinc oxide has many uses. By far the most important is in the *rubber industry*. Almost half the world's ZnO is used as an activator for vulcanization accelerators in natural and synthetic rubber. The reactivity of the ZnO is a function of its specific surface area, but is also influenced by the presence of impurities such as lead and sulfates. The ZnO also ensures good durability of the vulcanized rubber, and increases its thermal conductivity. The ZnO content is usually 2–5%.

In *paints and coatings*, zinc oxide is no longer the principal white pigment, although its superb white color is used by artists. It is used as an additive in exterior paints for wood preservation. It is also utilized in antifouling and anticorrosion paints (see Section 5.2.10.4) [2.79]. It improves film formation, durability, and resistance to mildew (having a synergistic effect with other fungicides) because it reacts with acidic products of oxidation and can absorb UV radiation.

The *pharmaceutical and cosmetic industries* use ZnO in powders and ointments because of its bactericidal properties. It is also used to form dental cements by its reaction with eugenol.

In the field of *glass, ceramics, and enamels*, ZnO is used for its ability to reduce thermal expansion, to lower the melting point, and to increase chemical resistance. It can also be used to modify gloss or to improve opacity.

Zinc oxide is used as a *raw material* for many products: stearates, phosphates, chromates, bromates, organic dithiophosphates, and ferrites (ZnO, MnO, Fe_2O_3). It is used as a source of zinc in animal feeds and in electrogalvanization. It is also used for desulfurizing gases.

Zinc oxide is used as a *catalyst* in organic syntheses (e.g., of methanol), often in conjunction with other oxides. It is present in some adhesive compositions.

The highest purity material is calcined with additives such as Bi_2O_3 and used in the manufacture of varistors [2.80]. The photoconducting properties of ZnO are used in photoreproduction processes. Doping with alumina causes a reduction in electrical resistance; hence, it can be used in the coatings on the master papers for offset reproduction [2.81].

2.3.6. Economic Aspects

The consumption of zinc oxide in Western Europe in 1990 was estimated to be 160 000 t. Annual world consumption is in the region of 500 000 t, representing ca. 10% of the total world zinc production. The rubber industry consumes ca. 45% of the total and the remainder is divided among a large number of industries.

2.3.7. Toxicology and Occupational Health

Unlike other heavy metals, zinc is not considered to be toxic or dangerous. It is an essential element for humans, animals, and plants. The human body contains ca. 2 g, and it is recommended that 10–20 mg should be ingested per day [2.82]. The oral LD_{50} value for rats is 630 mg/kg. The permitted concentration of the dust in air at the workplace is 5 mg/m^3 (MAK), 10 mg/m^3 (TLV-TWA). Values for zinc oxide fumes are 5 mg/m^3(TLV-TWA) and 10 mg/m^3 (TLV-STEL).

If large quantities of ZnO are accidentally ingested or inhaled, fever, nausea, and irritation of the respiratory tract ensue after several hours. These symptoms rapidly disappear without long-term consequences.

3. Colored Pigments

Colored pigments differ from black and white pigments in that their absorption and scattering coefficients are wavelength dependent with widely varying absolute values. The dependence of these coefficients on wavelength, particle size, particle shape, and their distributions determine the color, tinting strength, and hiding power of the pigments (see Section 1.3.1).

3.1. Oxides and Hydroxides

In many inorganic pigments, lanthanides and transition elements are responsible for color. Metal oxides and oxide hydroxides are, however, also important as colored pigments because of their optical properties, low price, and ready availability. Colored pigments based on oxides and oxide hydroxides are either composed of a single component or mixed phases. In the latter, color is obtained by incorporation of appropriate cations.

3.1.1. Iron Oxide Pigments

The continually increasing importance of iron oxide pigments is based on their nontoxicity; chemical stability; wide variety of colors ranging from yellow, orange, red, brown, to black; and low price. Natural and synthetic iron oxide pigments consist of well-defined compounds with known crystal structures [3.1]:

1) α-FeOOH, goethite [*1310-14-1*], diaspore structure, color changes with increasing particle size from green-yellow to brown-yellow
2) γ-FeOOH, lepidocrocite [*12022-37-6*], boehmite structure, color changes with increasing particle size from yellow to orange
3) α-Fe_2O_3, hematite [*1317-60-8*], corundum structure, color changes with increasing particle size from light red to dark violet
4) γ-Fe_2O_3, maghemite [*12134-66-6*], spinel super structure, ferrimagnetic, color: brown
5) Fe_3O_4, magnetite [*1309-38-2*], spinel structure, ferrimagnetic, color: black

Mixed metal oxide pigments containing iron oxide are also used (see Section 3.1.3). Magnetic iron oxide pigments are discussed in Sections 5.1.1 and 5.1.2. Transparent iron oxide pigments are described in Section 5.4.1. Methods of analysis and specifications of iron oxide pigments are listed in the standards given in Table 1.

3.1.1.1. Natural Iron Oxide Pigments

Naturally occurring iron oxides and iron oxide hydroxides were used as pigments in prehistoric times (Altamira cave paintings) [3.2]. They were also used as coloring materials by the Egyptians, Greeks, and ancient Romans.

Hematite (α-Fe_2O_3) has attained economic importance as a red pigment, geothite (α-FeOOH) as yellow, and the umbers and siennas as brown pigments. Deposits with high iron oxide contents are exploited preferentially. Naturally occurring magnetite (Fe_3O_4) has poor tinting strength as a black pigment, and has found little application in the pigment industry.

Hematite is found in large quantities in the vicinity of Malaga in Spain (Spanish red) and near the Persian Gulf (Persian red). The Spanish reds have a brown undertone. Their water-soluble salt content is very low and their Fe_2O_3 content often exceeds 90%. The Persian reds have a pure hue, but their water-soluble salt content is disadvantageous for some applications. Other natural hematite deposits are of only local importance. A special variety occurs in the form of platelets and is extracted in large quantities in Kärnten (Austria). This micaceous iron oxide, is mainly used in corrosion protection coatings.

Goethite is the colored component of yellow ocher; a weathering product mainly of siderite, sulfidic ores, and feldspar. It occurs in workable amounts mainly in the Republic of South Africa and France. The Fe_2O_3 content gives an indication of the iron oxide hydroxide content of the ocher, and is ca. 20% in the French deposits and ca. 55% in the South African.

Umbers are mainly found in Cyprus. In addition to Fe_2O_3 (45–70%), they contain considerable amounts of manganese dioxide (5–20%). In the raw state, they are deep brown to greenish brown and when calcined are dark brown with a red undertone (burnt umbers).

Siennas, mainly found in Tuscany, have an average Fe_2O_3 content of ca. 50%, and contain $< 1\%$ manganese dioxide. They are yellow-brown in the natural state and red-brown when calcined [3.3].

The processing of natural iron oxide pigments depends on their composition. They are either washed, slurried, dried, ground, or dried immediately and then ground in ball mills, or more often in disintegrators or impact mills.

Siennas and umbers are calcined in a directly fired furnace, and water is driven off. The hue of the products is determined by the calcination period, temperature, and raw material composition [3.4].

Natural iron oxide pigments are mostly used as inexpensive marine coatings or in coatings with a glue, oil, or lime base. They are also employed to color cement, artificial stone, and wallpaper. Ocher and sienna pigments are used in the production of crayons, drawing pastels, and chalks [3.5].

For standards, see Table 1 (Iron, manganese oxide pigments: "Methods of analysis"; "Natural, specification"; "Sienna, specification"; and "Umber, specification").

The economic importance of the natural iron oxide pigments has decreased in recent years in comparison with the synthetic materials.

3.1.1.2. Synthetic Iron Oxide Pigments

Synthetic iron oxide pigments have become increasingly important due to their pure hue, consistent properties, and tinting strength. Single-component forms are mainly produced with red, yellow, orange, and black colors. Their composition corresponds to that of the minerals hematite, geothite, lepidocrocite, and magnetite. Brown pigments usually consist of mixtures of red and/or yellow and/or black; homogeneous brown phases are also produced, e.g., $(Fe, Mn)_2O_3$ and γ-Fe_2O_3, but quantities are small in comparison to the mixed materials. Ferrimagnetic γ-Fe_2O_3 is of great importance for magnetic recording materials (see Sections 5.1.1 and 5.1.2).

Several processes are available for producing high-quality iron oxide pigments with controlled mean particle size, particle size distribution, particle shape, etc. (Table 23):

1) Solid-state reactions (red, black, brown)
2) Precipitation and hydrolysis of solutions of iron salts (yellow, red, orange, black)
3) Laux process involving reduction of nitrobenzene (black, yellow, red)

The raw materials are mainly byproducts from other industries: steel scrap obtained from deep drawing, grindings from cast iron, $FeSO_4 \cdot 7\,H_2O$ from TiO_2 production or from steel pickling, and $FeCl_2$ also from steel pickling.

Iron oxides obtained after flame spraying of spent hydrochloric acid pickle liquor, red mud from bauxite processing, and the product of pyrites combustion are no longer of importance. They yield pigments with inferior color properties that contain considerable amounts of water-soluble salts. They can therefore only be used in low-grade applications.

Solid-State Reactions of Iron Compounds. Black iron oxides obtained from the Laux process (see below) or other processes may be calcined in rotary kilns with an oxidizing atmosphere under countercurrent flow to produce a wide range of different red colors, depending on the starting material. The pigments are ground to the desired particle size in pendular mills, pin mills, or jet mills, depending on their hardness and intended use.

The calcination of yellow iron oxide produces pure red iron oxide pigments with a high tinting strength. Further processing is similar to that of calcined black pigments.

High-quality pigments called copperas reds are obtained by the thermal decomposition of $FeSO_4 \cdot 7\,H_2O$ in a multistage process (Fig. 22). If an alkaline-earth oxide or carbonate is included during calcination, the sulfate can be reduced with coal or carbon-containing compounds to produce sulfur dioxide, which is oxidized with air

Table 23. Reaction equations for the production of iron oxide pigments

Color	Reaction			Process
Red	$6\ FeSO_4 \cdot x\ H_2O + 1\frac{1}{2}\ O_2$	$\longrightarrow$	$Fe_2O_3 + 2\ Fe_2(SO_4)_3 + 6\ H_2O$	copperas process
	$2\ Fe_2(SO_4)_3$	$\longrightarrow$	$2\ Fe_2O_3 + 6\ SO_3$	(copperas process)
	$2\ Fe_3O_4 + \frac{1}{2}\ O_2$	$\longrightarrow$	$3\ Fe_2O_3$	calcination
	$2\ FeOOH$	$\longrightarrow$	$Fe_2O_3 + H_2O$	calcination
	$2\ FeCl_2 + 2\ H_2O + \frac{1}{2}\ O_2$	$\longrightarrow$	$Fe_2O_3 + 4\ HCl$	Ruthner process
	$2\ FeSO_4 + \frac{1}{2}\ O_2 + 4\ NaOH$	$\longrightarrow$	$Fe_2O_3 + 2\ Na_2SO_4 + 2\ H_2O$	precipitation
Yellow	$2\ FeSO_4 + 4\ NaOH + \frac{1}{2}\ O_2$	$\longrightarrow$	$2\ \alpha\text{-}FeOOH + 2\ Na_2SO_4 + H_2O$	precipitation
	$2\ Fe + 2\ H_2SO_4$	$\longrightarrow$	$2\ FeSO_4 + 2\ H_2$	Penniman process
	$2\ FeSO_4 + \frac{1}{2}\ O_2 + 3\ H_2O$	$\longrightarrow$	$2\ \alpha\text{-}FeOOH + 2\ H_2SO_4$	(Penniman process)
	$2\ Fe + \frac{1}{2}\ O_2 + 3\ H_2O$	$\longrightarrow$	$2\ \alpha\text{-}FeOOH + 2\ H_2$	(Penniman process)
	$2\ Fe + C_6H_5NO_2 + 2\ H_2O$	$\longrightarrow$	$2\ \alpha\text{-}FeOOH + C_6H_5NH_2$	Laux process
Orange	$2\ FeSO_4 + 4\ NaOH + \frac{1}{2}\ O_2$	$\longrightarrow$	$2\ \gamma\text{-}FeOOH + 2\ Na_2SO_4 + H_2O$	precipitation
Black	$3\ FeSO_4 + 6\ NaOH + \frac{1}{2}\ O_2$	$\longrightarrow$	$Fe_3O_4 + 3\ Na_2SO_4 + 3\ H_2O$	1-step precipitation
	$2\ FeOOH + FeSO_4 + 2\ NaOH$	$\longrightarrow$	$Fe_3O_4 + Na_2SO_4 + 2\ H_2O$	2-step precipitation
	$9\ Fe + 4\ C_6H_5NO_2 + 4\ H_2O$	$\longrightarrow$	$3\ Fe_3O_4 + 4\ C_6H_5NH_2$	Laux process
	$3\ Fe_2O_3 + H_2$	$\longrightarrow$	$2\ Fe_3O_4 + H_2O$	reduction
Brown	$2\ Fe_3O_4 + \frac{1}{2}\ O_2$	$\longrightarrow$	$3\ \gamma\text{-}Fe_2O_3$	calcination
	$3\ Fe_3O_4 + Fe_2O_3 + MnO_2 + \frac{1}{2}\ O_2$	$\longrightarrow$	$(Fe_{11},Mn)O_{18}$	calcination

to give sulfuric acid [3.6]–[3.9]. However, the waste gases and the dissolved impurities that are leached out in the final stage present ecological problems.

Lower quality products can be obtained by single-stage calcination of iron(II) sulfate heptahydrate in an oxidizing atmosphere. The pigments have a relatively poor tinting strength and a blue tinge. Decomposition of iron(II) chloride monohydrate in air at high temperatures also yields a low-quality red iron oxide pigment [3.10].

In a new process, micaceous iron oxide is obtained in high yield by reacting iron(III) chloride and iron at 500–1000 °C in an oxidizing atmosphere in a tubular reactor [3.11].

Black Fe_3O_4 pigments with a high tinting strength can be prepared by calcining iron salts under reducing conditions [3.12]. This process is not used industrially because of the furnace gases produced.

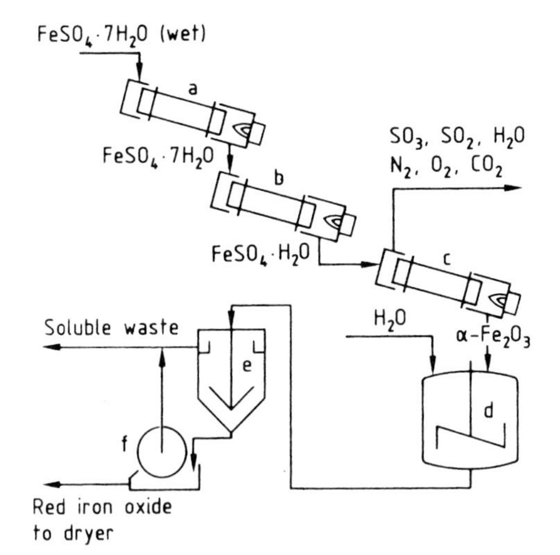

Figure 22. Production of copperas red
a) Dryer; b) Rotary kiln (dewatering); c) Rotary kiln; d) Tank; e) Thickener; f) Filter

Controlled oxidation of Fe_3O_4 at ca. 500 °C produces a single-phase brown γ-Fe_2O_3 with a neutral hue [3.13].

Calcination of α-FeOOH with small quantities of manganese compounds gives homogeneous brown pigments with the composition $(Fe, Mn)_2O_3$ [3.14]. Calcination of iron and chromium compounds that decompose at elevated temperatures yields corresponding pigments with the composition $(Fe, Cr)_2O_3$ [3.15].

Precipitation Processes. In principle, all iron oxide hydroxide phases can be prepared from aqueous solutions of iron salts (see Table 23). However, precipitation with alkali produces neutral salts (e.g., Na_2SO_4, NaCl) as byproducts which enter the wastewater.

Precipitation is especially suitable for producing soft pigments with a pure, bright hue. The manufacture of α-FeOOH yellow is described as an example. The raw materials are iron(II) sulfate ($FeSO_4 \cdot 7\,H_2O$) or liquors from the pickling of iron and steel, and alkali [NaOH, $Ca(OH)_2$, ammonia, or magnesite]. The pickle liquors usually contain appreciable quantities of free acid, and are, therefore, first optionally neutralized by reaction with scrap iron. Other metallic ions should not be present in large amounts, because they have an adverse effect on the hue of the iron oxide pigments.

The solutions of the iron salts are first mixed with alkali in open reaction vessels (Fig. 23, Route A) and oxidized, usually with air. The quantity of alkali used is such that the pH remains acidic. The reaction time (ca. 10–100 h) depends on the temperature (10–90 °C) and on the desired particle size of the pigment. This method yields yellow pigments (α-FeOOH) [3.16], [3.17]. If yellow nuclei are produced in a separate reaction (Fig. 23, Route A, tank c), highly consistent yellow iron oxide pigments with a pure color can be obtained [3.18].

If precipitation is carried out at ca. 90 °C while air is passed into the mixture at ca. pH $\geq$ 7, black iron oxide pigments with a magnetite structure and a good tinting strength are obtained when the reaction is stopped at a $FeO:Fe_2O_3$ ratio of ca. 1:1.

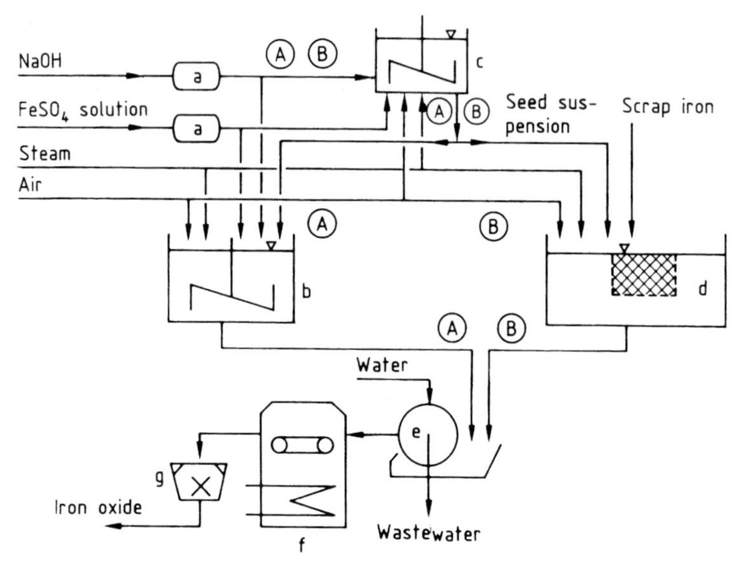

Figure 23. Production of yellow iron oxide by the precipitation (A) and Penniman (B) processes
a) Tank; b) Pigment reactor; c) Seed reactor; d) Pigment reactor with scrap basket; e) Filter; f) Dryer; g) Mill

The process can be accelerated by operating at 150 °C under pressure; this technique also improves pigment quality [3.19]. Rapid heating of a suspension of iron oxide hydroxide with the necessary quantity of $Fe(OH)_2$ to ca. 90 °C also produces black iron oxide of pigment quality [3.20], [3.21].

Orange iron oxide with the lepidocrocite structure (γ-FeOOH) is obtained if dilute solutions of the iron(II) salt are precipitated with sodium hydroxide solution or other alkalis until almost neutral. The suspension is then heated for a short period, rapidly cooled, and oxidized [3.22], [3.23].

Very soft iron oxide pigments with a pure red color may be obtained by first preparing α-Fe_2O_3 nuclei, and then continuously adding solutions of iron(II) salt with atmospheric oxidation at 80 °C. The hydrogen ions liberated by oxidation and hydrolysis are neutralized by adding alkali and keeping the pH constant [3.24]. Pigment-quality α-Fe_2O_3 is also obtained when solutions of an iron(II) salt, preferably in the presence of small amounts of other cations, are reacted at 60–95 °C with excess sodium hydroxide and oxidized with air [3.25].

The *Penniman process* is probably the most widely used production method for yellow iron oxide pigments [3.26], [3.27]. This method considerably reduces the quantity of neutral salts formed as byproducts. The raw materials are iron(II) sulfate, sodium hydroxide solution, and scrap iron. If the sulfate contains appreciable quantities of salt impurities, these must be removed by partial precipitation. The iron must be free of alloying components. The process usually consists of two stages (Fig. 23, Route B).

In the first stage, nuclei are prepared by precipitating iron(II) sulfate with alkali (e.g., sodium hydroxide solution) at 20–50 °C with aeration (c). Depending on the conditions, yellow, orange, or red nuclei may be obtained. The suspension of nuclei is pumped into vessels charged with scrap iron (d) and diluted with water. Here, the process is completed by growing the iron oxide hydroxide or oxide onto the nuclei. The residual iron(II) sulfate in the nuclei suspension is oxidized to iron(III) sulfate by blasting with air at 75–90 °C. The iron(III) sulfate is then hydrolyzed to form FeOOH or α-Fe_2O_3. The liberated sulfuric acid reacts with the scrap iron to form iron(II) sulfate, which is also oxidized with air. The reaction time can vary from ca. two days to several weeks, depending on the conditions chosen and the desired pigment. At the end of the reaction, metallic impurities and coarse particles are removed from the solid with sieves or hydrocyclones; water-soluble salts are removed by washing. Drying is carried out with band or spray dryers (f) and disintegrators or jet mills are used for grinding (g). The main advantage of this process over the precipitation process lies in the small quantity of alkali and iron(II) sulfate required. The bases are only used to form the nuclei and the relatively small amount of iron(II) sulfate required initially is continually renewed by dissolving the iron by reaction with the sulfuric acid liberated by hydrolysis. The process is thus considered environmentally friendly. The iron oxide pigments produced by the Penniman process are soft, have good wetting properties, and a very low flocculation tendency [3.26]–[3.34].

Under suitable conditions the Penniman process can also be used to produce reds directly. The residual scrap iron and coarse particles are removed from the pigment, which is then dried [3.35] and ground using disintegrators or jet mills. These pigments have unsurpassed softness. They usually have purer color than the harder red pigments produced by calcination.

The Laux Process. The Béchamp reaction (i.e., the reduction of aromatic nitro compounds with antimony or iron) which has been known since 1854, normally yields a black-gray iron oxide that is unsuitable as an inorganic pigment. By adding iron(II) chloride or aluminum chloride solutions, sulfuric acid, and phosphoric acid, LAUX modified the process to yield high-quality iron oxide pigments [3.36]. Many types of pigments can be obtained by varying the reaction conditions. The range extends from yellow to brown (mixtures of α-FeOOH and/or α-Fe_2O_3 and/or Fe_3O_4) and from red to black. If, for example, iron(II) chloride is added, a black pigment with very high tinting strength is produced [3.36]. However, if the nitro compounds are reduced in the presence of aluminum chloride, high-quality yellow pigments are obtained [3.37]. Addition of phosphoric acid leads to the formation of light to dark brown pigments with good tinting strength [3.38]. Calcination of these products (e.g., in rotary kilns) gives light red to dark violet pigments. The processes are illustrated in Figure 24.

The type and quality of the pigment are determined not only by the nature and concentration of the additives, but also by the reaction rate. The rate depends on the grades of iron used, their particle size, the rates of addition of the iron and nitrobenzene (or another nitro compound), and the pH value. No bases are required to precipitate the iron compounds. Only ca. 3% of the theoretical amount of acid is required to dissolve all of the iron. The aromatic nitro compound oxidizes the Fe^{2+} to Fe^{3+} ions, acid is liberated during hydrolysis and pigment formation, and more metallic iron is dissolved by the liberated acid to form iron(II) salts; consequently, no additional acid is necessary.

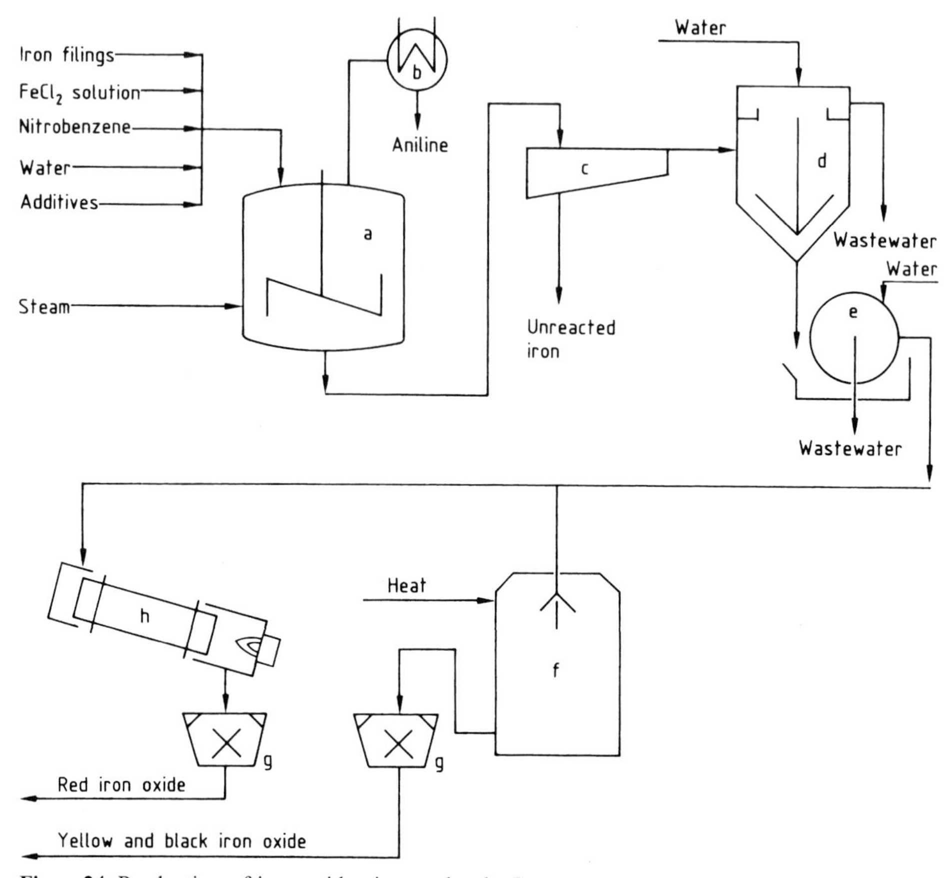

Figure 24. Production of iron oxide pigment by the Laux process
a) Reactor; b) Condenser; c) Classifier; d) Thickener; e) Filter; f) Dryer; g) Mill; h) Rotary kiln

The iron raw materials used are grindings from iron casting or forging that must be virtually free of oil and grease. The required fineness is obtained by size reduction in edge runner mills and classification with vibratory sieves. The iron and the nitro compound are added gradually via a metering device to a stirred tank (a) containing the other reactants (e.g., iron(II) chloride, aluminum chloride, sulfuric acid, and phosphoric acid). The system rapidly heats up to ca. 100 °C and remains at this temperature for the reaction period. The nitro compound is reduced to form an amine (e.g., aniline from nitrobenzene) which is removed by steam distillation. Unreacted iron is also removed (e.g., in shaking tables, c). The pigment slurry is diluted with water in settling tanks (d) and the pigment is washed to remove salts, and filtered on rotary filters (e). It may then be dried on band, pneumatic conveyor, or spray dryers to form yellow or black pigments, or calcined in rotary kilns (h) in an oxidizing atmosphere to give red or brown pigments. Calcination in a nonoxidizing atmosphere at 500–700 °C improves the tinting strength [3.39]. The pigments are then ground to the desired fineness in pendular mills, pin mills, or jet mills, depending on their hardness and application.

The Laux process is a very important method for producing iron oxide because of the coproduction of aniline; it does not generate byproducts that harm the environment.

Other Production Processes. The three processes already described are the only ones that are used on a large scale. The following processes are used on a small scale for special applications:

1) Thermal decomposition of $Fe(CO)_5$ to form transparent iron oxides (see Section 5.4) [3.40]
2) Hydrothermal crystallization for the production of α-Fe_2O_3 in platelet form [3.41]

3.1.1.3. Toxicology and Environmental Aspects

The Berufsgenossenschaft der Chemischen Industrie (Germany) has recommended that all iron oxide pigments should be classified as inert fine dusts with an MAK value of 6 mg/m^3. This is the highest value proposed for fine dusts.

Iron oxide pigments produced from pure starting materials may be used as colorants for food and pharmaceutical products [3.42]. Synthetic iron oxides do not contain crystalline silica and therefore are not considered to be toxic, even under strict Californian regulations.

3.1.1.4. Quality

The red and black iron oxide pigments produced by the methods described have an Fe_2O_3 content of 92–96 wt%. For special applications (e.g., ferrites) analytically pure pigments with Fe_2O_3 contents of 99.5–99.8 wt% are produced. The Fe_2O_3 content of yellow and orange pigments lies between 85 and 87 wt% corresponding to FeOOH contents of 96–97 wt%. For standards, see Table 1 (Iron oxide pigments: "Black, specification"; "Red, specification"; "Yellow, specification"; and "FeO content"). Variations of 1–2% are of no importance with respect to the quality of the pigments. Pigment quality is mainly determined by the quantity and nature of the water-soluble salts, the particle size distribution (hue and tinting strength are effected), and the average particle size of the ground product. The hue of red iron oxide is determined by the particle diameter, which is ca. 0.1 μm for red oxides with a yellow tinge and ca. 1.0 μm for violet hues (see Figure 24a).

The optical properties of the yellow, usually needle-shaped, iron oxide pigments depend not only on the particle size, but also on the length to width ratio (e.g., length = 0.3–0.8 μm, diameter = 0.05–0.2 μm, length: diameter ratio = ca. 1.5–8). In applications for which needle-shaped particles are unsuitable, spheroidal pigments are available [3.43] (see Figure 24b). Black iron oxide pigments (Fe_3O_4) have a particle diameter of ca. 0.1–0.6 μm.

Some iron oxide pigments have a limited stability on heating. Red iron oxide is stable up to 1200 °C in air. In the presence of oxygen, black iron oxide changes into

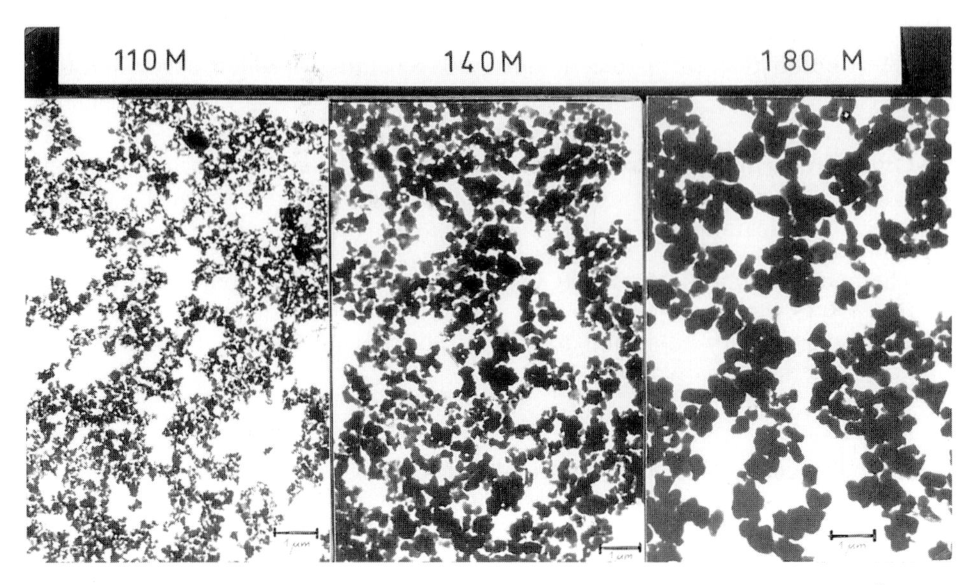

Figure 24a. Electronmicrographs of iron oxide red of different particle size. Bayferrox® 110M shows a yellow tinge while Bayferrox® 180M shows a bordeaux tinge

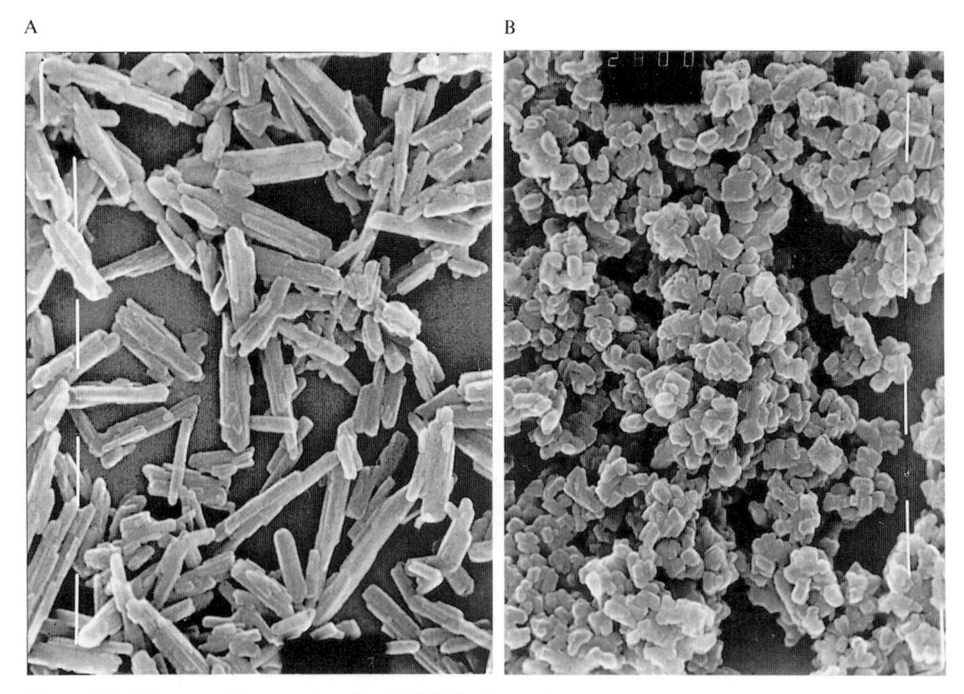

Figure 24b. Electronmicrographs of α-FeOOH pigment
A) Needle-like pigment (Bayferrox® 420);
B) Spheroidal pigment (Bayferrox® 915)

brown γ-Fe_2O_3 at ca. 180 °C and then into red α-Fe_2O_3 above 350 °C. Yellow iron oxide decomposes above ca. 180 °C to form red α-Fe_2O_3 with liberation of water. This temperature limit can be increased to ca. 260 °C by stabilization with basic aluminum compounds. The thermal behavior of brown iron oxides produced by mixing depends on their composition.

3.1.1.5. Uses

All synthetic iron oxides possess good tinting strength and excellent hiding power. They are also lightfast and resistant to alkalis. These properties are responsible for their versatility. The principle areas of use are shown in Table 24.

Iron oxide pigments have long been used for coloring construction materials. Concrete roof tiles, paving bricks, fibrous cement, bitumen, mortar, rendering, etc., can be colored with small amounts of pigment that do not affect the setting time, compression strength, or tensile strength of the construction materials. Synthetic pigments are superior to the natural pigments due to their better tinting power and purer hue.

Natural rubber can only be colored with iron oxides that contain very low levels of copper and manganese (Cu < 0.005 %, Mn < 0.02 %). Synthetic rubber is less sensitive.

In the paint and coating industries, iron oxide pigments can be incorporated in many types of binders. Some reasons for their wide applicability in this sector are pure hue, good hiding power, good abrasion resistance, and low settling tendency. Their high temperature resistance allows them to be used in enamels.

The use of iron oxide as a polishing medium for plate glass manufacture has decreased now that other methods of glass production are available.

3.1.1.6. Economic Aspects

Accurate production figures for natural and synthetic iron oxide pigments are difficult to obtain, because statistics also include nonpigmentary oxides (e.g., red mud from bauxite treatment, intermediate products used in ferrite production). World production of synthetic iron oxides in 1995 was estimated to be 600 000 t;

Table 24. Main areas of use for natural and synthetic iron oxide pigments

Use	Amount, %		
	Europe	United States	World
Coloring construction materials	64	37	60
Paints and coatings	30	48	29
Plastics and rubber	4	14	6
Miscellaneous	2	1	5

production of natural oxides was ca. 100000 t. The most important producing countries for synthetic pigments are Germany, the United States, the United Kingdom, Italy, Brazil, the People's Republic of China, and Japan. The natural oxides are mainly produced in France, Spain, Cyprus, India, Iran, Italy, and Austria.

The most important *manufacturers* are Bayer, Harcross, Laporte and Toda (Japan).

3.1.2. Chromium Oxide Pigments

Chromium oxide pigments, also called chromium oxide green pigments, consist of chromium(III) oxide [*1308-38-9*], Cr_2O_3, M_r 151.99. Chromium oxide green is one of the few single-component pigments with green coloration. Chrome green is a blend of chrome yellow and iron blue pigments; phthalochrome green is a blend of chrome yellow and blue phthalocyanine pigments.

Natural, minable deposits of chromium oxide are not known. In addition to pigment grade, chromium oxide producers usually also offer a technical grade for applications based on properties other than coloration. These include:

1) Metallurgy: aluminothermic production of chromium metal by reaction of aluminum powder and Cr_2O_3
2) Refractory industry: production of thermally and chemically resistant bricks and lining materials
3) Ceramic industry: coloring of porcelain enamels, ceramic frits, and glazes
4) Pigment industry: raw material for the production of chromium-containing stains and pigments based on mixed metal oxide phases
5) Grinding and polishing agent: chromium(III) oxide is used in brake linings and polishing agents due to its high hardness

Chromium oxide hydroxide and hydrated chromium oxide pigments (Guignets Green) have a very attractive blue-green color. They are of low opacity, but provide excellent lightfastness and good chemical resistance. Loss of water on heating limits the application temperature. These pigments are no longer of industrial importance [3.43 a].

3.1.2.1. Properties

Chromium(III) oxide crystallizes in the rhombohedral structure of the corundum type; space group D^6_{3d}-$R\bar{3}c$, ϱ 5.2 g/cm^3. Because of its high hardness (ca. 9 on the Mohs scale) the abrasive properties of the pigment must be taken into account in certain applications [3.44]. It melts at 2435 °C but starts to evaporate at 2000 °C. Depending on the manufacturing conditions, the particle sizes of chromium oxide pigments are in the range 0.1–3 μm with mean values of 0.3–0.6 μm. Most of the particles are isometric. Coarser chromium oxides are produced for special applications, e.g., for applications in the refractory area.

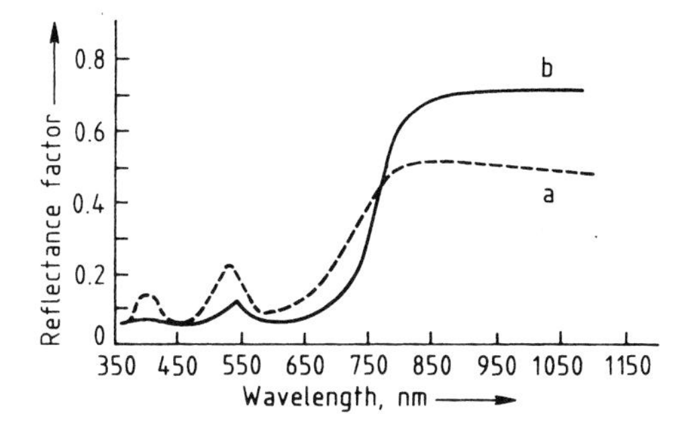

Figure 25. Dependence of the reflectance of chromium oxide on the wavelength
a) Regular pigment; b) Special product with larger particle size and high IR reflectance

Chromium oxide has a refractive index of ca. 2.5. Chromium oxide green pigments have an olive green tint. Lighter greens with yellowish hues are obtained with finely divided pigments, and darker, bluish tints with larger particle diameters; the darker pigments are weaker colorants. The maximum of the reflectance curve lies in the green region of the spectrum at ca. 535 nm (Fig. 25, curve a). A weaker maximum in the violet region (ca. 410 nm) is caused by Cr–Cr interactions in the crystal lattice. Chromium oxide green pigments are used in IR-reflecting camouflage coatings because of their relatively high reflectance in the near infrared (Fig. 25, curve b).

Since chromium(III) oxide is virtually inert, chromium oxide green pigments are remarkably stable. They are insoluble in water, acid, and alkali and are thus extremely stable to sulfur dioxide and in concrete. They are light, weather, and temperature resistant. A change of the tint only occurs above 1000 °C due to particle growth.

3.1.2.2. Production

Alkali dichromates are used as starting materials for the production of chromium(III) oxide pigments. They are available as bulk industrial products in the required purity. High impurity levels have an unfavorable effect on the hue.

Reduction of Alkali Dichromates. In industrial processes, solid alkali dichromates are reacted with reducing agents such as sulfur or carbon compounds. The reaction is strongly exothermic, and with sulfur proceeds as follows:

$$Na_2Cr_2O_7 + S \longrightarrow Cr_2O_3 + Na_2SO_4$$

Sodium sulfate can then easily be separated by washing, because it is water soluble. The use of sulfur was first described in 1820 [3.45]. Roth described the use of $K_2Cr_2O_7$ in 1927 [3.46]. If charcoal is used in place of sulfur, Na_2CO_3 is formed as byproduct [3.47].

Finely divided sodium dichromate (dihydrate or anhydrous) is mixed homogeneously with sulfur. This mixture is then reacted in a furnace lined with refractory bricks at 750–900 °C. An excess of sulfur is used to ensure completion of the reaction. The reaction mass is leached with water to remove water-soluble components such as sodium sulfate. The solid residue is then separated, dried, and ground.

If potassium dichromate is used instead of sodium dichromate, a green pigment with a more bluish hue is obtained.

If it is to be used as a pigment in paints and lacquers, chromium oxide green can be subjected to jet milling (micronization) to obtain the required properties (e.g., gloss).

Reduction of Ammonium Dichromate. Chromium(III) oxide can be obtained by thermal decomposition of ammonium dichromate. Above ca. 200 °C, a highly voluminous product is formed with elimination of nitrogen [3.48]. The pigment is obtained after addition of alkali salts (e.g., sodium sulfate) and subsequent calcination [3.49].

In the industrial process, a mixture of ammonium sulfate or chloride and sodium dichromate is calcined [3.50]:

$$Na_2Cr_2O_7 \cdot 2\,H_2O + (NH_4)_2SO_4 \longrightarrow Cr_2O_3 + Na_2SO_4 + 6\,H_2O + N_2$$

The workup is then carried out as described above. A chromium oxide pigment obtained by this process typically contains (wt%):

Cr_2O_3	99.0–99.5
SiO_2	0.05 (max.)
Al_2O_3	0.1 (max.)
Fe_2O_3	0.05 (max.)
S	ca. 0.02
Water	ca. 0.3

Chromium oxides with a minimal sulfur content are preferred for metallurgical applications. These are obtained by reacting sodium dichromate with ammonium chloride or sulfate in a deficiency of 10 mol% [3.51]. Chromium (III) oxides with a low sulfur content can also be obtained by thermal aftertreatment [3.52]. Thermal decomposition of chromic acid anhydride (CrO_3) yields high-purity chromium(III) oxide [3.53].

The pigment properties of chromium oxides can be modified by precipitation of hydroxides (e.g., of titanium or aluminum), and subsequent calcining. This treatment changes the color to yellow-green, and decreases the flocculation tendency [3.54]. Aftertreatment with organic compounds (e.g., alkoxylated alkylsulfonamides) is also used [3.55].

Other Processes. Other production processes are suggested in the patent literature, but have not so far gained industrial importance. For instance, sodium dichromate can be mixed with heating oil and reacted at 300 °C. The soda formed must be washed out prior to calcining at 800 °C to avoid reoxidation in the alkaline melt [3.56].

In alkaline solution, sodium chromate can be reduced with sulfur at atmospheric pressure with formation of sodium thiosulfate. After neutralization, more sodium chromate is added to exhaust the reducing capacity of the thiosulfate. The mixture is calcined at 900–1070 °C [3.57].

Another process involves the shock heating of sodium dichromate in a flame at 900–1600 °C in the presence of excess hydrogen and chlorine to bind the alkali as sodium chloride [3.58]. This method is suitable for the preparation of pigment-grade chromium oxide of high purity, with an especially low sulfur content.

Environmental Protection. Since alkali dichromates or chromic acid anhydride are used as starting materials for the production of chromium(III) oxides, occupational health requirements for the handling of hexavalent chromium compounds must be observed [3.59]. The sulfur dioxide formed on reduction with excess sulfur must be removed from the flue gases according to national regulations, e.g., by oxidation to H_2SO_4.

Process wastewater may contain small amounts of unreacted chromates; recovery is uneconomical. Prior to release into drainage systems, the chromates in these wastewater streams must be reduced (e.g., with SO_2 or $NaHSO_3$) and precipitated as chromium hydroxide [3.60]. In Germany, for example, the minimum requirements for wastewater in the production of chromium oxide pigments are specified in [3.61].

3.1.2.3. Quality Specifications and Analysis

International, technical specifications for chromium oxide pigments are defined in ISO 4621 (1986), they must have a minimum Cr_2O_3 content of 96 wt %.

Various grades are defined according to their particle fineness as measured by the residue on a 45-µm sieve: grade 1, 0.01 % residue (max.); grade 2, 0.1 % (max.); and grade 3, 0.5 % (max.).

ISO 4621 (1986) also specifies analytical methods. Usually, analysis of chromium and the byproducts is preceded by melting with soda and sodium peroxide. The content of water-soluble or acid-soluble chromium is becoming important from the toxicological and ecological point of view. It is determined according to DIN 53 780 with water, or according to ISO 3856, part 1 with 0.1 mol/L hydrochloric acid.

3.1.2.4. Storage and Transportation

Chromium(III) oxide pigments are thermally stable and insoluble in water. They are not classified as harzardous materials and are not subject to international transport regulations. As long as they are kept dry their utility as a pigment is practically unlimited.

3.1.2.5. Uses

The use of chromium(III) oxide as a pigment for toys, cosmetics, and in plastics and paints that come in contact with food is permitted in national and international regulations [3.62]–[3.68]. Maximum limits for heavy metals or their soluble fractions are usually a prerequisite. Because pure starting materials are used, these limits are satisfied by most types of chromium oxide.

Chromium oxide is equally important as a colorant and in its other industrial applications. As a pigment, it is used predominantly in the paint and coatings industry for high quality green paints with special requirements, especially for steel constructions (coil coating), facade coatings (emulsion paints), and automotive coatings.

A series of RAL (Reichs-Ausschuß für Lieferbedingungen) tints (e.g., Nos. 6003, 6006, 6011, 6014, and 6015) can be formulated based on chromium oxide. As mentioned previously, chromium oxide is also an important pigment for the formulation of green camouflage coatings (e.g., RAL 6031-F 9, Natogreen 285, Stanag 2338, and Forestgreen MIL-C-46168 C).

Except for the expensive cobalt green, chromium oxide is the only green pigment that meets the high color stability requirements for building materials based on lime and cement [3.69]. In plastics, however, chromium oxide green is only of minor importance because of its dull tint.

The industrial significance of chromium oxide is due to its chemical and physical properties. Its high purity makes it suitable as a starting material for the aluminothermic production of very pure chromium metal.

Since the late 1970s chromium oxide has gained significance as a raw material in the refractory industry. The addition of chromium oxide to bricks and refractory concrete based on alumina significantly improves their stability against slag in the production and processing of pig iron. Chromium oxide bricks containing ca. 95 wt% Cr_2O_3 have become important in the production of E-glass reinforcement fibers for lining melting tanks. These linings have substantially improved furnace stability (i.e., prolonged furnace life).

The high hardness of chromium oxide resulting from its crystal structure is exploited in polishing agents for metals and in brake linings. Addition of a small amount of chromium oxide to magnetic materials of audio and video tapes imparts a self-cleaning effect to the sound heads.

3.1.2.6. Economic Aspects

Important producers are American Chrome and Chemicals (USA), Bayer (Germany), British Chrome and Chemicals (UK), and Nihon Denko (Japan).

Statistical data on the consumption of chromium oxide have not been published recently. However, it can be assumed that the world capacity 1996 is about 72000 t/y, thereof 43000 t/y in Eastern Europe and GUS. The consumption as pigments is estimated to be 24000 t in 1996.

3.1.2.7. Toxicology and Occupational Health

Toxicological or carcinogenic effects have not been detected in rats receiving up to 5% chromium(III) oxide in their feed [3.70] nor in medical studies performed in chemical plants producing chromium(III) oxide and chromium(III) sulfate [3.71]. The oral LD_{50} for chromium(III) oxide in the rat is >10000 mg/kg; it does not irritate the skin or mucous membranes.

Chromium(III) oxide is not included in the MAK list (Germany), the TLV list (USA), or in the list of hazardous occupational materials of the EC [3.72]. In practice, this means that chromium(III) oxide can be regarded as an inert fine dust with a MAK value of 6 mg/m^3.

3.1.3. Mixed Metal Oxide Pigments

The term mixed metal oxide pigment denotes a pigment that crystallizes in a stable oxide lattice, and in which the color is due to the incorporation of colored cations in this lattice. Such compounds are regarded as solid solutions. The American "Dry Color Manufacturers' Association" has abandoned the term "mixed metal oxide inorganic colored pigments", it denotes these colorants as "complex inorganic color pigments" [3.73].

According to V. M. GOLDSCHMIDT, two substances with the same basic formula and crystal structure form solid solutions in a concentration range that depends on the degree of similarity of their ionic radii. A large range of solid solutions may be expected if the radius of the larger ion does not exceed that of the smaller by more than 15% [3.73 a].

Solid solutions can also be formed to a limited extent with substances that have a similar formula but a different crystal structure. One of the substances imposes its structure upon the other. In lattices which are stable even if some sites in the unit cell are unoccupied, substances with different formulae can form solid solutions (anomalous mixed crystals). The incorporated ions must be of similar size, but do not need to have the same valency because electroneutrality is ensured by the presence of a compensating valency elsewhere in the lattice. Changes in the relative concentrations in the solid solutions lead to continuous variation of their properties.

In principle, the number of possible mixed metal oxide phases that can be produced from different substituent elements and oxide lattices is extremely large. Systematic investigations have greatly increased the knowledge about mixed metal oxide phases [3.74].

The only stable oxide lattices that have so far been of value as mixed metal oxide pigments are those with spinel, rutile, and hematite structures. These lattices possess not only good thermal and chemical stability, but also have a high refractive index which is important for good optical pigment properties.

3.1.3.1. Properties

Rutile Mixed Metal Oxide Pigments. The rutile (TiO_2) lattice offers considerable scope for variation (see Section 2.1.1) [3.74]. In this lattice, titanium ions are surrounded by six neighboring oxygen atoms at the corners of a regular, but slightly distorted octahedron. Colored pigments may be obtained by substituting transition-metal ions for the titanium ions [3.75]. If the colored cation substituent is not tetravalent, another cation of appropriate valency must be substituted into the rutile lattice to maintain an average cation valency of four.

H. H. SCHAUMANN prepared the first mixed-phase rutile pigments by incorporating nickel oxide and antimony oxide into rutile to give a yellow pigment, and cobalt and antimony oxides to give an ocher pigment [3.76]. Other rutile mixed metal oxide phases were first used as ceramic colorants (e.g., with chromium–tungsten, brown [3.77]) or for coloring porcelain enamels (e.g., with copper–antimony, lemon yellow; manganese–antimony, dark brown; iron–antimony, gray [3.78]; and vanadium–antimony, dark gray). In commercial mixed-phase rutile pigments, between 10 and 20 mol% of the titanium ions are replaced by substituent metal ions.

Nickel and chromium mixed-phase rutile pigments are industrially important. When nickel and chromium are substituted into the rutile lattice, higher valency metals (e.g., antimony, niobium, or tungsten) must also be substituted to maintain an average valency of four.

Nickel rutile yellow [*8007-18-9*], C.I. Pigment Yellow 53:77788, is a light lemon yellow pigment with the approximate composition $(Ti_{0.85}Sb_{0.10}Ni_{0.05})O_2$. Antimony can be replaced by niobium without any appreciable change in color [3.79].

Chromium rutile yellow pigments [*68186-90-3*], C.I. Pigment Brown 24:77310, usually contain antimony to balance the valency. Their composition is approximately $(Ti_{0.90}Sb_{0.05}Cr_{0.05})O_2$. Depending on particle size, the color varies from light to medium ocher (buff). Stability towards plastics at high temperature is considerably improved by the incorporation of small quantities of lithium [3.80] or magnesium [3.81]. If antimony is replaced by tungsten, the products become darker in color, whereas replacement by niobium leaves the color unchanged.

Spinel Mixed Metal Oxide Pigments. The crystal structure of spinel, $MgAl_2O_4$, is made up of an approximately cubic closely packed array of oxygen ions (Fig. 26). The unit cell contains 32 oxygen ions. Of the 64 possible tetrahedral sites, 8 are occupied by magnesium ions; of the 32 possible octahedral sites, 16 are occupied by aluminum ions.

The inverse spinel structure differs in that one type of cation occupies the tetrahedral and half of the octahedral sites of the spinel lattice, and the other cations occupy the remaining octahedral sites. This is indicated by writing the formula of, for example, zinc titanium spinel as $Zn(TiZn)O_4$. Spinels containing di- and tetravalent cations are mostly of the inverse type. Normal and inverse spinels should be regarded as idealized limiting structures, intermediate forms are often observed in practice. A

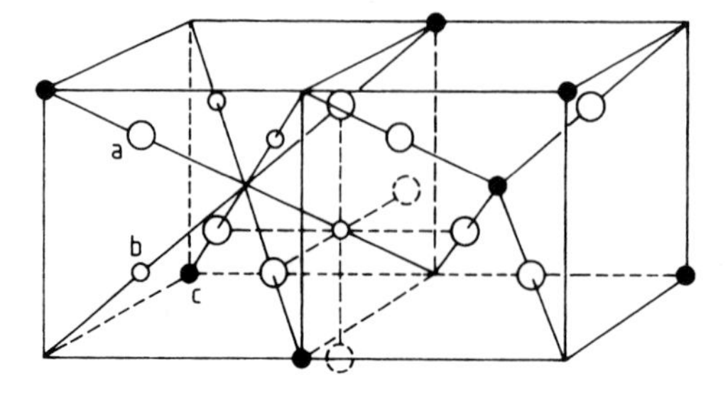

Figure 26. Structure of spinel showing two octants of the unit cell
Two oxygen ions of neighboring octants are included (dashed circles) to show the arrangement of oxygen around the octahedral sites.
a) Oxygen; b) Cations occupying octahedral sites; c) Cations occupying tetrahedral sites

large range of spinel phases can be formed due to the numerous combinations of cations in the oxidation states 1–6, especially from elements of the first transition-metal series. This range is increased if some of the cation sites remain unoccupied. Further details are given in [3.82].

Many spinel phases have been used as colorants for ceramics and porcelain enamels. The most important pigments based on spinel phases used in paints, plastics, and building materials are given in Table 25.

Cobalt Blue. This was formerly known as Thénard's blue, and is a cobalt aluminum spinel. It usually has a deficiency of cobalt (ca. 0.6–0.7 mol CoO/mol Al_2O_3) compared with the ideal formula $CoAl_2O_4$. It has a brilliant blue hue with a red tinge. The tetrahedral holes can be filled by incorporating zinc or magnesium ions in the lattice. If chromium is used in place of aluminum, the hue is shifted via neutral blue (molar ratio Al:Cr $\approx$ 7:3) to a greenish blue (molar ratio Al:Cr $\approx$ 3:7). Brilliance decreases with increasing chromium content, while the hiding power increases.

Cobalt Green. Incorporation of cobalt in the inverse titanium spinels Mg_2TiO_4 and Zn_2TiO_4 produces blue to green hues. The green color results when cobalt ions occupy the octahedral sites [3.83]. Both the blue and the green pigments have patent protection [3.84], and correspond to inverse spinels with the general formula $Mg_aCo_bZn_cTiO_4$ (where $a+b+c=2$). A brilliant cobalt green is obtained if magnesium is replaced by nickel, giving the approximate stoichiometry $NiCo_{0.5}Zn_{0.5}TiO_4$ [3.85]. Turquoise spinels are obtained by including lithium, $Li_2(Co,Ni,Zn)Ti_3O_8$ [3.86].

Zinc Iron Brown. Variation of the zinc to iron ratio in red-brown iron spinel ($ZnFe_2O_4$), or replacement of some of the iron ions by aluminum and titanium ions, gives light to medium brown pigments. Inclusion of a small proportion of lithium ions considerably improves stability towards reducing agents, e.g., when the pigments are used to color plastics [3.87]. The partial replacement of iron ions by chromium(III) ions yields dark brown products [3.88].

Spinel Black. Iron–chrome spinels of composition $CuFe_{0.5}Cr_{1.5}O_4$ provide black pigments with good tinting strength. Other commercial black pigments are copper chromites, $CuCr_2O_4$, in which part of the chromium may be replaced by manganese.

Mixed Metal Oxide Pigments with Hematite and Bixbyite Structures. Manganese and chromium can be incorporated in the hematite lattice of α-Fe_2O_3 which has a higher refractive index than rutile. Brown, temperature-resistant iron oxide pigments with the hematite structure are described in Section 3.1.1.2, e.g., α-$(Fe,Mn)_2O_3$ [3.89] and α-$(Fe,Cr)_2O_3$ [3.90]. The bixbyite lattice of α-Mn_2O_3 can take up > 50% Fe_2O_3 in solid solution while retaining its structure [3.91]. Heat-stable black pigments with good tinting strength are obtained.

3.1.3.2. Production

The mixed metal oxide pigments are usually produced by reaction of the components in the solid state at 800–1400 °C. The reactions proceed more readily if the components are reactive, finely divided, and intimately mixed. The reactants are

Table 25. Mixed-phase spinel pigments

Trivial name	Formula	CAS registry no.	C.I. name and no.	Color	Crystal structure
Cobalt blue	$CoAl_2O_4$	[*1345-16-0*]	Pigment Blue 28:77 346	reddish blue	spinel
	$Co(Al,Cr)_2O_4$	[*68187-11-1*]	Pigment Blue 36:77 343	greenish blue	spinel
Cobalt green	$(Co,Ni,Zn)_2TiO_4$	[*68186-85-6*]	Pigment Green 50:77 377	green	inverse spinel
Zinc iron brown	$ZnFe_2O_4$	[*68187-51-9*]	Pigment Yellow 119:77 496	light to medium brown	spinel
Spinel black	$Cu(Cr,Mn)_2O_4$	[*68186-91-4*]	Pigment Black 28:77 428	black	spinel
	$Cu(Fe,Cr)_2O_4$	[*55353-02-1*]	Pigment Black 22:77 429		intermediate between normal and inverse spinel

therefore usually extremely finely divided oxides, hydroxides, carbonates, or nitrates that are mixed in the form of aqueous suspensions. As these raw materials are commercially available with great purity and fineness, coprecipitation of hydroxides or carbonates from aqueous salt solutions is generally not required. Rutile mixed metal oxide phases based on titanium dioxide are prepared from the reactive anatase modification or titanium dioxide hydrolysate, an intermediate in titanium dioxide production.

Wet mixes are usually dried before calcination. Calcination is performed continuously in rotary or tunnel kilns, or batchwise in directly fired drum or box furnaces. The temperature at which the mixed metal oxide pigments are formed can be reduced by adding mineralizing agents [3.75]. In the case of chromium rutile pigments, addition of magnesium compounds [3.81] or lithium compounds [3.80] before calcination improves thermal stability in plastics.

The formation of rutile mixed phases in the system Ti–Ni–Sb–O is described in detail in [3.92]. When nickel ions are to be included in the pigment and antimony trioxide is used to provide the antimony ions, oxygen must be present to ensure that nickel and antimony occupy the titanium sites in the lattice. If oxygen is excluded, only small amounts of nickel and antimony are taken up in solid solution, and no useful products are formed. The presence of oxidizing agents (e.g., nitrates) is necessary for the production of antimony-containing mixed metal oxide pigments.

After calcination, the products are ground to the desired fineness. Calcined products containing salts are subjected to wet grinding followed by washing, drying, and dry grinding. With salt-free calcination products, washing of the pigments can be omitted, and dry grinding is possible.

When mixed-phase rutile pigments are used in special paint systems, (e.g., stoving or acid-catalyzed lacquers), inorganic surface treatment in an aqueous medium can improve the gloss and flocculation properties. For example, an aqueous pigment suspension is first treated with a surfactant, and then coated with metal hydroxides or oxide hydrates [3.93].

Chromium rutile pigments with extremely high resistance to flocculation are obtained by applying a double-layer coating of oxides or oxide hydrates formed from tetravalent metals and oxides or oxide hydrates of aluminum [3.94].

3.1.3.3. Quality Specifications and Analysis

There are as yet no national or international (ISO) standards for mixed metal oxide pigments, and none are expected in the foreseeable future. However, the general tests of chemical, physical, and optical properties described in ASTM, DIN, or ISO standards for pigments may be utilized, such as DIN 6174, DIN 55986, and ISO 787 Part 16 for the optical properties, and ISO 787 Parts 1–20 for the physical properties (see also Table 1).

There are no standard methods for the chemical analysis of mixed metal oxide pigments. They are first decomposed (e.g., with peroxide in the case of rutile mixed phases, or soda/borax for the spinels) and the appropriate methods are then used to determine the elements.

3.1.3.4. Storage and Transportation

Mixed metal oxide pigments are thermally stable, water-insoluble materials. They are not classified as hazardous substances, and are therefore not subject to international transport regulations. When stored under dry conditions their pigment properties do not deteriorate.

3.1.3.5. Legal Aspects

Mixed metal oxide pigments are permitted in almost all countries for coloring paints and plastics in contact with foods, or for toy manufacture [3.95]–[3.99]. The legal limits for a number of elements (sometimes for the soluble fraction only) must be observed. Owing to the large number of raw materials and processes used, general purity requirements cannot be specified. Each manufacturer must supply relevant data.

In the EC guidelines for coloring agents for foods [3.100], cosmetics [3.101], or pharmaceuticals [3.102], there are no references to mixed metal oxide pigments, and these are therefore not permitted.

3.1.3.6. Uses

The outstanding properties of the mixed metal oxide pigments are their lightfastness and resistance to temperature, chemicals (including acid and alkali), and weathering.

Owing to their high scattering power, their hiding power is much better than that of organic pigments. They are therefore used in combination with organic pigments as base pigments for colored coatings, e.g., for lead-free and chromate-free automobile topcoats (nickel rutile yellow, chromium rutile yellow).

The stability of these pigments makes them suitable for all types of coatings, especially those which have to meet high standards of lightfastness and resistance to weathering or chemicals. They are especially suitable for baking enamels, coil coatings, powder coatings, as well as for coloring plastics, masonry coatings based on resin emulsions or waterglass, and building materials.

A very important use for rutile yellow is in coil coating of aluminum and steel, especially in the building industry, but also for containers, vehicles, and machinery.

Rutile yellow is used in rigid poly(vinyl chloride) for outdoor use (USA: vinyl siding); cobalt blue and cobalt green are employed to color high-density polyethylene (bottle crates). Rutile yellow is being used increasingly in engineering plastics to replace cadmium pigments, which have unfavorable toxicological and ecological properties. Rutile yellow pigments can be used cheaply and advantageously for this purpose instead of a combination of titanium dioxide and expensive organic colorants. Nickel and chromium rutiles can replace cadmium yellow as toning pigments. Rutile yellow pigments can be safely used in all plastics. The color resistance of chromium rutiles in some plastics is limited up to 280 °C, though this disadvantage can be overcome by lattice stabilization (see Section 3.1.3.1).

Trade names of some mixed metal oxide pigments include Heucodur (Dr. H. Heubach, Germany); Irgacolor (Ciba-Geigy, Switzerland); Lichtechtpigmente (Bayer, Germany); Sicotan-Gelb, Sicopal-Blau (BASF, Germany); Tipaque Yellow (Ishihara, Japan); Kerafast (Blythe Colours, UK); Meteor Colors (Engelhard, USA); and Ferro (USA).

3.1.3.7. Economic Aspects

No statistical information concerning the use of mixed metal oxide pigments has been published. However, it is estimated that world consumption, excluding the former Eastern bloc and China, was ca. 15 000 t in 1996. There is no definite information available on whether or not production of mixed metal oxide pigments takes place in the former Eastern bloc countries or China.

3.1.3.8. Toxicology and Occupational Health

Mixed metal oxide pigments contain one or more heavy metals, whose oxides can have toxicological effects. However, these heavy metals in the mixed metal oxide phase occur as different compounds, even if the respective oxides are used as raw materials for pigment production. They have an extremely low solubility in water and dilute acids. Toxicological investigations with nickel and chromium rutile yellow have shown that, even at high dose rates (up to 1 % in animal feeds), these pigments are not toxic or biologically available [3.103]. Cobalt spinels, unlike cobalt oxide, do not have carcinogenic properties [3.104]. Also, the waste disposal of paints and plastics colored with mixed metal oxide pigments is regarded as safe [3.105].

Only nickel rutile yellow has so far been included in the MAK list in Class II b edition 1996; toxicological data available for other pigments are inadequate. The mixed metal oxide pigments are not included in the EC guidelines for hazardous materials (67/548/EC) or in the TLV list for 1995.

3.2. Cadmium Pigments

Among the inorganic pigments, cadmium pigments have particularly brilliant red and yellow colors, as well as a high durability. All cadmium pigments are based on cadmium sulfide and crystallize with a hexagonal wurtzite lattice (Fig. 27). The sulfur atoms have a hexagonal, tightly packed arrangement, in which the cadmium atoms occupy half of the tetrahedral holes. In this wurtzite lattice the cations and anions can be replaced within certain limits by chemically similar elements; nevertheless, the only substitutions which are used in practice are zinc and mercury for the cations and selenium for the anions.

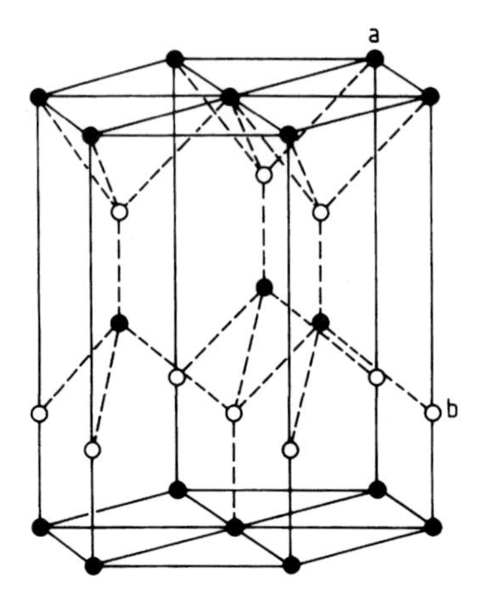

Figure 27. Crystal lattice of cadmium pigments (wurtzite structure)
a) Sulfur (selenium); b) Cadmium (zinc, mercury)

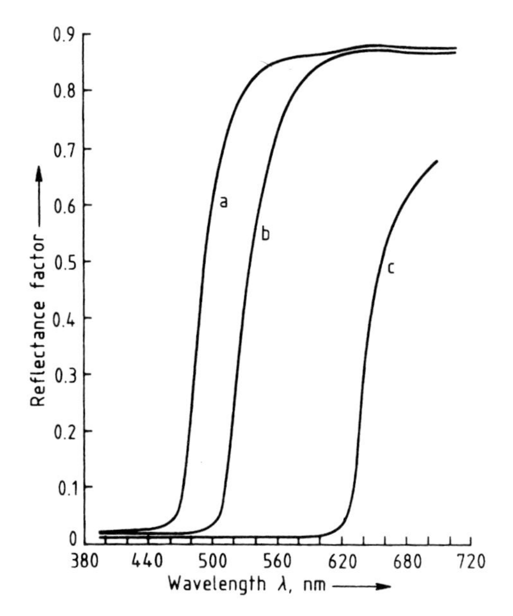

Figure 28. Reflectance curves of cadmium pigments (pigment volume concentration 10%)
a) Cadmium yellow; b) Cadmium golden yellow; c) Cadmium red (bordeaux)

The use of zinc yields greenish yellow pigments due to the lower lattice constants; mercury and selenium lead to expansion of the lattice. With an increasing content of selenium, or especially mercury, the shades of the pigments change to orange, red, and ultimately to deep red (bordeaux). The brilliant colors of cadmium pigments are primarily due to their almost ideal reflectance curves with a steep ascent (Fig. 28).

3.2.1. Cadmium Sulfide

Cadmium sulfide [*68859-25-6*], CdS, M_r 144.48, occurs as greenockite or cadmium-blende in several natural deposits, which are, however, of no importance as pigments. The mineral crystallizes hexagonally in the wurtzite lattice (α-form).

When hydrogen sulfide is passed into cadmium salt solutions, cadmium sulfide is formed as a yellow precipitate with a zinc blende structure (cubic, β-form). The β-form can be converted into the α-form (e.g., by heating). α-Cadmium sulfide shows photoconductivity due to defects in the crystal lattice (usage in photovoltaic cells) [3.106]. The solubility in water at 25 °C is 1.46×10^{-10} mol/L [3.107]. Cadmium sulfide forms the basis for all cadmium pigments.

3.2.2. Cadmium Yellow

Cadmium yellow consists of pure cadmium sulfide (golden yellow color) or mixed crystals of zinc and cadmium sulfide [*8048-07-5*], (Cd, Zn)S, in which up to one-third of the cadmium can be replaced by zinc. The density of this pigment is 4.5–4.8 g/cm^3 and its refractive index is 2.4–2.5. The prevalent parcticle size is approx. 0.2 µm with cubic to spheroidal habits. Cadmium yellow is practically insoluble in water and alkali, and of low solubility in dilute mineral acid. It dissolves in concentrated mineral acid with generation of hydrogen sulfide.

Production. The raw material for the production of cadmium yellow pigments is high-purity cadmium metal (99.99 %), cadmium oxide, or cadmium carbonate. If the metal is used it is first dissolved in mineral acid. A zinc salt is then added to the solution; the amount added depends on the desired shade. The zinc salt is followed by addition of sodium sulfide solution. An extremely finely divided cadmium sulfide or cadmium zinc sulfide precipitate is formed, which does not possess any pigment properties. This intermediate product can also be obtained by mixing the cadmium or cadmium–zinc salt solution with sodium carbonate solution. An alkaline cadmium carbonate or cadmium zinc carbonate precipitate is formed which reacts in suspension with added sodium sulfide solution.

The crude precipitated cadmium yellow is washed and then calcined at approx. 600 °C, at which temperature the cubic crystal form changes to the hexagonal form. This process determines the particle size distribution, which is essential for the pigment properties.

If the starting material is cadmium oxide or cadmium carbonate, it is mixed with sulfur and calcined at approx. 600 °C. After calcination, the pigment product is washed with dilute mineral acid to remove any remaining soluble salts, dried, and ground.

3.2.3. Cadmium Sulfoselenide (Cadmium Red)

Pure cadmium selenide [*1306-24-7*], CdSe, is brownish black and has no pigment properties. Like cadmium sulfide, it is dimorphous and occurs in hexagonal and cubic modifications. Cadmium selenide is insoluble in dilute acid. It readily liberates hydrogen selenide in concentrated hydrochloric acid. It dissolves completely in fuming nitric acid, the Se^{2-} ions being converted to SeO_4^{2-} ions. Cadmium selenide is an *n*-type semiconductor.

Cadmium red consists of cadmium sulfoselenide [*12656-57-4*], [*58339-34-7*], Cd(S,Se), and is formed when sulfur is replaced by selenium in the cadmium sulfide lattice. With increasing selenium content, the color changes to orange, red, and finally dark red. The density of these pigments increases correspondingly from 4.6 to 5.6 g/cm^3 and the refractive index from 2.5 to 2.8. The crystals have cubic or spheroidal habits, the prevalent particle size is 0.3–0.4 μm.

Production. Cadmium red pigments are produced in a similar way to the cadmium yellow pigments. The cadmium salt solution is prepared by dissolving the metal in mineral acid and then sodium sulfide is added. A certain amount of selenium powder is dissolved in the sodium sulfide solution to obtain the desired color shade. In an alternative procedure, the cadmium solution is mixed with sodium carbonate solution to precipitate cadmium carbonate which is reacted with the selenium-containing sodium sulfide solution.

The cadmium red pigment intermediate is obtained as a precipitate which is filtered off, washed, and calcined at approx. 600 °C. As with cadmium yellow, calcination yields the red pigment and determines the particle size, particle size distribution, and color shade. Analogously to the cadmium yellow process, cadmium red can be produced by direct reaction of cadmium oxide or cadmium carbonate with sulfur and the required amount of selenium at approx. 600 °C.

Products in which selenium is totally or partially replaced by tellurium [3.108], [3.109] have not had any commercial application owing to their poor coloristic properties.

3.2.4. Cadmium Mercury Sulfide (Cadmium Cinnabar)

The cadmium in the wurtzite lattice of cadmium sulfide can be replaced by divalent mercury to give cadmium mercury sulfide (cadmium cinnabar). As the quantity of mercury increases, the lattice expands, and the color deepens, changing from yellow to orange and finally dark red. The coloring properties of cadmium cinnabar resemble those of the selenium-containing cadmium red pigments. Use of these mercury-containing pigments can be justified for economic reasons. However, their use is no longer justifiable from the ecological point of view because adequate nontoxic substitutes are available.

3.2.5. Properties and Uses

Cadmium pigments are mainly used to color plastics. They have brilliant, pure shades (yellow, orange, red, and bordeaux), good hiding power, and moderate tinting strength.

Their process-relevant properties are determined by their high thermal stability and chemical resistance to aggressive additives or to molten plastics that have a reducing action (e.g., polyamides). Other advantages for use in plastics are good lightfastness, weather resistance, and migration resistance. The cadmium pigments also protect polyalkenes against ageing because they absorb UV light. Premature embrittlement is prevented and the polymer can be recycled [3.110].

A very useful feature of plastics colored with cadmium pigments is the dimensional stability of injection-molded parts with a large surface area. The combination of these properties is not matched by any other class of colorant.

For economic reasons, cadmium pigments are no longer used for coloring plastics where the pigment property requirements are less stringent. In poly(vinyl chloride) and to a large extent in low-density polyethylene, they have been replaced by less expensive inorganic and organic pigments.

For ecological reasons, cadmium pigments are increasingly being replaced in plastics where high demands are made on the colorants, sometimes with concessions as regards product quality. Plastics processed at high temperature are now colored with high-quality organic pigments (e.g., perylene, quinacridone, and high-quality azo pigments) often in combination with inorganic pigments, that provide for the required covering power. At increasing processing temperatures and outdoor uses the number of substitution product declines. Partly compromises in processability and product quality inclusive shade and brilliance have to be accepted.

A further use of cadmium pigments is in paints and coatings (powder paints, silicone resins, and automotive topcoats) but this is declining.

Cadmium pigments are irreplaceable for coloring enamels, ceramic glazes, and glass to achieve highest brilliance. Brilliant transparent glasses are obtained by adding a small amount of cadmium red, while 10% cadmium red with a cadmium oxide stabilizer produces dark decorative glasses. Other areas of application are ceramics (wall and floor tiles, household and decorative ceramic ware). Very small amounts of cadmium pigments are used for artists' colors. The terms cadmium yellow and cadmium red have become synonyms for very brilliant red and yellow shades.

Cadmium pigments, especially cadmium red, are very sensitive to intensive grinding, which causes loss of brilliance due to an increase in the number of irregular lattice defects. A brilliant red shade may become a dirty brownish red.

The cadmium pigments are lightfast but, like all sulfide pigments, are slowly oxidized to soluble sulfates by UV light, air, and water. This photooxidation is more pronounced with cadmium yellow than with cadmium red and can still be detected in the powder pigment which normally contains 0.1% moisture.

Regulations Affecting Foods. The cadmium pigments fulfill the legal requirements of the EC countries for colorants used in plastics which come in contact with food.

Except for polyamide 6, only microgram quantities of cadmium can be extracted from colored plastics into simulants [3.111], [3.112].

To prevent even minimal intake of cadmium, the European Commission has passed a resolution which states that cadmium pigments should only be used for these purposes if adequate substitutes are not available [3.113].

The use of cadmium pigments in ceramics is controlled in the EC Guideline No. 84/500/EC. International standards [Part 1 (Methods) and Part II (Limit Values)] are as follows: ceramic surfaces ISO 6486, enamel ISO 4531, and glass ISO 7086.

3.2.6. Quality Specifications

The quality requirements and conditions for the supply of cadmium pigments given in ISO 4620 have lost their significance in practice. According to this standard, cadmium pigments must be free from organic colorants, inorganic colored pigments, and other additives. The acid-soluble components (0.1 mol/L HCl) should not exceed 0.2 % Cd (determined according to ISO 4620 and ISO 3856/4); in practice, they are usually < 0.01 % Cd.

Cadmium pigments are sold as homogeneous powders and as preparations mixed with barium sulfate to give the required tinting strength. To reduce the risk of inhalation (see Section 3.2.8), they are supplied as low-dust powders and fine granules.

In dust-free and dispersed form, they are supplied as concentrated plastic granules (masterbatch pellets), as concentrated pastes, and as liquid colors. These products are added at different stages in the processing of plastics.

3.2.7. Economic Aspects

The output of cadmium pigments in the industrialized countries in 1995 was approximately 3000 t, of which approx. 2000 t were produced in Europe and 500 t in the United States. Consumption is distributed approximately as follows: plastics 90 %, ceramics 8 % and others 2 %.

Manufacturers and trade names of cadmium pigments include Cerdec AG, Frankfurt, Germany; Orkem, Narbonne, France (Langdopec Pigments); J. M. Brown, Stoke-on-Trent, U.K.; Cookson-Matthey, Kidsgrove, U.K.; Ferro Corporation, Color Division, Cleveland, USA and Harshaw Chemical Co., Cleveland, USA.

3.2.8. Toxicology and Environmental Protection

Production. Since soluble cadmium compounds have a toxic effect on human beings and the environment, all wastewater originating from the production of cadmium pigments must be treated to remove cadmium. In the EU, the following limits for the pigment production must be observed: 0.2 mg Cd/L in wastewater and 0.3 kg Cd per ton of used cadmium compounds calculated as metal [3.114].

The exhaust air from the production plants and the ventilation equipment must be cleaned. Tubular filters and, more recently, absolute filters have been used to remove cadmium compounds in dust form. With these high-performance filters, it is possible to comply with the limits specified in Germany (0.2 mg Cd/m^3 air, 1 g Cd/h, emission limit; 5 µg Cd/m^2 per day, immission limit). During pigment production, occupational hygiene and safety measures for toxic materials must be observed. Exposure to cadmium can be determined by measurements of the workplace concentration and by examination of the cadmium levels in blood and urine. In Germany, the following maximum biological tolerance levels are allowed at the workplace (BAT values): 15 µg Cd/L urine and 1.5 µg Cd/100 mL blood. The TRK value for cadmium pigments is 0.015 mg/m^3. In the United States, the TLV value for cadmium and its compounds is 0.01 mg Cd/m^3. The OSHA Cadmium Standard has a PEL value of 0.005 mg Cd/m^3.

Toxicity. Cadmium pigments are cadmium compounds with a low solubility, however, small quantities of cadmium dissolve in dilute hydrochloric acid (concentration equivalent to stomach acid), and in cases of long-term oral intake of cadmium pigments, they can accumulate in the human body. On inhalation of subchronic amounts of cadmium pigments, a small proportion of cadmium is biologically available [3.115], [3.116].

Cadmium pigments have no acute toxic effect (oral LD_{50}, rat, > 10 g/kg). The pigments do not have any adverse effects on the skin and mucous membranes.

Genotoxic Effects and Carcinogenicity. Tests for damages of the DNA [3.117] and cell transformation caused by crystalline cadmium sulfide were positive [3.118]. Cadmium sulfide also proved to be carcinogenic by intraperitoneal and after intratracheal administration [3.119]. The significance of such animal studies is being controversially discussed by toxicologists.

Long-term animal feeding studies with various cadmium compounds show no carcinogenic risks. However, inhalation studies with rats, mice and hamsters (with cadmium oxide, cadmium chloride, cadmium sulfate and cadmium sulfide) showed, that all four compounds produced a significant increase of lung cancer in rats [3.120]. The results for mice were inconclusive and no carcinogenicity in hamsters was observed [3.121]. Reinvestigation of the inhalation studies has shown that the test was not applicable to cadmium sulfide. The inhaled liquid cadmium sulfide suspension contained $\leq 40\%$ cadmium sulfate formed by light-induced oxidation.

In 1989 the German Senate Commission for the Assessment of Dangerous Substances classified cadmium and its compounds (including cadmium sulfide) as substances that have carcinogenic properties as found in animal studies. According to MAK classification cadmium pigments and preparations which contain cadmium pigments in concentrations above 0.1 % must therefore be regarded as carcinogenic if they occur in an inhalable form.

In the United States, the National Toxicology Program (NTP) has shown that cadmium and different cadmium compounds (including cadmium sulfide) have a carcinogenic effect in animal experiments.

In 1991 the EU reported the above-mentioned fault in the inhalation studies and classified cadmium sulfide in the EU Cancer List in Category III (suspected carcinogen).

In the EU pure cadmium sulfide is labelled with the symbol T (Toxic) and the Risk phrases 22-40-48/23/25 – Harmful if swallowed. Possible risks of irreversible effects. Toxic: Danger of serious damage to health by prolonged exposure by inhalation and if swallowed.

Cadmium compounds have to be labelled with the symbol Xn (Harmful) and the Risk Phrases 20/21/22 – Harmful by inhalation, in contact with skin and if swallowed –.

The cadmium pigments (cadmium-/zinc sulfide and cadmium sulfide/-selenide) are exempted from labelling. The cadmium pigments are slightly soluble in diluted acids and the soluble cadmium is bioavailable. Therefore some cadmium pigment producers are labelling their cadmium-/zinc sulfide and cadmium sulfide/-selenide pigments like the cadmium compounds.

Limitations of Use. In 1981 Sweden prohibited the use of cadmium pigments (with some exceptions) for ecological reasons. In 1987 Switzerland prohibited the use of cadmium pigments in plastics. Exceptions are possible if valid reasons can be given.

In 1991 the EU passed a directive to prohibit the use of cadmium pigments as colorants for certain plastics that can easily be colored with other pigments [3.122]. A transition period was granted for further series of plastics, these must not be colored with cadmium pigments since 01.01.96. The maximum cadmium content in plastics is limited to 0.01 %, because in this range only technical caused impurities are possible and a technical convenient coloring is not practicable.

The use in coating media is also prohibited since 01.01.96. Use in artists' colors and ceramic products is still permitted.

In the directive was determined, that the prohibited lists shall be revised and perhaps extended after 5 years. The revision takes place in 1997. It will be expected that the limit value of 0.01 % is lowered to 0.005 % cadmium.

In 1989 the CONEG Model (Coalition of North-Eastern Governors) was passed by 14 US States to reduce the environmental pollution by environmentally relevant heavy metals. They decided to limit the total content of the environmentally relevant heavy metals lead, cadmium, mercury and hexavalent chromium. The other States of the U.S.A. accepted the model in the meantime. In the U.S.A. the sum content of the four elements must not exceed 100 ppm in packaging since 1994.

These four heavy metals have been taken over in the EU directive for packaging and packaging waste (94/62/EEC). The limitation of certain heavy metals shall be effective in 1998. The sum content of the four elements shall be reduced in three steps.

	Cd, Pb, Hg, Cr(VI)		Cd, Pb, Hg, Cr(VI)		Cd, Pb, Hg, Cr(VI)
from 01.07.1998	600 ppm	from 01.07.1999	250 ppm	from 01.07.2001	100 ppm

For packaging colored with cadmium pigments the furthermore production, handling or putting into circulation in the EU will be no longer permitted. For cadmium containing returnable crates which are in circulation an exception rule is in discussion in 1997.

3.3. Bismuth Pigments

Bismuth-containing special effect pigments based on platelet-shaped crystals of bismuth oxide chloride (bismuth oxychloride, BiOCl) have been known for a long time (see Section 5.3.1.4.). More recently, greenish yellow pigments based on bismuth orthovanadate [14059-33-7], $BiVO_4$, M_r 324, have attracted increasing interest. They represent a new class of pigments with interesting colouristic properties, extending the familiar range of yellow inorganic pigments (iron yellow, chrome yellow, cadmium yellow, nickel titanium yellow, and chromium titan yellow). In particular they are able to substitute the greenish yellow shades of the lead chromate and cadmium sulphide pigments.

Historical Aspects. Bismuth vanadate occurs naturally as the brown mineral pucherite (orthorhombic), as Clinobisvanite (monoclinic) and as Deyerite (tetragonal). Its syntheses was first reported in 1924 in a patent for pharmaceutical uses [3.123]. The development of pigments based on $BiVO_4$ began in the mid 1970s. In 1976, Du Pont described the preparation and properties of "brilliant primrose yellow" monoclinic bismuth vanadate [3.124]. Montedison developed numerous pigment combinations based on $BiVO_4$ [3.125]. Pigments containing other phases besides $BiVO_4$, e.g. Bi_2XO_6 (X = Mo or W), have been reported by BASF [3.126] and became the first commercial product (trade name Sicopal® Yellow L 1110). Since then Bayer [3.127], Ciba-Geigy [3.128], BASF [3.129] and others have published further methods for the manufacture of pigments based on $BiVO_4$. On the market are now pigments for the use in paints and plastics of the following suppliers: BASF, Ciba-Geigy, Bayer, Capelle, Bruchsaler Farbenfabrik, and Heubach.

Recently, BASF placed a new reddish yellow bismuth vanadate pigment on the market (trade name Sicopal® Yellow L 1600).

3.3.1. Properties

Most of the commercial bismuth vanadate pigments are now based on pure bismuth vanadate with monoclinic structure, though there are still pigments based on the two phase system bismuth vanadate molybdate, $4\,BiVO_4 \cdot 3\,Bi_2MoO_6$. The pure bismuth vanadate pigments show a higher chroma than the two phase systems. In the following the physical and colouristic properties of a pure bismuth vanadate pigment are given (Sicopal® Yellow L 1100, BASF):

Density	6.1 g/cm^3
Refractive index n	2.45
Specific surface area (BET)	13 m^2/g
Oil absorption	33 g/100 g of pigment

Composition:		
	Bi	64.51 wt%
	V	15.73 wt%
	O	19.76 wt%

Bismuth vanadate, C. I. Pigment Yellow 184, is a pigment with a greenish yellow colour, having high tinting strength, high chroma, and very good hiding power. When compared with other yellow inorganic pigments, it most closely resembles cadmium yellow and chrome yellow in its colouristic properties (Fig. 29). Bismuth vanadate shows a sharp increase in reflection at 450 nm and considerably higher chroma than iron yellow or nickel titanium yellow. It has very good weather resistance both in full shade and in combination with TiO_2. Pigment properties follow:

Hue angle H° (HGD, hue grade; CIELAB)	93.7
Chroma *C*ab* (CIELAB)	95.5
Hiding power	At 42.5% by weight in dry film about 70 μm over black/white
Weather resistance (DIN 54002)	alkyd/melamine
Full shade	4–5
Mixed with TiO_2, 1:10	4–5
Chemical resistance in crosslinked paint films	
acid	5 (2% HCl)
alkali	5 (2% NaOH)
Heat stability	> 200 °C

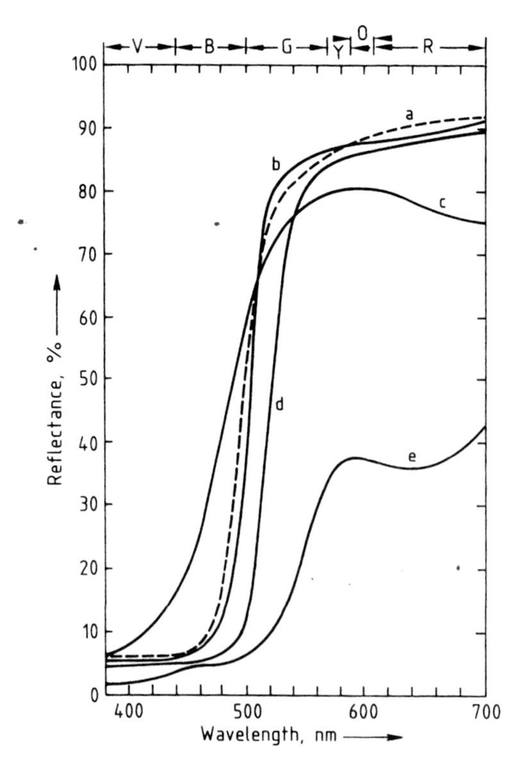

Figure 29. Reflectance curves of yellow pigments
a) $BiVO_4$; b) CdS;
c) (Ti,Ni,Sb)O_2; d) $PbCrO_4$–$PbSO_4$;
e) FeOOH

3.3.2. Production

The production process for bismuth vanadate pigments consists usually of a precipitation reaction followed by a calcination step. This calcination leads to an appreciable increase in particle size as demonstrated in Figure 30. The calcination step is different from producer to producer and can be completely absent depending on the precipitation process and the desired product properties.

$$Bi(NO_3)_3 + NaVO_3 \xrightarrow{1.\ H_3PO_4,\ \ 2.\ NaOH} BiVO_4$$

$$BiVO_4 \xrightarrow{\text{stabilisation}} \xrightarrow{\text{calcination, } 300-700\,^{\circ}C} \text{Pigment}$$

In this process first a fine precipitate is formed by adding an alkaline solution of sodium or ammonium vanadate to an acidic bismuth nitrate solution in presence of a considerable amount of phosphate or in the inverse process by adding the bismuth nitrate solution to a sodium or ammonium vanadate solution. Thereafter the pH is set between 5 and 8 with sodium hydroxide and the precipitate crystallizes to monoclinic bismuth orthovanadate usually by heating to reflux. After this crystallization the pigment is often coated with calcium or zinc phosphate or aluminum oxide to improve the lightfastness and weather-resistance. The final process is usually the calcination followed by wet milling and spray drying. The spray drying process produces a fine granulated dust-free pigment. For the use in plastics the pigment is additionally coated with silica and other components to increase the heat stability in certain polymers, like polyamide, up to 320 °C.

3.3.3. Uses

Bismuth vanadate pigments are used in the manufacture of lead-free, weather resistant, brilliant yellow colors for automobile finishes and industrial paints. They are suitable for the pigmentation of solvent-containing paints, water-based paints, powder coatings, and coil-coating systems. It can be mixed with other pigments

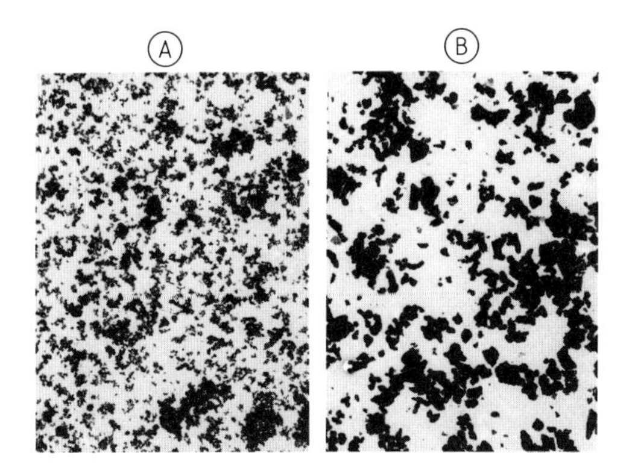

Figure 30. Particle size after precipitation (A) and subsequent heating (B) (magnification × 6000)

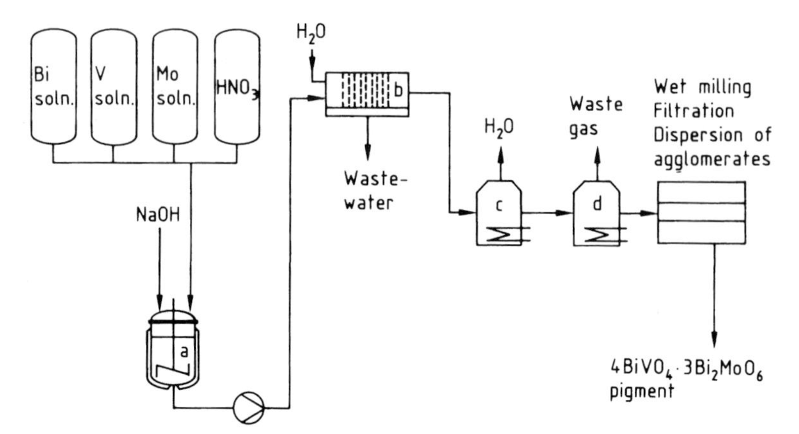

Figure 31. Flow diagram for the production of bismuth vanadate molybdate pigment
a) Reaction vessel; b) Filter; c) Dryer; d) Furnace treatment

especially to produce brilliant colors in the orange, red and green regions like the German standard colors RAL 1021, 1028, 2004, 3020, 6018, and 6029.

Recently, different producers have developed bismuth vanadate pigments for the use in plastics. These pigments are used in every type of polymer, especially in polyethylene and technical plastics.

The annual production of bismuth vanadate pigments is now about 500 t. In the near future an increasing market of 1000 to 2000 t/a is expected.

3.3.4. Toxicology

Bismuth vanadate pigments are not acute toxic whether on inhalative or oral incorporation. They show chronic toxicity on inhalation due to the vanadium content. The "no effect level" for rats is 0.1 mg/m^3 (exposure: 3 months, 6 hrs./day, 5 days/week). The critical factor for the inhalation toxicity is the real amount of pigment in the lungs, and not just the concentration in air. Because of the high density of the pigment and because the supply form is a fine granulate the risk of inhalation is very low. Therefore the dust-free pigments can be handled under usual hygienic working conditions [3.130].

3.4. Chromate Pigments

The most important chromate pigments include the lead chromate (chrome yellow) and lead molybdate pigments (molybdate orange and molybdate red) whose colors range from light lemon yellow to reds with a blue hue. Chrome yellow, molybdate orange, and molybdate red are used in the production of paints, coatings, and plastics, and are characterized by brilliant hues, good tinting strength, and good hiding power. Special treatment of the pigments has allowed continual improvement of their resistance to light, weathering, chemicals, and temperature.

The chromate pigments are also combined with blue pigments (e.g., iron blue or phthalocyanine blue) to obtain high-quality chrome green and fast chrome green pigments. Molybdate orange and molybdate red pigments are often combined with red organic pigments, giving a considerable extension of the color range.

Lead chromates, lead molybdates, chrome greens, and fast chrome greens are supplied as pigment powders, low-dust or dust-free preparations, or as pastes. For standard specifications, see Table 1.

The anticorrosive pigments zinc chromate, zinc tetraoxychromate, and strontium chromate are described in Section 5.2.6

3.4.1. Chrome Yellow

The chrome yellow pigments [*1344-37-2*], C.I. Pigment Yellow 34:77600 and 77603, are pure lead chromate or mixed-phase pigments with the general formula $Pb(Cr,S)O_4$ [3.131] (refractive index 2.3–2.65, density ca. 6 g/cm^3). Chrome yellow is insoluble in water. Solubility in acids and alkalis and discoloration by hydrogen sulfide and sulfur dioxide can be reduced to a minimum by precipitating inert metal oxides on the pigment particles.

Both lead chromate and lead sulfochromate (the latter is a mixed-phase pigment) can be orthorhombic or monoclinic; the monoclinic structure is the more stable [3.132]. The greenish yellow orthorhombic modification of lead chromate is metastable at room temperature, and is readily transformed to the monoclinic modification under certain conditions (e.g., concentration, pH, temperature). The latter modification occurs naturally as crocoite.

Partial replacement of chromate by sulfate in the mixed-phase crystals causes a gradual reduction of tinting strength and hiding power, but allows production of the important chrome yellows with a greenish yellow hue.

Production. In large-scale production, lead or lead oxide is reacted with nitric acid to give lead nitrate solutions, which are then mixed with sodium dichromate solution. If the precipitation solutions contain sulfate, lead sulfochromate is formed as a mixed-phase pigment. After stabilization the pigment is filtered off, washed until free of electrolyte, dried, and ground.

The color of the pigment depends on the ratio of the precipitating components and other factors during and after precipitation (e.g., concentration, pH, temperature, and time). According to WAGNER [3.133], the precipitated crystals are orthorhombic, but change very readily to the monoclinic form on standing; higher temperatures accelerate this conversion. Almost isometric particles that do not show any dichroism can be obtained by appropriate control of the process conditions. Needle-shaped monoclinic crystals should be avoided because they lead to disadvantages such as low bulk density, high oil absorption, and iridescence in the coating film.

Unstabilized chrome yellow pigments have poor lightfastness, and darken due to redox reactions. Recent developments have led to improvements in the fastness properties of chrome yellow pigments, especially toward sulfur dioxide and temperature. This has been achieved by coating the pigment particles with compounds of titanium, cerium, aluminum, antimony, and silicon [3.134]–[3.142].

Carefully controlled precipitation and stabilization provide chrome yellow pigments with exceptional fastness to light and weathering, and very high resistance to chemical attack and temperature, enabling them to be used in a wide field of applications. The following qualities are commercially available:

1) Unstabilized chrome yellows (limited importance)
2) Stabilized chrome yellows with high color brilliance, stable to light and weathering
3) Highly stabilized chrome yellow pigments
   - very stable to light and weathering
   - very stable to light and weathering, and resistant to sulfur dioxide
   - very stable to high temperature, light, and weathering
   - very stable to high temperature, sulfur dioxide, light, and weathering
4) Low-dust products (pastes or powders)

Lightfast chrome yellow pigments that are coated with metal oxides (e.g., of aluminum, titanium, manganese) are produced by Du Pont [3.135].

A chrome yellow that is coated with large amounts of silicate and alumina and which shows improved stability to temperature, light, and chemicals is also produced by Du Pont [3.136].

Bayer describes pigments containing lead chromate stabilized in aqueous slurry with silicate-containing solutions and antimony(III), tin(II), or zinc compounds [3.137].

ICI produces light- and weatherfast chrome yellow pigments stabilized with antimony compounds and silicates in the presence of polyhydric alcohols and hydroxyalkylamines [3.138].

Ten Horn describes a process for the production of lead sulfochromate containing at least 50% lead chromate [3.139]. This has a low acid-soluble lead content (< 5% expressed as PbO, by BS 3900, Part B3, 1965).

BASF produces temperature-stable lead chromate pigments with a silicate coating obtained by hydrolysis of magnesium silicofluoride [3.140]

Heubach has developed a process for the alternate precipitation of metal oxides and silicates [3.141], [3.142]. A homogenizer is used to disperse the pigment particles during stabilization. Products obtained have a very good temperature resistance and very low lead solubility in acid (< 1% Pb by DIN 55770, 1986 or DIN/ISO 6713, 1985).

Continuous processes for the production of chromate pigments have been developed in the United States and Hungary [3.143], [3.144].

Uses. Chrome yellow pigments are mainly used for paints, coil coatings, and plastics. They have a low binder demand and good dispersibility, hiding power, tinting strength, gloss, and gloss stability. Chrome yellows are used in a wide range of applications not only for economic reasons, but also on account of their valuable pigment properties. They are important base pigments for yellow colors in the production of automotive and industrial paints.

Chrome yellow pigments stabilized with a large amount of silicate play a major role in the production of colored plastics (e.g., PVC, polyethylene, or polyesters) with high temperature resistance. Incorporation into plastics also improves their chemical resistance to alkali, acid, sulfur dioxide, and hydrogen sulfide.

Chrome green and fast chrome green mixed pigments are produced by combining chrome yellow with iron blue or phthalocyanine blue (see p. 121).

World production of chrome yellow in 1996 was 37000 t.

3.4.2. Molybdate Red and Molybdate Orange

Molybdate red and molybdate orange [*12656-85-8*], C.I. Pigment Red 104:77605, are mixed-phase pigments with the general formula $Pb(Cr,Mo,S)O_4$ [3.131]. Most commercial products have a MoO_3 content of 4–6 % (refractive index 2.3–2.65, density 5.4–6.3 g/cm^3). Their hue depends on the proportion of molybdate, crystal form, and particle size.

Pure tetragonal lead molybdate, which is colorless, forms orange to red tetragonal mixed-phase pigments with lead sulfochromate. The composition of molybdate red and molybdate orange pigments can be varied to give the required coloristic properties; commercial products usually contain ca. 10% lead molybdate. Lead molybdate pigments have a thermodynamically unstable tetragonal crystal modification that can be transformed into the undesirable stable yellow modification merely by dispersing [3.145]. This is especially true of the bluish varieties of molybdate red which have larger particles whose color can be changed to yellow by shear forces. The tetragonal modification of the lead molybdate pigments must therefore be stabilized after precipitation [3.146], [3.147].

The fastness properties of the molybdate orange and molybdate red pigments are comparable with those of the chrome yellows. As with the chrome yellows, the pigment particles can be coated with metal oxides, metal phosphates, silicates, etc., to give stabilized pigments with high color brilliance and good fastness properties, as well as highly stabilized grades with very good resistance to light, weathering, sulfur dioxide, and temperature, and with a very low content of acid-soluble lead (DIN 55770, 1986 or DIN/ISO 6713, 1985).

The colors of lead molybdate pigments vary from red with a yellow hue to red with a blue hue. Since chrome orange is no longer available (see Section 3.4.3), molybdate orange has become much more important.

Production. In the Sherwin–Williams process, a lead nitrate solution is reacted with a solution of sodium dichromate, ammonium molybdate, and sulfuric acid [3.148]. Instead of ammonium molybdate, the corresponding tungsten salt can be used, giving a pigment based on lead tungstate. The pigment is stabilized by adding sodium silicate (25% SiO_2) and aluminum sulfate ($Al_2(SO_4)_3 \cdot 18\ H_2O$) to the suspension, which is then neutralized with sodium hydroxide or sodium carbonate. The pigment is filtered off, washed until free of electrolyte, dried, and ground. Treatment with silicate increases the oil absorption; it also improves light fastness and working properties.

In the Bayer process molybdate red is formed from lead nitrate, potassium chromate, sodium sulfate, and ammonium molybdate [3.137]. The pigment is then stabilized by adding water glass (28 % SiO_2, 8.3 % Na_2O) to the suspension with stirring, followed by solid antimony trifluoride, stirring for 10 min, and further addition of water glass. The pH is adjusted to 7 with dilute sulfuric acid and the pigment is filtered off, washed free of electrolyte, dried, and ground.

To give the lead molybdate pigments very good stability to light, weathering, chemical attack, and temperature, the same methods are used as those for the stabilization of chrome yellow pigments (see Section 3.4.1) [3.134]–[3.142].

Uses. Molybdate orange and molybdate red are mainly used in paints, coil coatings, and for coloring plastics (e.g., polyethylene, polyesters, polystyrene). The temperature-stable grades are the most suitable for coil coatings and plastics.

Molybdate orange and molybdate red have a low binder demand; good dispersibility, hiding power, and tinting strength; and very high lightfastness and weather resistance. Stabilization (see "Production") also gives high-grade pigments with good resistance to sulfur dioxide and high temperature.

Like the chrome yellows, the molybdate reds are used to produce mixed pigments. Combinations with organic red pigments give a considerably extended color range. Such combinations have very good stability properties because the lightfastness and weather resistance of many organic red pigments are not adversely affected by molybdate pigments.

Total world production of molybdate orange and molybdate red in 1996 was 13000 t.

3.4.3. Chrome Orange

Chrome orange [*1344-38-3*], C.I. Pigment Orange 21:77601, is a basic lead chromate with the composition $PbCrO_4 \cdot PbO$ but is no longer of technical or economic importance.

This product was obtained by precipitating lead salts with alkali chromates in the alkaline pH range. By controlling the pH and temperature, the particle size and thus the hue could be varied between orange and red.

3.4.4. Chrome Green and Fast Chrome Green

Chrome greens, C.I. Pigment Green 15:77510 and 77600, are combined or mixed pigments of chrome yellow (see Section 3.4.1 and iron blue (see Section 3.6) with the formula

$$Pb(S,Cr)O_4/Fe^{III}_4[Fe^{II}(CN)_6]_3 \cdot x\ H_2O$$

Fast chrome greens, C.I. Pigment Green 48:77600, 74160, and 74260, are combinations of chrome yellow (see Section 3.4.1) and phthalocyanine blue or phthalocyanine green. For high-grade fast chrome greens, stabilized and highly stabilized chrome yellows are usually used.

The density and refractive index of the chrome greens and fast chrome greens depend on the ratio of the components of the mixture. Their hues vary from light green to dark blue-green, again depending on the ratio of the components.

Production. Chrome green and fast chrome green pigments can be prepared by dry or wet mixing.

Dry Mixing. The yellow and blue or green pigments are mixed and ground in edge runner mills, high-performance mixers, or mills giving intimate contact of the pigment particles. Excessive increase of temperature must be avoided, because this can lead to spontaneous combustion [3.149]. Differences in the density and particle size of the components can lead to segregation and floating of the pigment components in the coating. Wetting agents are therefore added to avoid these effects [3.150].

Wet Mixing. Pigments with brilliant colors, high color stability, very good hiding power, and good resistance to floating and flocculation are obtained by precipitating one component onto the other. Solutions of sodium silicate and aluminum sulfate or magnesium sulfate are then added for further stabilization [3.148].

Alternatively, the components are wet milled or mixed in suspension and then filtered. The pigment slurry is dried, and the pigment is ground.

Uses. Chrome greens have very good dispersibility, resistance to flocculation, bleeding, and floating and very good fastness properties. This is especially true of the fast chrome greens that are based on high-grade phthalocyanine and highly stabilized chrome yellows. They are therefore used in the same applications as chrome yellow and molybdate red pigments (i.e., for the pigmentation of coating media and plastics).

Pigments consisting of zinc potassium chromates combined with blue pigments are no longer of importance.

3.4.5. Toxicology and Occupational Health

Occupational Health. Precautions have to be taken and workplace concentration limits have to be observed when handling lead- and lead chromate-containing pigments. General regulations exist for all lead-containing materials [3.151]. Concentration limits are as follows:

MAK value (lead)	< 0.1 mg/m^3
BAT values	
Lead (blood)	< 70 µg/dL
Lead (blood– women < 45 years)	< 30 µg/dL

δ-Aminolevulinic acid	
(urine, Davies method)	<15 mg/L
(women $<$ 45 years)	<6 mg/L
TLV-TWA value (lead)	<0.15 mg/m^3

It is accepted that the BAT limit has been complied with if the blood lead level does not exceed 50 μg/dL (or for women of $<$ 45 years, 30 μg/dL).

The EEC Directive EEC 82/605 specifies maximum lead concentrations in the air of $<$ 150 μg/m^3 and permitted blood lead levels of 70–80 μg/dL, with δ-aminolevulinic acid values of 20 mg/g creatinine [3.152].

MAK limits for lead chromates and lead chromate pigments are not given. They are classified as substances suspected of having carcinogenic potential (MAK: Group III B; TLV-TWA: 0.05 mg Cr/m^3, A 2). However, extensive epidemiological investigations have given no indication that the practically insoluble lead chromate pigments have any carcinogenic properties [3.153], [3.154]. Such properties have been reported for the more soluble zinc chromate and strontium chromate pigments.

These chromate pigments can be safely handled if the various rules and regulations regarding concentration limits, safe working practices, hygiene and industrial medicine are adhered to.

Environmental Aspects. Dust emissions from approved manufacturing plants must not exceed 5 mg/m^3 for lead and chromium with a total mass flow exceeding 25 g/h (TA-Luft) [3.155].

According to latest German wastewater legislation [3.156] for inorganic pigment manufacturing processes discharging directly into public stretches of water, mass limits for lead and chromium related to tonnes of average output (t_{prod}) are:

Lead	0.04 kg/t_{prod}
Chromium (total)	0.03 kg/t_{prod}

These requirements recently replaced earlier legislation dating from 1984 [3.157]. Lower limits might be set by local or regional authorities, even for "nondirect" discharges into municipal sewer systems.

Waste containing lead and lead chromate that cannot be recycled must be taken to a special waste disposal site under proper control.

Labelling. In the EC lead chromate and lead chromate pigments must be appropriately labelled. Such substances must be marked with a skull and crossbone (T) [3.158], [3.159]. Additionally, the following risk (R) and safety phrases (S) must be used:

R61	May cause harm to the unborn child.
R62	Possible risk of impaired fertility.
R33	Danger of cumulative effects.
R40	Possible risks of irreversible effects.
S53	Avoid exposure – obtain special instructions before use.
S45	In case of accident or if you feel unwell, seek medical advice immediately (show the label where possible).

In the 21st adaption to EEC Council Directive 67/548, the lead chromate pigments C.I. Pigment Yellow 34 and C.I. Pigment Red 104 have been added individually [3.160]. These pigments are classified in the same manner as lead chromate and must be labelled with a skull and crossbone (T) and the above mentioned risk (R) and safety phrases (S).

According to EEC Council Directive for the labelling of preparations [3.161] in conjunction with the 21st adaption to EEC Council Directive 67/548, Nota 1, [3.160] such materials containing more than 0.5% lead, are labelled in the same way as the pure lead pigment, with a skull and crossbone (T) and the corresponding R and S phrases.

With respect to improved protection of public health, special restrictions on cancerogenic and teratogenic substances and their corresponding preparations have been established by the 14th amendment of EEC Council Directive EEC 76/769 [3.162].

In accordance with the 14th amendment of EEC Council Directive EEC 76/769 [3.162] and revised ChemVerbV [3.163], lead chromate based pigments and preparations are no longer permitted to be used by private consumers and have to be labelled with the phrase "Only for industrial purposes".

Lead containing coatings and paints with a total lead content exceeding 0.15% of the total weight of the preparation must carry the phrase: "Contains lead. Should not be used on surfaces liable to be chewed or sucked by children" in accordance with EEC Council Directive 89/178 [3.164] and German GefStoffV [3.158].

Lead chromate pigments are not permitted for use as coloring materials for plastics for consumer goods [3.165], or for coatings for toys, according to European Standard EN 71 part 3 [3.166].

For transition, the labelling to be used (GGVS/GGVE, ADR/RID) for the pigment as well as for its preparations is class 6.1, No. 62c, hazard symbol 6.1A, and a skull and crossbone, if the lead content soluble in hydrochloric acid [c(HCl) = 0.07 mol/L] exceeds 5%.

3.5. Ultramarine Pigments

Blue ultramarine – blue from over the sea – is the name which European artists of the Middle Ages gave to the pigment derived from lapis lazuli, a semiprecious stone imported mainly from Afghanistan. Ultramarine was the supreme blue of medieval times, but eventually it became scarce and very expensive.

In 1828, J. B. Guimet in France and Christian Gmelin in Germany independently devised similar processes for synthetic preparation. Relatively abundant supplies soon became available, the price fell dramatically, and ultramarine was adopted as a general-purpose color.

Synthetic ultramarines are inorganic powder pigments, commercially available in three colors:

1) Reddish blue, C.I. Pigment Blue 29:77007 [*57455-37-5*]
2) Violet, C.I. Pigment Violet 15:77007 [*12769-96-9*]
3) Pink, C.I. Pigment Red 259:77007 [*12769-96-9*]

A green variety, once produced in small quantities, is no longer available.

Within limits set by stability considerations, the proportions of the chemical constituents can vary. The typical lattice repeat unit of a blue ultramarine is $Na_{6.9}Al_{5.6}Si_{6.4}O_{24}S_{4.2}$. The violet and pink variants differ from the blue mainly in the oxidation state of the sulfur groups. This is reflected in somewhat lower sodium and sulfur contents.

3.5.1. Chemical Structure

Reviews of work on the structure of ultramarine are given in [3.167], [3.168].

Ultramarine is essentially a three-dimensional aluminosilicate lattice with entrapped sodium ions and ionic sulfur groups (Fig. 32). The lattice has the sodalite structure, with a cubic unit cell dimension of ca. 0.9 nm. In synthetic ultramarine derived from china clay by calcination (see Section 3.5.3), the lattice distribution of silicon and aluminum ions is disordered. This contrasts with the ordered array in natural ultramarines.

In the simplest ultramarine structure, equal numbers of silicon and aluminum ions are present and the basic lattice unit is $Na_6Al_6Si_6O_{24}$ or $(Na^+)_6(Al^{3+})_6(Si^{4+})_6(O^{2-})_{24}$ with a net ionic charge of zero as required for structural stability.

The nature of the sulfur groups responsible for the color is reviewed in [3.169]–[3.171]. There are two types of sulfur group in blue ultramarine, S_3^- and S_2^-, both

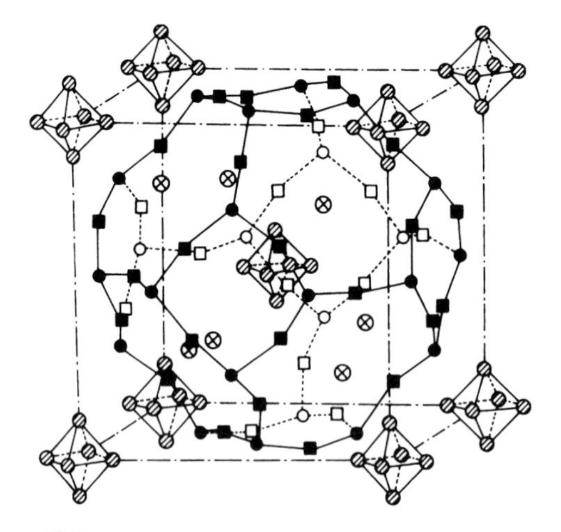

Figure 32. Schematic drawing of the basic structure of standard ultramarine showing the available sites for sulfur and sodium

being free radicals stabilized by lattice entrapment. In the predominant S_3^- species, the spacing between the three sulfurs is 0.2 nm and the angle between them is 103°. S_3^- absorbs a broad energy band in the visible green–yellow–orange region centered at 600 nm, while S_2^- absorbs in the violet–ultraviolet region at 380 nm (Fig. 33).

The basic lattice $(Na^+)_6(Al^{3+})_6(Si^{4+})_6(O^{2-})_{24}$ is derived from $Si_{12}O_{24}$ by substituting six of the silicon ions by aluminum. Every Al^{3+} must be accompanied by a Na^+ so that the overall ionic charge for the structure is zero. Hence, six of the eight sodium sites are always filled by sodiums required for lattice stability and the remaining two sites are filled with sodiums associated with ionic sulfur groups. This means that only one S_3^{2-} polysulfide ion can be inserted into the lattice (as Na_2S_3) even though subsequent oxidation to S_3^- (Section 3.5.3) leads to loss of one of the accompanying sodium ions. This gives a basic ultramarine lattice formula of $Na_7Al_6Si_6O_{24}S_3$.

To increase the sulfur content and thereby improve color quality, the lattice aluminum content can be decreased by including a high-silicon felspar in the manufacturing recipe (Section 3.5.3). This reduces the number of sodium ions needed for lattice stabilization and leaves more for sulfur group equivalence. A typical product would be $Na_{6.9}Al_{5.6}Si_{6.4}O_{24}S_{4.2}$ with a stronger, redder shade of blue than the simpler type.

In violet (Fig. 34) and pink (Fig. 35) ultramarines, the lattice structure is little changed, but the sulfur chromophores are further oxidized, possibly to S_3Cl^-, S_4, or S_4^-.

Ultramarines are zeolites, though lattice paths are restricted by 0.4 nm diameter channels. The sodium ions can be exchanged for other metal ions (e.g., silver, potassium, lithium, copper). Although this produces marked color change, none of the products have commercial value.

3.5.2. Properties

The basic ultramarine color is a rich, bright reddish blue, the red-green tone varying with chemical composition. The violet and pink derivatives have weaker, less saturated colors (see Figs. 33–35, for reflectance spectra).

The color quality of commercial pigments is developed by grinding to reduce particle size and thus enhance tinting strength. Mean particle size ranges typically from 0.7 to 5.0 μm. Figure 36 shows an electron micrograph of a 1.0 μm grade. Fine pigments are lighter in shade and rather greener than coarser grades, but when reduced with white, their color is brighter and more saturated.

With a refractive index close to 1.5, similar to that of paint and plastics media, ultramarine gives a transparent blue in gloss paints and clear plastics. Opacity is obtained by adding a small quantity of a white pigment. Increasing quantities of white give paler shades, and a trace of ultramarine added to a white enhances the whiteness and acceptability.

In many applications ultramarine blue is stable to around 400 °C, violet to 250 °C, and pink to 200 °C. All have excellent lightfastness with a 7–8 rating (full and

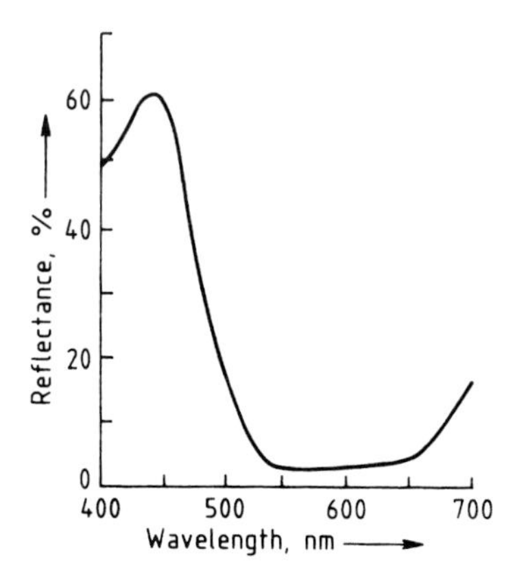

Figure 33. Spectral reflectance distribution of blue ultramarine

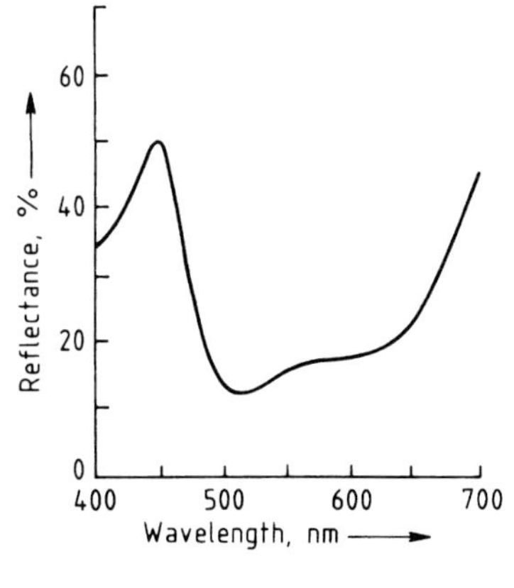

Figure 34. Spectral reflectance distribution of violet ultramarine

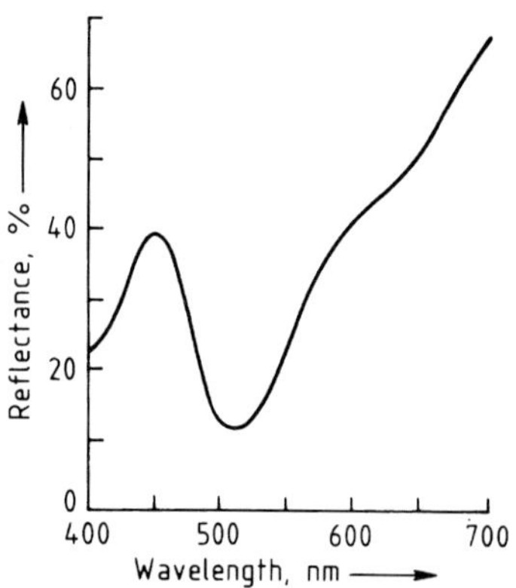

Figure 35. Spectral reflectance distribution of pink ultramarine

reduced shades) on the International Blue Wool Scale. Color fade attributed to light exposure or moderate heat is almost always caused by acid attack. Ultramarines react with all acids, and if there is sufficient acid, the pigment is completely decomposed losing all color to form silica, sodium and aluminum salts, sulfur, and hydrogen sulfide. Evolution of hydrogen sulfide with acids is a useful test for ultramarine.

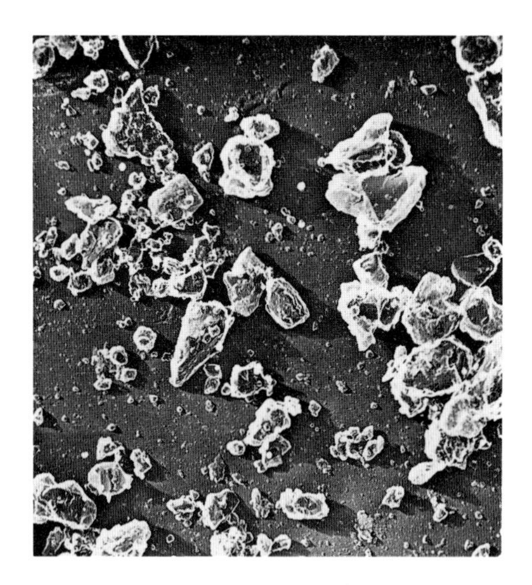

Figure 36. Electron micrograph of blue ultramarine with a mean particle size of 1.0 μm (magnification × 5 800)

Grades resistant to transient acidity are available, in which the pigment particles are protected by a coating of impervious silica. Blue and violet grades are stable in mildly alkaline conditions, but pink tends to revert to a violet shade.

Ultramarines are insoluble in water and organic solvents, so the color does not bleed or migrate from paints or polymers.

Being macromolecular, the fine ultramarine particles have a high surface energy and are cohesive. The finer grades, with their greater surface area, are therefore less easy to disperse than the coarser types, and some are available with pigment surfaces treated to reduce energy and improve dispersability.

Ultramarines adsorb moisture on the external particle surfaces and at the internal surfaces of the zeolite structure. External surface moisture (1–2% according to particle size) is driven off at 100–105 °C, but the additional 1% of internal moisture needs 235 °C for complete removal.

Ultramarine particles are hard and can cause abrasion in equipment handling either dry or slurried pigment.

Specific gravity is 2.35, but bulk density of the pigment powder is much lower, varying with particle size between 0.5 and 0.9 g/cm^3.

Oil absorption also varies with particle size (usually 30–40 g%). The pH lies between 6 and 9. Ultramarine pigments are largely odorless, nonflammable, and do not support combustion.

3.5.3. Production

Ultramarine is made from simple, relatively cheap materials, typically china clay, felspar, anhydrous sodium carbonate, sulfur, and a reducing agent (oil, pitch, coal etc.). Use of a synthetic zeolite has been proposed [3.172], [3.173], but this method is not known to be in commercial use.

Clay Activation. The clay is heated to about 700 °C to destabilize the kaolinite structure by removing hydroxyl ions as water. This can be either a batch process with the clay in crucibles in a directly fired kiln, or a continuous process in a tunnel kiln, rotary kiln, or other furnace.

Blending and Heating Raw Materials. The activated clay is blended with the other raw materials and dry-ground, usually in batch or continuous ball mills, to a mean size approaching 15 μm. Typical recipes (in wt%) are:

	Green-tone	Red-tone
Calcined clay	32.0	30.0
Felspar		7.0
Sodium carbonate	29.0	27.0
Sulfur	34.5	33.0
Reducing agent	4.5	3.0

The ground mixture is heated to about 750 °C under reducing conditions, normally in a batch process. This can be done in directly fired kilns with the blend in lidded crucibles of controlled porosity, or muffle kilns. The heating medium can be solid fuel, oil, or gas. The sodium carbonate reacts with the sulfur and reducing agent at 300 °C to form sodium polysulfide. At higher temperatures the clay lattice reforms into a three-dimensional framework, which at 700 °C is transformed to the sodalite structure, with entrapped sodium and polysulfide ions.

Oxidation. The furnace is allowed to cool to ca. 500 °C when air is admitted in controlled amounts. The oxygen reacts with excess sulfur to form sulfur dioxide, which exothermically oxidizes the di- and triatomic polysulfide ions to S_2^- and S_3^- free radicals, leaving sodium sulfoxides and sulfur as byproducts. When oxidation is complete, the furnace cools and is unloaded – a full kiln cycle can take several weeks. The "raw" ultramarine product typically contains 75 wt% blue ultramarine, 23 wt% sodium sulfoxides, and 2 wt% free (uncombined) sulfur with some iron sulfide.

New Process. After a long period of reduction of capacities and shut downs nowadays new plants using new continuous processes are under construction [3.174], [3.175]. These new processes consume less energy, produce less byproducts conform to environmental regulations and better qualities. The raw materials are the same as in the above described batch process. The processing time is shortened to less than one day.

Purification and Refinement. The purification and refinement operations can be batch or continuous. The raw blue is crushed and ground, slurried in warm water, then filtered and washed to remove the sulfoxides. Reslurrying and wet grinding release the sulfurous impurities and reduce the ultramarine particle size, often to 0.1–10.0 μm. The impurities are floated off by boiling or cold froth flotation.

The liquor is then separated into discrete particle size fractions by gravity or centrifugal separation; residual fine particles are reclaimed by flocculation and filtration. The separated fractions are dried and disintegrated to give pigment grades differentiated by particle size. These are blended to sales-grade standard, adjusting hue, brightness, and strength to achieve specified color tolerance.

Violet ultramarine can be prepared by heating a mid-range blue grade with ammonium chloride at ca. 240 °C in the presence of air. Treating the violet with hydrogen chloride gas at 140 °C gives the pink derivative.

A good ultramarine pigment would meet the following specification:

Tinting strength/standard	± 2% max.
Reduced shade/standard	1 CIELAB unit max.
Free sulfur	0.05% max.
Water-soluble matter	1.0% max.
Sieve residue (45 μm)	0.1% max.
Moisture	
coarse grade	1.0% max.
fine grade	2.0% max.
Heavy metals	traces

For standards see Table 1 (“Ultramarine pigments: specification”).

3.5.4. Uses

The stability and safety of ultramarine pigments are the basis of their wide range of applications which include the following:

Plastics
Paints and powder coatings
Printing inks
Paper and paper coatings
Rubber and thermoplastic elastomers
Latex products
Detergents
Cosmetics and soaps
Artists' colors
Toys and educational equipment
Leather finishes
Powder markers
Roofing granules
Synthetic fibers
Theatrical paints and blue mattes
Cattle salt licks
White enhancement

Plastics. Blue ultramarine can be used in any polymer; violet ultramarine has a maximum processing temperature of 250 °C, and pink ultramarine has a maximum processing temperature of 200 °C. With PVC, acid-resistant grades must be used if color fades during processing. Surface-treated grades are available for enhanced dispersibility. Ultramarines do not cause shrinkage or warping in polyolefins. Ultramarine pigments are permitted worldwide for coloring food-contact plastics.

Paints. Ultramarine pigments are used in decorative paints, stoving finishes, transparent lacquers, industrial paints, and powder coatings. They are not recommended for colored, air-drying paints for outdoor use in urban atmospheres.

Printing Inks. Ultramarine pigments can be used in inks for most printing processes including hot-foil stamping. Letterpress, flexography, and gravure need high-strength grades; lithography needs water-repellent grades; any grade is suitable for screen inks, fabric printing, and hot-foil stamping.

Paper and Paper Coatings. Ultramarine pigments are used to enhance the hue of white paper or for colored paper. They can be added directly to the paper pulp, or used in applied coatings. When added to the pulp, acid-resistant grades must be used if acid-sizing is employed. They are particularly suitable for colored paper for children's use.

Detergents. Ultramarine pigments are widely used to enhance the effects of optical brightening agents in improving whiteness of laundered fabrics [3.176]. They do not stain or build up with repeated use.

Cosmetics and Soaps. Ultramarine pigments are widely used in cosmetics. Pink is not recommended for toilet soaps because of the color shift to violet. Advantages are complete safety, nonstaining, and conformance to all major regulations.

Artists' Colors. This traditional use for ultramarine in all types of media is still an important application. Unique color properties, stability, and safety are highly prized.

Toys and other Articles/Materials for Children's Use. Ultramarine pigments are widely used in plastics and surface coatings for toys, children's paints and finger paints, modeling compositions, colored paper, crayons, etc. They comply with major regulations and standards.

3.5.5. Toxicology and Environmental Aspects

Ultramarines are safe in both manufacture and use. Their only known hazard is the evolution of hydrogen sulfide on contact with acid. Massive exposure of workers in well over a century of manufacture, worldwide use for whitening clothes, and use in a number of countries for whitening sugar have all been without reported ill-effects.

Tests sponsored by Reckitt's Colours confirm that acute oral toxicity in rats and mice (LD_{50}) is greater than 10000 mg/kg. Fish toxicity (LC_{50} in rainbow trout) exceeds 32000 mg/L. Ultramarine is nonmutagenic, nonirritant, and nonsensitizing to skin.

There is no listed threshold limit value or maximum exposure limit for the pigment. Normal practice is to consider it a nuisance dust with TLV 10 mg/m^3. The pigment is not listed as a dangerous substance in the EC nor in any similar national or international classification; neither is it classified as hazardous for disposal.

The production process evolves close to 1 t of gaseous sulfur dioxide and 0.3 t of water-soluble sodium sulfoxides for every tonne of pigment produced. These must be disposed of in an environmentally acceptable manner. If the soluble salts are fully oxidized, they can be discharged safely into tidal waters. Future legislation in all producing countries may require removal of sulfur dioxide from the effluent gases before discharge to the atmosphere.

3.5.6. Economic Aspects

Ultramarines can be categorized as either laundry grades, which are low-strength and sometimes low-purity materials, or as industrial grades, which are high-strength, high-purity pigments.

Factories in several countries produce laundry-grade materials, including several in the People's Republic of China, India, Eastern European countries, and one in Pakistan. Neither the numbers of these units nor their outputs are accurately known.

There are only four major producers of high-grade ultramarine pigments – Dainichi Seika (Japan), Nubiola (Spain), Prayon (Belgium) and Reckitt's Colours (France, United Kingdom) – with two smaller producers in Austria and Colombia. In 1990 total worldwide production was ca. 20000 t/a.

3.6. Iron Blue Pigments

The term iron blue pigments as defined in ISO 2495 has largely replaced a great number of older names (e.g., Paris blue, Prussian blue, Berlin blue, Milori blue, Turnbull's blue, toning blue, and nonbronze blue). These names usually stood for insoluble pigments based on microcrystalline Fe(II)Fe(III) cyano complexes; many were associated with specific hues. A standardized naming system has been demanded by users and welcomed by manufacturers, and has led to a reduction in the number of varieties [3.177].

Iron blue [*14038-43-8*], C.I. Pigment Blue 27:77510 (soluble blue is C.I. Pigment Blue 27:77520), was discovered in 1704 by Diesbach in Berlin by a precipitation reaction, and can be regarded as the oldest synthetic coordination compound. Milori was the first to produce it as a pigment on an industrial scale in the early nineteenth century [3.178].

3.6.1. Structure

X-ray and infrared spectroscopy show that iron blue pigments have the formula $M^IFe^{II}Fe^{III}(CN)_6 \cdot H_2O$ [3.179]. M^I represents potassium, ammonium, or sodium, of which the potassium ion is preferred because it produces excellent hues in industrial manufacture.

The crystal structure of the $Fe^{II}Fe^{III}(CN)_6$ grouping is shown in Figure 37. A face-centered cubic lattice of Fe^{2+} is interlocked with another face-centered cubic lattice of Fe^{3+} to give a cubic lattice with the corners occupied by iron ions. The CN^- ions are located at the edges of the cubes between each Fe^{2+} ion and the neighboring Fe^{3+}; the carbon atom of the cyanide is bonded to the Fe^{2+} ion and the nitrogen atom is coordinatively bonded to the Fe^{3+} ion. The alkali-metal ions and water molecules are inside the cubes formed by the iron ions.

The presence of coordinative water is essential for stabilization of the crystal structure. Removal of this water, however carefully carried out, destroys the pigment properties.

Many investigations helped to elucidate the structure of iron blue [3.181]–[3.184].

3.6.2. Production

Iron blue pigments are produced by the precipitation of complex iron(II) cyanides by iron(II) salts in aqueous solution. The product is a whitish precipitate of iron(II) hexacyanoferrate(II) $M^I_2Fe^{II}[Fe^{II}(CN)_6]$ or $M^{II}Fe^{II}[Fe^{II}(CN)_6]$, (Berlin white), which is aged and then oxidized to the blue pigment [3.180].

Potassium hexacyanoferrate(II) or sodium hexacyanoferrate(II) or mixtures of these salts are usually used. When the pure sodium salt or a calcium hexacyanofer-

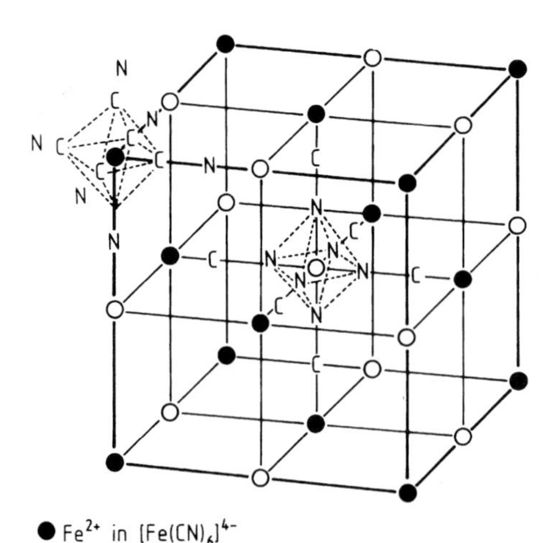

Figure 37. Crystal structure of iron blue [3.180]

rate(II) solution is used, the pigment properties are obtained by adding a potassium or ammonium salt during the precipitation of the white paste product or prior to the oxidation stage.

The iron(II) salt used is crystalline iron(II) sulfate or iron(II) chloride solution. The oxidizing agent can be hydrogen peroxide, alkali chlorates, or alkali dichromates. Industrial precipitation is carried out batchwise in large stirred tanks by simultaneous or sequential addition of aqueous solutions of potassium hexacyanoferrate(II) and iron(II) sulfate to a dilute acid. The filtrate from the white paste product must contain a slight excess of iron. Temperature and concentration of the starting solutions have a decisive influence on the size and shape of the precipitated particles. The suspension of white paste is aged by heating. The ageing period varies in length and temperature depending on the required properties of the finished pigment. This is followed by the oxidation to form the blue pigment by adding hydrochloric acid and sodium or potassium chlorate [3.185]. Finally, the suspension of the blue pigment is pumped into filter presses either immediately or after having been washed with cold water and decanted. After filtering, it is washed until free of acid and salt. The washed filter cake (30–60 % solids) is carefully dried in tunnel or belt dryers to form irregularly shaped pieces which are ground, packed into bags, and stored in silos. Another possibility is to form rods or pellets by extruding the washed filter cake with a granulator. On drying, this gives a dust-free iron blue pigment.

Dispersibility can be improved by adding organic compounds to the pigment suspension before filtering to prevent the particles from agglomerating too strongly on drying, for example [3.186], [3.187]. In another method (the Flushing process), the water in the wet pigment paste is replaced by a hydrophobic binder [3.188]. Although these and other methods of pigment preparation produce fully dispersible products consisting mainly of iron blue and a binder [3.189]–[3.191], they have not become established on the market.

A "water-soluble" blue can be manufactured by adding pepticizing agents (the latter improve the water solubility via an emulsifying action). This forms a transparent colloidal solution in water without the use of high shear forces [3.180].

For quality standards, see Table 1 ("Iron blue pigments: Methods of analysis" and "Specification").

3.6.3. Properties

Hue, relative tinting strength, dispersibility, and rheological behavior are the properties of iron blue pigments with the most practical significance. Other important properties are the volatiles content at 60 °C, the water-soluble fraction, and acidity (ISO 2495). Pure blue pigments are usually used singly (e.g., in printing inks) and do not need any additives to improve them. Finely divided iron blue pigments impart a pure black tone to printing ink.

Due to their small particle size (see Table 26 and Figs. 38 and 39), iron blue pigments are very difficult to disperse. A graph of cumulative particle size distribu-

Table 26. Physical and chemical properties of iron blue pigments (VOSSEN BLAU® and MANOX® Blue grades)

Type	VOSSEN BLAU® 705	VOSSEN BLAU® 705 LS[g]	VOSSEN BLAU® 724	MANOX® Blue 460 D	MANOX EASISPERSE® HSB 2
Color Index Number	77510	77510	77510	77510	77510
Color Index Pigment	27	27	27	27	27
Tinting strength[a]	100 pure blue	100 pure blue	100 pure blue	115 pure blue	95[h] pure blue
Oil absorption[b], g/100 g	36–42	40–50	36–42	53–63	22–28
Weight loss on drying[c], %	2–6	2–6	2–6	2–6	2–6
Tamped density[d], g/L	500	200	500	500	550
Density[e], g/cm^3	1.9	1.9	1.9	1.8	1.8
Mean diameter of primary particles, nm	70	70	70	40	80
Specific surface area[f], m^2/g	35	35	35	80	30
Thermal stability, °C	150	150	150	150	150
Resistance to acids	very good	very good	very good	very good	very good
Resistance to alkalines	poor	poor	poor	poor	poor
Resistance to solvents	very good	very good	very good	very good	very good
Resistance to bleeding	very good	very good	very good	very good	very good

[a] DIN ISO 787/XVI and DIN ISO 787/XXIV. [b] DIN ISO 787/V, ASTM D 281, or JIS K 5101/19 (JIS: Japanese Industrial Standard). [c] DIN ISO 787/II, ASTM D 280, or JIS K 5101/21. [d] DIN ISO 787/XI or JIS K 5101/18. [e] DIN ISO 787/X or JIS K 5101/17. [f] DIN 66131. [g] LS = Luftstrahlmühle (air jet mill). [h] Surface-treated easily dispersible type.

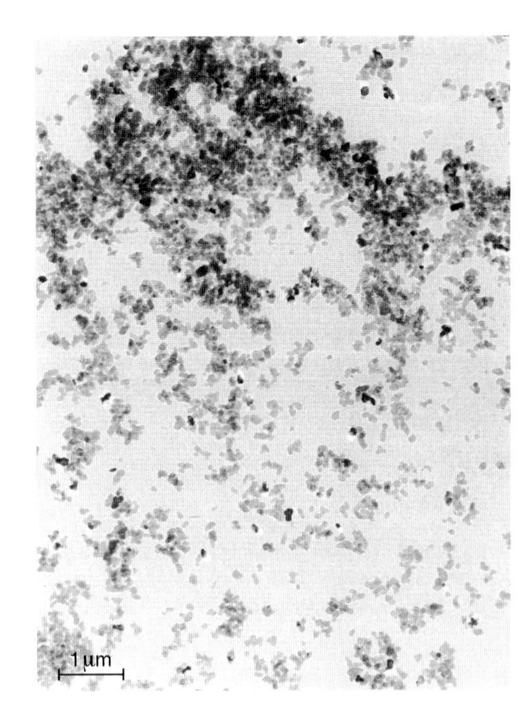

Figure 38. Electron micrograph of an iron blue pigment of small particle size (Manox Blue® 460 D)

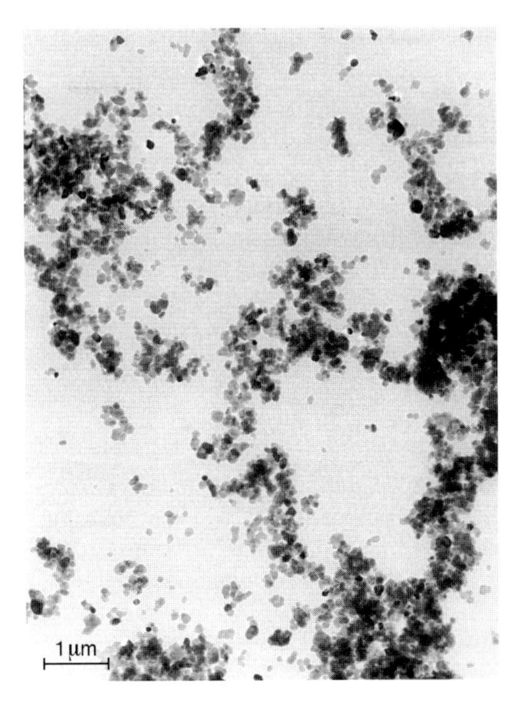

Figure 39. Electron micrograph of an iron blue pigment of normal particle size (Vossen-Blau® 705)

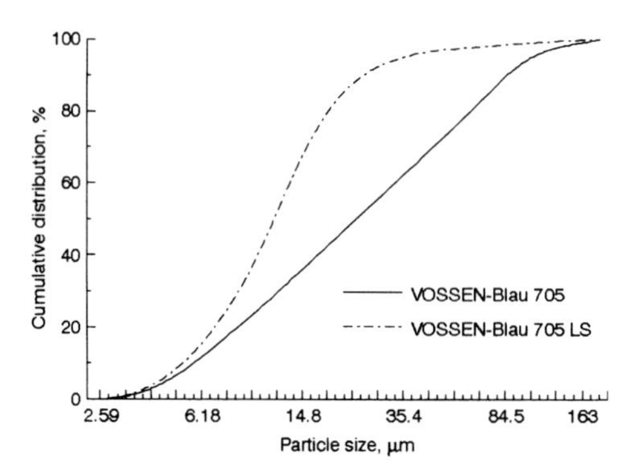

Figure 40. Cumulative particle size distribution curve of a normal (705) and a micronized (705 LS) iron blue pigment of equal primary particle size
LS = Luftstrahlmühle (air jet mill)

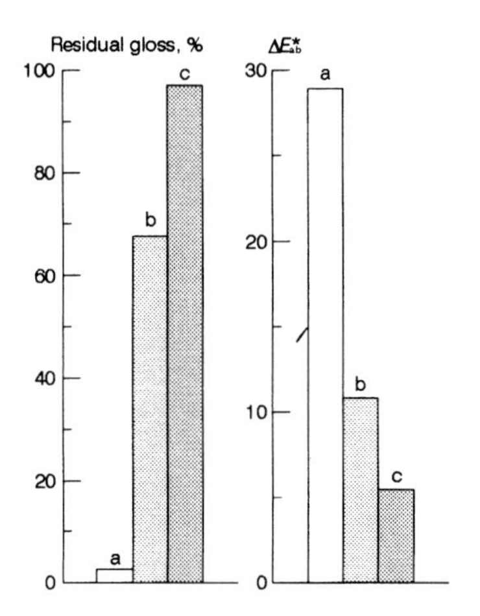

Figure 41. Residual gloss and ΔE^*_{ab} values for isocyanate-cross-linked polyacrylate resins that contain 15 wt% Vossen Blau® 2000 (older pigment type which has been replaced by Manox® Blue 460 D) relative to the binder and 15 wt% TiO_2 (rutile) relative to the iron blue pigment after 1000 h fast exposure to UV [3.193]
a) Without clearcoat; b) With clearcoat but without UV protection; c) With clearcoat and UV protection

tion is given in Figure 40 for a commercial quality iron blue and for a micronized grade with similar primary particle size. The micronized grade gives greater tinting strength in dry mixtures than the blues obtained from standard grinding. The average size of the aggregates in the micronized material is ca. 10 µm compared with ca. 35 µm for the normal quality product.

Iron blue pigments are thermally stable for short periods at temperatures up to 180 °C, and therefore can be used in stoving finishes. The powdered material presents an explosion hazard, the ignition point is 600–625 °C (ASTM D 93-52). The pigments are combustible in powder form, ignition in air being possible above 140 °C [3.180].

Iron blue pigments used alone have excellent light- and weatherfastness. When mixed with white pigments, these properties can disappear [3.192]. Recent investigations have shown that a topcoat (as commonly applied in automobile manufacture) overcomes this problem [3.193]. Figure 41 shows changes in residual gloss and color after a short weathering period. The pigments are resistant to dilute mineral acid and oxidizing agents, and do not bleed. They are decomposed by hot, concentrated acid and alkali. Other properties are listed in Table 26.

3.6.4. Uses

Total production of iron blue in 1975 was ca. 25000 t/a, but in 1995 it was only ca. 13000–15000 t/a. The main consumer in Europe and USA is the printing ink industry. The second largest use in Europe, especially of micronized iron blue pigments, is for coloring fungicides, but use in the paint industry is decreasing.

Printing Ink Industry. Iron blue pigments are very important in printing, especially rotogravure, because of their deep hue, good hiding power and economic cost/performance-basis. Iron blue is often mixed with phthalocyanine pigments for multicolor printing. Another important use is in controlling the shade of black printing inks. Typical amounts used are 5 % – 8 % for full-shade rotogravure inks and 2 % – 8 % for toning black gravure and offset inks.

Iron blue pigments are used in the manufacture of single- and multiple-use carbon papers and blue copying papers, both for toning the carbon black and as blue pigments in their own right [3.180].

Toning of Black Gravure Inks. For the toning of black gravure inks, for example, 2 % to 6 % of Vossen-Blau 705 are used together with 6 % to 12 % of carbon black. Combinations with red pigments with a blue undertone are also common. When using organic pigments, resistance to solvents must be taken in account. Because of the poor dispersibility of iron blue compared with carbon black it is both economical and practical to disperse the blue pigment in a separate step.

While the visual judgement of black is influenced by the individual ability of the observer to distinguish small color differences in deep black, it is possible, with the help of photometric measurements, to graphically interpret objective evaluations by means of physical data [3.194].

Figure 42 illustrates the color changes of a low structure LCF-type carbon black by addition of Pigment Blue 27 (Vossen-Blau 705) and Pigment Violet 27 or by toning with a 4:1 combination of Pigment Blue 27 and Pigment Red 57:1. A mixture of asphalt resin, calcium/zinc resinate and phenol resin was used as a binder. The pigment concentration for all was 13,2 %. The toner was added in 2,2 % steps up to 6,6 % with a simultaneous reduction from 13,3 % to 6,6 %.

Toning of Black Offset Printing Inks. The basic requirements for the successful use of iron blue as a toning agent in offset printing inks are resistance to damping or "fountain" solutions and good dispersibility. "Resistance" is understood here as the hydrophobic characteristics of the pigment.

This property prevents wetting of the pigment by water and therefore its peptisation. Non-resistant iron blue can render the ink useless by adsorbing water to above the normal content. A negative side-effect of peptisation is the "dissolution" of the blue pigment from the printing ink and the resulting blue coloration of the fountain solution with the familiar problems of printing-plaste contamination, as known as scumming or toning.

The combined dispersion of pigments is only practical with colorants of similar dispersibility. Toning agents with a considerable higher resistance to dispersion than

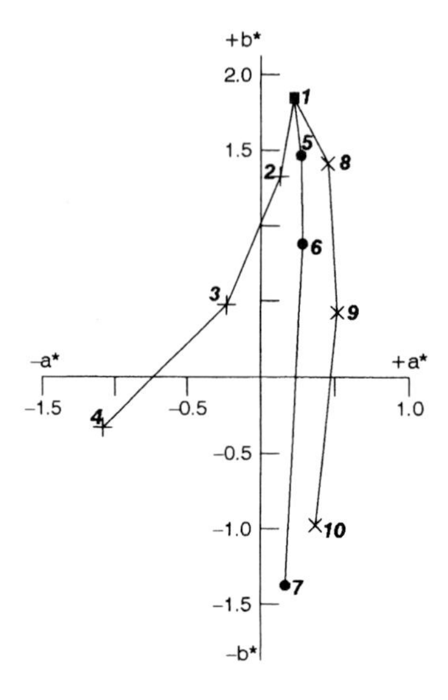

Figure 42. Color coordinates of black gravure inks with different toning + 2–4: Pigment Black LCF*/Vossen-Blau 705; • 5–7: Pigment Black LCF/[VB 705–Pigment Red 57:1 (4:1)]; × 1–10: Pigment Black LCF*/[VB 705 Pigment Violet 27 (2:1)]; ■ 1: Pigment Black LCF = Printex 35

carbon black are therefore delivered by the manufacturer in the form of a predispersed paste or must be ground separately by the user.

Developments in the fields of iron blue technology have overcome these problems. A new generation of pigments has been generated which covers both the demand for a sufficient resistance against damping solutions and the request for a good dispersibility. Easily dispersible iron blue is prefered to be used for combined dispersion with carbon black in the so-called "Co-Grinding" process.

In the following section, the coloristic effects of those iron blue pigments are described, as obtained in toning experiments involving a LCF-type carbon black, in comparison with Pigment Blue 15:3 and Pigment Blue 61.

The pigment concentration of the inks is 24%, i.e. the amount of carbon black was reduced correspondingly with the addition of 3%, 6%, or 9% of blue pigment. Black inks containing 15% to 24% of carbon black are excluded from this experiment, and are presented to give an additional coloristic description of "pigment" black as regards the development of the hue when used as a self-color pigment in increasing concentrations.

In the visual and the colorimetric evaluation the color location change of the carbon black when used as a self-color pigment is noticeable since there is a tendency to a black with a blue untertone at higher pigment concentration, even without the addition of toning agent. In Figure 43 these are the color locations 1, 2, 3, and 4, starting with 15% carbon black and in increasing additions in steps of 3% to a maximum concentration of 24%.

However, it is also clear that without the addition of a blue pigment the achromatic point cannot be achieved.

By adding various toning agents the desired color hue is achieved – although with varying red/green spread. Iron blue (Manox Easisperse 154) brings about a clear shift towards green with a relatively small shift in the blue direction (see color locations 5, 6, and 7). The required target is more successfully achieved by the use

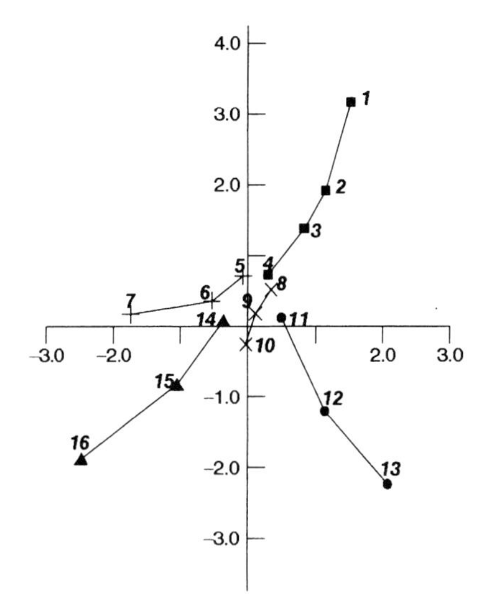

Figure. 43. Color locations of black offset printing inks with different toning ■ 1–4: Pigment Black LCF (= Printex 35); + 5–7: Pigment Black LCF/Manox Easisperse; × 8–10: Pigment Black LCF/[Manox Easisperse/Pigment Red 57:1 (4:1)]; ● 11–13: Pigment Black LCF/Pigment Blue 61; ▲ 14–16: Pigment Black LCF/Pigment Blue 15:3

of a mixture of iron blue and Pigment Red 57:1 in the ratio 4:1. The color locations 8, 9, and 10 illustrate useful ways of approaching the achromatic point with the addition of 3%, 6%, and 9% of the mixture.

The addition of 3% of Pigment Blue 61 already results in a significant step towards blue/red and shows almost identical incremental changes with further additions (see color locations 11, 12, and 13). In addition, the bronze effect occurs which intensifies with increasing distance from the achromatic point. With Pigment Blue 15:3 in numerically equivalent increments, the hue of the black ink moves towards blue/green in the opposite direction from the achromatic point and with a negative shift (color locations 14–16).

Agriculture. Since ca. 1935, and especially in Mediterranean countries, blue inorganic fungicides based on copper and used for treating vines have largely been replaced by colorless organic compounds. Micronized iron blue pigments are used to color these fungicides (normally at a concentration of 3–8 wt%), so that even small amounts become visible due to the high color intensity, and precise control is possible. The fungicide is usually milled or mixed with a micronized iron blue pigment [3.195].

A welcome side effect of treating fungi (e.g. peronospora plasmopara viticola) with iron blue is the fertilizing of vines in soils that give rise to chlorosis. Leaf color is intensified, ageing of the leaves is retarded, and wood quality ("ripeness") is also improved [3.196]–[3.198]. Iron is necessary for chlorophyll synthesis, which improves grape quality and yield. Other iron salts do not have this effect [3.197].

Paints and Coatings. Iron blue pigments are used in the paint industry, especially for full, dark blue colors for automotive finishes. A full shade with good hiding power is produced by 4–8% iron blue pigments.

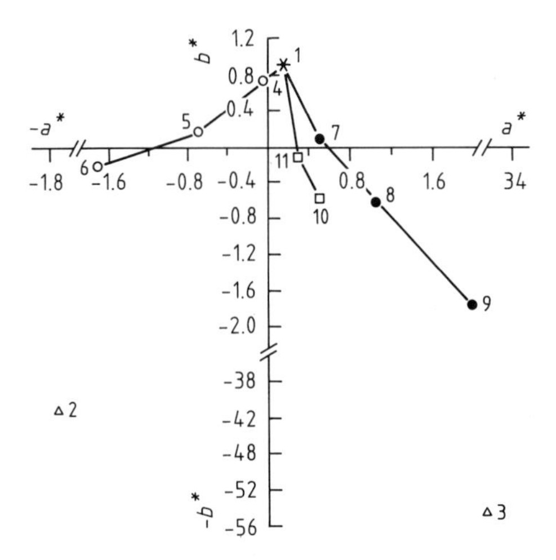

Figure 44. Stepwise approach to the achromatic point of a Printex 35 offset printing ink by using Manox Easisperse and/or Pigment Blue 61 as toning pigment [3.199]
1) Printex 35; 2) Manox Easisperse; 3) Pigment Blue 61; 4)–6) Printex 35/Manox Easisperse; 7)–9) Printex 35/Pigment Blue 61; 10), 11) Printex 35/Manox Easisperse/Pigment Blue 61
Points 4 and 7 contain 3% blue pigment; points 5 and 8 6%; points 6 and 9 9%. Point 10 results by mixing 2% Manox Easisperse and 4% Pigment Blue 61; point 11 by mixing in the ratio 3:3.

Paper. Blue paper can be produced by adding "water-soluble" iron blue pigment directly to the aqueous phase. Alternatively, a suitable iron blue pigment can be ground together with a water-soluble binder, applied to the paper, dried, and glazed (quantity applied: ca. 8% in the dispersion).

Pigment Industry. The importance of iron blue in the production of chrome green and zinc green pigments has greatly increased worldwide (see Section 3.4.4).

Medical Applications. Iron blue has become important as an agent for decontaminating persons who have ingested radioactive material. The isotope ^{137}Cs which would otherwise be freely absorbed via the human or animal digestive tract exchanges with the iron(II) of the iron blue [3.200], [3.201] and is then excreted in the feces [3.202]. Gelatin capsules containing 500 mg iron blue are marketed as Radiogardase-Cs (Heyl). Thallium ions have been found to behave similarly [3.203]–[3.205]. The gelatin capsules for this purpose are sold as "Antidotum Thalii" (Heyl) [3.206].

3.6.5. Toxicology and Environmental Aspects

Blue pigment compounds show no toxicity in animal studies therefore it is not expected to cause any adverse effects on human health. No toxic effects were reported in humans when blue pigment compounds were used experimentally or therapeutically.

Toxikokinetic studies showed, that the adsorption of iron blue pigments is very low. Following intravenous injection of a ^{59}Fe-radio labelled iron blue pigment, the $[^{59}Fe(CN)_6]^{4-}$ ion was rapidly and virtually completely excreted with the urine. After oral administration of ferric cyanoferrate (^{59}Fe) approx. 2% of the labelled hexacyanoferrate ion was adsorbed by the gastro-intestinal tract [3.207]. Most of the substance is excreted with the feces [3.208] and there was no evidence of its decomposition.

The decomposition of blue pigment salts to toxic cyanide in aqueous systems is very low. The CN-release of $KFe[Fe(CN)_6]$ in artificial gastric or intestinal juice was 141 or 26 µg/g/5 hours respectively and in water 37 µg/g/5 hours. The corresponding figure of $Fe[4Fe(CN)_6]$ were 64, 15 and 22 µg/g/5 hours [3.209].

In the breath of rats after i.p. injection of ^{14}C-labelld $KFe[Fe(CN)_6]$ less than 0.01% (detection limit) was found, whereas in another study 0.04–0.08% of the orally administered dose was found in the exhaled air [3.210]. It can be concluded, that the hexacyanoferrate(II) complex disintegrates only to a small extent in the intestinal tract after oral administration. This is confirmed by the results of acute oral toxicity studies which show in high doses no clinical symptoms or lethality. The LD_{50} values are above 5000–15000 mg/kg (limit tests) [3.211]–[3.213].

In primary irritation tests no or only slight effects were seen at the skin or in the eyes of the treated rabbits respectively [3.213], [3.214]. No skin sensitisation occurred in a Guinea pig maximisation test [3.211].

The subchronic (90–120 days) consumption of iron blue pigments at concentrations of 1–2% in the food or drinking water influenced slightly the body weight grain, but no other clinical signs or histopathological changes were observed [3.215]–[3.218]. After the administration of daily doses of 200 or 400 mg/kg for ten days to dogs the body weight gained and the general condition remained unaffected [3.219].

In a bacterial test system [ames test] no increase of mutagenicity was detected without or in presence of a metabolic system [3.213].

In human volunteers who received 1.5 or 3.0 g ferric cyanoferrate for up to 22 days apart from a slight obstipation no effects were reported [3.219], [3.220].

$Fe_4[Fe(CN)_6]$ can bind caesium therefore iron blue pigments are used in clinical practice as an antidote for the treatment of humans contaminated with radioactive caesium (see also Section 3.6.4). Clinical use of iron(III) ferrocyanide in doses up to 20 g/d for decontaminations of persons exposed to radio caesium has not been associated with any reported toxicity [3.209].

Blue pigment salts are also used as an effective antidote for thallium intoxications. Ferric cyanoferrate interferes with the enterosystemic circulation of thallium ions and enhances their fecal excretion [3.221].

In a semistatic acute fish toxicity test (Leuciscus idus, melanotus, fresh water fish) a saturated solution with different blue pigment compounds (with unsolved material on the bottom or filtrated solution) no death occurred within 96 hours. Based on the quantity weighed the No Observed Effect Level (NOEL) is greater than 1000 mg/l (nominal concentrations) [3.222].

The bacterial toxicity was measured according DEV, DIN 38412, L3 (TTC(2,3,5-triphenyl-2*H*-tetrazolium chlorid) test). The result gives an EC_{50} (effective concentration) varying between 2290 and 14700 mg/L, and estimated NOEC values in the range of < 10 to 100 mg/L [3.223].

There are no harmful effects on fish, but the toxic effects on bacteria constitute a slight hazard when iron blue pigments are present in water.

4. Black Pigments (Carbon Black)

The light absorption coefficient of black pigments determines their optical quality. Their color intensity and hiding power depend on the particle size and particle size distribution. The most important black pigments are carbon blacks, iron oxides (Section 3.1.1), and mixed metal oxides (Section 3.1.3).

Carbon black [*1333-86-4*] and soot are formed either by pyrolysis or by partial combustion of vapors containing carbon. Soot as an unwanted byproduct of combustion (e.g., in chimneys or diesel engines) is a poorly defined material. Besides carbon black particles, it often contains significant amounts of ash and high amounts of polycyclic aromatic hydrocarbons (PAH) [4.1]. Residual hydrocarbons, which can be determined by extraction with solvents (e.g., toluene), can account for 30 wt%.

On the other hand, the term "carbon black" is used for a group of well-defined, industrially manufactured products. They are produced under carefully controlled conditions. The physical and chemical properties of each type of carbon black are kept within narrow specifications. Carbon black is one form of highly dispersed elemental carbon with extremely small particles. Depending on the production process and the raw materials, carbon black also contains chemically bound hydrogen, oxygen, nitrogen, and sulfur.

Due to its excellent pigmentation properties, especially its light stability and universal insolubility, carbon black has been used as a black pigment since early times. It was produced for this purpose by burning oils, fats, or resinous materials. The flame was either quenched on a cool surface (impingement black) or cooled in special chimneys (lamp black), where the carbon black was deposited.

Both processes are still used in industry for the production of carbon black. The channel black process, a process for making impingement blacks, has been used in the United States since the end of the 19th century. This process, which has now been abandoned because of economic and environmental considerations, used natural gas as raw material. A similar process for the production of impingement blacks, the "Degussa gas black process," is still used today.

The increasing demand led to new production processes. The most important process today is the furnace black process. It was developed in the United States in the 1930s and substantially improved after World War II. It is a continuous process, which allows the production of a variety of carbon black types under carefully controlled conditions. Nearly all rubber grades and a significant part of pigment-grade carbon blacks are now manufactured by the furnace black process. Nevertheless, other production processes, such as gas black, lamp black, thermal black, and acetylene black processes, are still used for the production of specialties.

While carbon black was exclusively used as a pigment until the beginning of this century, its use as an active filler in rubber was the starting point for a new rapidly expanding application. In the production of automobile tires, it was found that treads filled with carbon black had a markedly higher abrasion resistance than those filled with zinc oxide. This discovery, together with increasing use of motor vehicles, was the basis for the present importance of carbon black as a filler in rubber.

Today at least 35 different types of carbon black are used as fillers in rubber, and about 80 types of carbon black are used in pigments or special applications. The total world production in 1994 was 6×10^6 t. More than 90% of the carbon black was produced for the use in the rubber industry.

4.1. Physical Properties

Morphology. Electron micrographs show that the primary particles of carbon black are almost spherically shaped. In general, a larger number of such primary particles builds aggregates in the form of chains or clusters. In practice, the degree of aggregation is called the "structure" of carbon black. These aggregates tend to agglomerate.

The mean primary particle diameter, the width of the particle size distribution, and the degree of aggregation can be varied within relatively wide ranges (Figs. 45–47) by varying the production process and several process parameters.

The diameter of the primary particles ranges from 5 to 500 nm. Diffraction patterns produced by the so-called phase-contrast method in high-resolution electron microscopy show that the spherical primary particles are not amorphous (Fig. 48). They consist of relatively disordered nuclei surrounded by concentrically deposited carbon layers [4.2]. The degree of order increases from the center to the periphery of each particle, a phenomenon important to the understanding of the chemical reactivity of carbon black.

The carbon atoms within each layer are arranged in almost the same manner as in graphite. The layers are nearly parallel to each other; however, the relative position of these layers is random, so that there is no order as in the *c* direction of graphite ("turbostratic structure") [4.3]. X-ray diffraction permits the determination of "crystalline" regions within the carbon black primary particle. These regions are

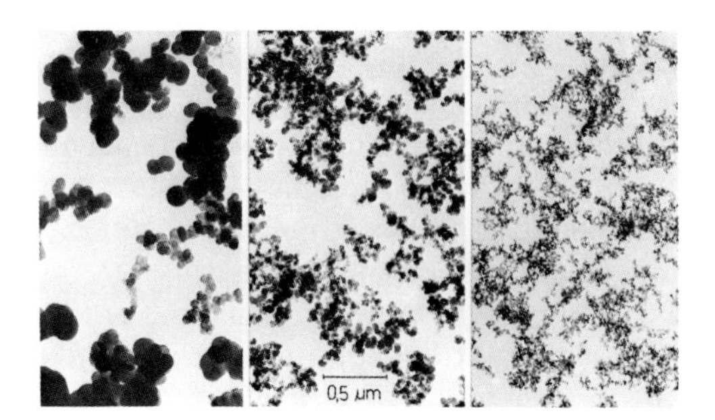

Figure 45. Carbon blacks of different primary particle sizes and specific surface areas
From left to right: lamp black (mean primary particle diameter 95 nm, BET surface area 21 m^2/g), furnace black (27 nm, 90 m^2/g), finely divided gas black (13 nm, 320 m^2/g)

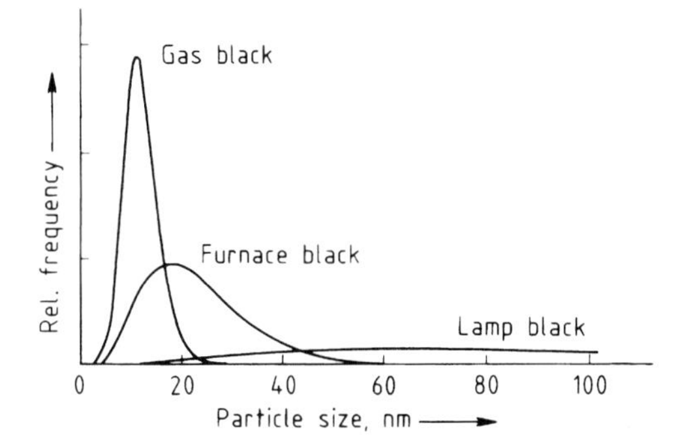

Figure 46. Particle distribution curves for the carbon blacks of Figure 45

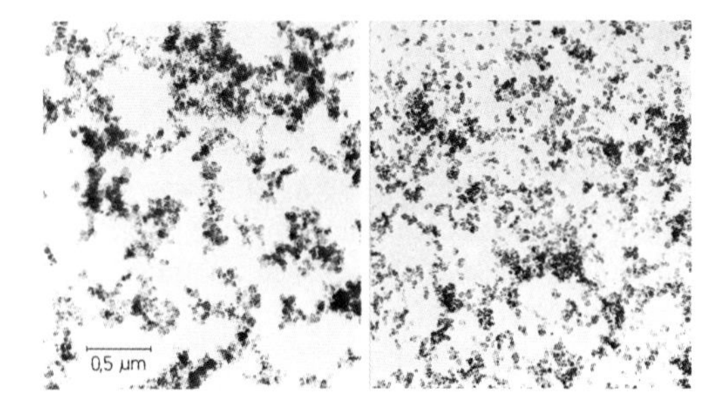

Figure 47. Furnace blacks of different aggregation degrees

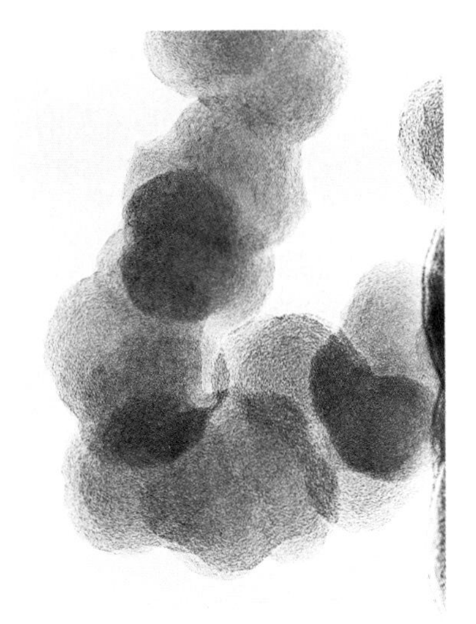

Figure 48. Phase-contrast electron micrograph of a carbon black aggregate

parts of more extended layers, not isolated crystallites. X-ray diffraction reflexes are observed wherever parts of at least three layers are parallel and equidistant. For most carbon blacks, these "crystalline" regions are 1.5–2.0 nm in length and 1.2–1.5 nm in height, corresponding to 4–5 carbon layers [4.4]. The "crystalline" or well-ordered carbon fraction in carbon blacks varies according to oxidation kinetic studies between 60 and 90%.

The morphology of carbon black primary particles indicates that during formation of carbon black, the first nuclei of pyrolyzed hydrocarbons condensate from the gas phase. Thereafter, further carbon layers or their precursors are adsorbed onto the surface of the growing particle. Due to this adsorption, the new layers are always orientated parallel to the existing surface. In the case of high-structure carbon blacks, several particles are joined by collision while they grow. Aggregates are formed by further carbon deposits on these initially loose agglomerates. Polyacetylene seems to play a role in the formation of precursors when aliphatic hydrocarbons are used as starting materials. With aromatic raw materials, however, it is more likely that remaining aromatic degradation products are the intermediates.

The carbon layers of carbon black rearrange to a graphitic order, beginning at the particle surface at temperatures above 1200 °C. At 3000 °C, graphite crystallites are formed and the carbon black particles assume polyhedral shape.

Specific Surface Area. The specific surface area of industrial carbon blacks varies widely. While coarse thermal blacks have specific surface areas as small as 8 m^2/g, the finest pigment grades can have specific surface areas as large as 1000 m^2/g. The specific surface areas of carbon blacks used as reinforcing fillers in tire treads lie between 80 and 150 m^2/g. In general, carbon blacks with specific surface areas > 150 m^2/g are porous with pore diameters of less than 1.0 nm. The area within the pores of high-surface-area carbon blacks can exceed the outer (geometrical) surface area of the particles.

Adsorption Properties. Due to their large specific surface areas, carbon blacks have a remarkable adsorption capacity for water, solvents, binders, and polymers, depending on their surface chemistry. Adsorption capacity increases with a higher specific surface area and porosity. Chemical and physical adsorption not only determine wettability and dispersibility to a great extent, but are also most important factors in the use of carbon blacks as fillers in rubber as well as in their use as pigments. Carbon blacks with high specific surface areas can adsorb up to 20 wt% of water when exposed to humid air. In some cases, the adsorption of stabilizers or accelerators can pose a problem in polymer systems.

Density. Density measurements using the helium displacement method yield values between 1.8 and 2.1 g/cm^3 for different types of carbon black. A mean density value of 1.86 g/cm^3 is commonly used for the calculation of electron microscopic surface areas. Graphitization raises the density to 2.18 g/cm^3. The lower density with respect to graphite (2.266 g/cm^3) is due to slightly greater layer distances.

Electrical Conductivity. The electrical conductivity of carbon blacks is inferior to that of graphite, and is dependent on the type of production process, as well as on the specific surface area and structure. Since the limiting factor in electrical conductivity is generally the transition resistance between neighboring particles, compression or concentration of pure or dispersed carbon black, respectively, plays an important role. Special grades of carbon black are used to donate to polymers antistatic or electrically conductive properties. Carbon blacks with a high conductivity and high adsorption capacity for electrolyte solutions are used in dry-cell batteries.

Light Absorption. The wide use of carbon blacks as black pigments is due to their absorption of visible light. The absorption rate can reach 99.8 %. The black color can take on a bluish or brownish tone, depending on light scattering, wavelength, the type of carbon black, and the system into which the carbon black is incorporated. Infrared and ultraviolet light is also absorbed. Therefore, some carbon blacks are used as UV stabilizers in plastics.

4.2. Chemical Properties

Chemical Composition and Surface Chemistry. The global chemical composition according to elemental analysis is within the following limits:

carbon	80.0–99.5 wt%
hydrogen	0.3– 1.3 wt%
oxygen	0.5–15.0 wt%
nitrogen	0.1– 0.7 wt%
sulfur	0.1– 0.7 wt%

depending on the manufacturing process, raw material, and eventual chemical aftertreatment. The ash content of most furnace blacks is < 1 wt%. The ash components can result from the raw material, the salts which are injected to control the structure, and the salts of the process water. The ash content of gas blacks is less than 0.02%.

The surface of carbon blacks contains certain amounts of polynuclear aromatic substances. These are strongly adsorbed and can only be isolated by continuous extraction with solvents, e.g., boiling toluene. For most industrial carbon blacks, the amount of extractable material is below the limit defined by the food laws.

The hydrogen in carbon black is bound as CH groups at the edge of the carbon layers. Nitrogen seems to be primarily integrated into the aromatic layer system as heteroatoms.

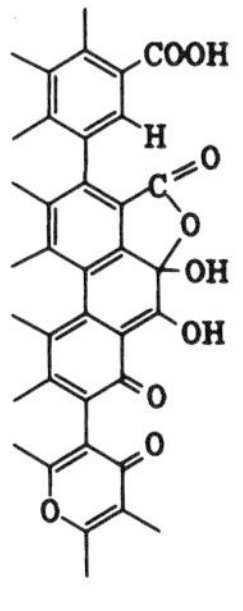

The oxygen content of carbon blacks is of greatest importance for their application. Oxygen is bound to the surface in the form of acidic or basic functional groups. The amount of surface oxides and their composition depend on the production process and an eventual aftertreatment. Furnace blacks and thermal blacks, which have been produced in a reducing atmosphere, contain about 0.2–2.0 wt% oxygen in the form of almost pure, basic surface oxides. Gas and channel blacks, which are manufactured in the presence of air, contain up to 8 wt% oxygen. In this case, the greater part of the oxygen is contained in acidic surface oxides and only a small portion in basic oxides. The amount of acidic surface oxides can be increased by oxidative aftertreatment. Oxygen contents of up to 15 wt% may be obtained.

The surface oxides are destroyed at high temperatures. Due to this fact the weight loss at 950 °C ("volatiles") is a rough indication of the oxygen content of a carbon black.

The pH measured in an aqueous slurry is another indication of the degree of oxidation. In general, the pH is > 7 for furnace blacks (low oxygen content, basic surface oxides), 4–6 for gas blacks, and 2–4 for oxidized carbon blacks (high oxygen content, acidic surface oxides with a high amount of polar functional groups). Further organic reactions, e.g., alkylation, halogenation, esterification, can be carried out with the surface oxides to modify the surface properties.

Oxidation Behavior. Industrial carbon blacks do not spontaneously ignite when stored in air at 140 °C according to the IMDG Code [4.5]. When ignited in air, carbon black glows slowly. In contrast to coal, dust explosions are not observed under normal test conditions [4.6]. However, ignition sources of extremely high energy, e.g., a gas explosion, may induce a secondary dust explosion in air.

For modification of the application properties with regard to surface oxidation, see Section 4.4.7.

4.3. Raw Materials

Mixtures of gaseous or liquid hydrocarbons which can be vaporized represent the raw materials preferable for the industrial production of carbon black. Since aliphatic hydrocarbons give lower yields than aromatic hydrocarbons, the latter are primarily used. The best yields are given by unsubstituted polynuclear compounds with 3–4 rings. Certain fractions of coal tar oils and petrochemical oils from petroleum refinement or the production of ethylene from naphtha (aromatic concentrates and pyrolysis oils) are materials rich in these compounds. These aromatic oils, which are mixtures of a variety of substances, are the most important feedstocks today. Oil on a petrochemical basis is predominant. A typical petrochemical oil consists of 10–15% monocyclic, 50–60% bicyclic, 25–35% tricyclic, and 5–10% tetracyclic aromates.

Important characteristics determining the quality of a feedstock are the C/H ratio as determined by elemental analysis and the BMC Index [4.7] (Bureau of Mines Correlation Index), which is calculated from the density and the mid-boiling point resp. the viscosity. Both values give some information on the aromaticity and therefore the expected yield. Further characteristics are viscosity, pourpoint, alkaline content (due to its influence on the carbon black structure), and sulfur content, which should be low because of environmental and corrosion considerations.

Natural gas, which was previously the predominant feedstock for the production of channel blacks and furnace blacks in the United States, has lost its importance for economic reasons. Currently, only thermal blacks are produced with natural gas. However, natural gas is still the most important fuel in the furnace black process, even though the use of other gases and oils is also possible. In several patents recycled tail gas in combination with oxygen or oxygen-enriched air, has also been proposed as a fuel, but has not reached any commercial importance.

Acetylene, due to its high price, is used only for the production of highly specialized conductivity blacks and battery blacks.

4.4. Production Processes

A summary of the most important production processes is given in Table 27. In general, the processes are divided into two groups: those employing partial combustion and those based on pure pyrolysis. This nomenclature is somewhat misleading insofar as the carbon black resulting from the partial combustion process is also formed by pyrolysis. The two types of processes differ in that air is used in the one to burn part of the feedstock, thus producing the energy required to carry out the pyrolysis, whereas in the other heat is generated externally and introduced into the process.

The furnace black process is currently the most important production process. It accounts for more than 95% of the total worldwide production. The advantages of the furnace black process are its great flexibility in manufacturing various grades of carbon black and its better economy compared to elder processes. The following comparison makes this apparent: for similar grades of carbon black, the production rate of one flame is ca. 0.002 kg/h for channel black, ca. 0.2 kg/h for gas black, and ca. 2000 kg/h for a modern furnace black reactor. However, in spite of the more advantageous furnace black process, the production processes listed in Table 27 (except for the channel black process) are still in use for the production of special carbon blacks which cannot be obtained via the furnace black process.

Table 27. Summary of the manufacturing processes and feedstocks for the production of carbon black

Chemical process	Manufacturing process	Feedstock
Incomplete combustion	furnace black process	petrochemical oils and coal tar oils
	Degussa gas black process	coal tar oils
	channel black process	natural gas
	lamp black process	petrochemical and coal tar oils
Thermal cracking	thermal black process	natural gas
	acetylene black process	acetylene

4.4.1. Furnace Black Process

In the past decades the rapidly expanding automobile industry required increasing numbers of tires with various characteristics. This led not only to the development of new rubber grades, but also to the development of new carbon blacks required by the increasingly refined application processes and to the development of a new and better manufacturing process, the furnace black process. Unlike the old channel black process, this process allows the production of nearly all types of carbon black required by the rubber industry. It also meets the high economic and ecological requirements of our times.

The furnace black process was developed in the United States in the 1920s, and since then, it has been greatly refined. It is a continuous process, carried out in closed reactors, so that all inputs can be carefully controlled [4.8]. Today most semireinforcing rubber blacks (carcass or soft blacks) with specific surface areas of 20–60 m^2/g and the active reinforcing blacks (tread or hard blacks) (see Table 30, Section 4.7) with specific surface areas of 65–150 m^2/g are manufactured by this process, as well as to an increasing extent, pigment-grade carbon blacks with much greater specific surface areas and smaller particle sizes. In addition to the specific surface area, other quality specifications such as structure, measured as DBP absorption, and application properties of rubber such as abrasion resistance, modulus, and tear strength or jetness and tinting strength for color blacks can also be systematically varied in the furnace black process by adjusting the operating parameters. This flexibility is necessary to meet the very narrow specifications required by customers.

The heart of a furnace black production plant is the furnace in which the carbon black is formed. The feedstock is injected, usually as an atomized spray, into a high-temperature and high-energy density zone, which is achieved by burning a fuel (natural gas or oil) with air. The oxygen, which is in excess with respect to the fuel, is not sufficient for the complete combustion of the feedstock, which, therefore is for the most part pyrolyzed to form carbon black at temperatures of 1200–1900 °C.

After the reaction mixture is quenched with water and further cooled in heat exchangers, the carbon black is collected from the tail gas by using a filter system.

Figure 49 shows a schematic drawing of a furnace black plant. The feedstock, preferably petrochemical or carbochemical heavy oils, which usually begins to crystallize near the ambient temperature, is stored in heated tanks equipped with circulation pumps to maintain a homogeneous mixture. Oil is conducted to the reactor by means of rotary pumps via heated pipes and a heat exchanger, where it is heated to 150–250 °C to obtain a viscosity appropriate for atomization. Various types of spraying devices are used to introduce the feedstock into the reaction zone. An axial oil injector with a spraying nozzle at its tip, producing a hollow-cone spray pattern, is a frequently used device. One- and two-component atomizing nozzles [4.9] are in use, air and steam being the preferred atomizing agents in the latter case. However, the feedstock is injected into other reactors as a plurality of coherent or atomized streams into the accelerated combustion gases perpendicular to the direction of stream [4.10].

As the carbon black structure may be reduced by the presence of alkali metal ions in the reaction zone [4.11], alkali metal salts, preferably aqueous solutions of potassium hydroxide or potassium chloride, are often added to the make oil in the oil injector. Alternatively, the additives may be sprayed separately into the combustion chamber. In special cases, other additives, e.g., alkaline-earth metal compounds which increase the specific surface area are introduced in a similar manner.

The high temperature necessary for pyrolysis is obtained by burning fuel in excess air in a combustion chamber. Natural gas is still the fuel of choice, but other gases, e.g., coke oven gases or vaporized liquid gas, are occasionally used. Various oils including the feedstock are occasionally be used as fuel for economic reasons. Special burners, depending on the type of fuel, are used to obtain fast and complete combustion.

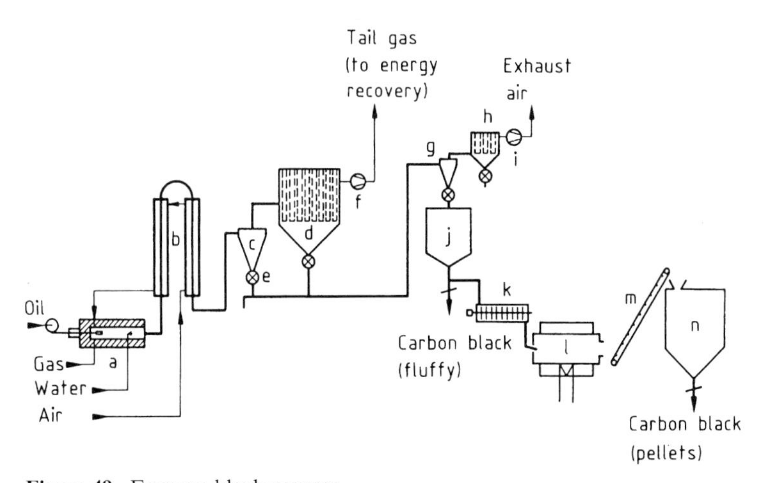

Figure 49. Furnace black process
a) Furnace black reactor; b) Heat exchanger; c) Collecting or agglomerating cyclone; d) Bag filter; e) Carbon black outlet to pneumatic conveying system; f) Tail gas blower; g) Collector; h) Exhaust air filter; i) Blower for the pneumatic conveying air; j) Fluffy black storage tank; k) Pelletizer; l) Dryer drum; m) Conveying belt; n) Storage tank for carbon black pellets

The air required for combustion is compressed by rotating piston compressors or turbo blowers. The air is preheated in heat exchangers by hot gases containing carbon black leaving the reactor. This conserves energy and thus improves the carbon black yield. Preheated air temperatures of 500–700 °C are common.

Important progress has been made on the reactor throughput: A production plant with a capacity of 20 000 t/a (2.5 t/h) was previously run with as many as 12 furnaces, which in the last decades have been replaced by only one high-performance reactor for the same capacity. Modern plants are one-stream units with only one aggregate for each process step (reactor, collecting system, beading device, dryer). From a technical point of view, even larger units could be built. However, due to the great variety of carbon black types required, the capacity of one unit is economically limited by the frequency of switching over to other types and the amount of off-grade carbon black produced during this procedure.

The reactors of modern furnace plants vary considerably in internal geometry, flow characteristics, and the manner in which fuel and feedstock are introduced. Nevertheless, they all have the same basic process steps in common: producing hot combustion gases in a combustion chamber, injecting the feedstock and rapidly mixing it with the combustion gases, vaporizing the oil, pyrolyzing it in the reaction zone, and rapidly cooling the reaction mixture in the quenching zone to temperatures of 500–800 °C.

Schematic drawings of some typical modern furnace black reactors are shown in Figure 50. They all have a gas-tight metal jacket. The reaction zone is coated with a ceramic inner liner, generally on an alumina base, which is stable to temperatures of ca. 1800 °C. Several quenching positions allow the changing of the effective volume of the reactor. This allows variation of the mean residence time of the carbon black at the high reaction temperature. Typical residence times for reinforcing blacks are 10–100 ms.

Most furnace black reactors are arranged horizontally. They can be up to 18 m long with an outer diameter of up to 2 m. Some vertical reactors are used especially for the manufacture of certain semireinforcing blacks [4.12] (Fig. 51). For further reactors, see [4.13].

The properties of carbon blacks are dependent on, e.g., the ratios of fuel, feedstock, and air, which therefore must be controlled carefully [4.14]. The particle size of the carbon black formed generally decreases with increasing amounts of excess air relative to the amount needed for the complete combustion of the fuel. Since the excess air reacts with the feedstock, a greater amount of air leads to higher oil combustion rates, resulting in rising temperatures in the reaction zone. As a consequence, the nucleation velocity and the number of particles formed increase, but the mass of each particle and the total yield decrease. This allows semireinforcing carbon blacks to be manufactured with better yields than active reinforcing carbon blacks. The yields, which depend on the carbon black type and the type of feedstock, range between 50 and 65% for semireinforcing blacks and 40 and 60% for reinforcing blacks. High-surface-area pigment black with markedly smaller particle size than rubber blacks gives lower yields.

Other parameters influencing carbon black quality are the manner in which the oil is injected, atomized, and mixed with the combustion gases, the type and amount of additives, the preheating temperature of the air, and the quench position. As long as

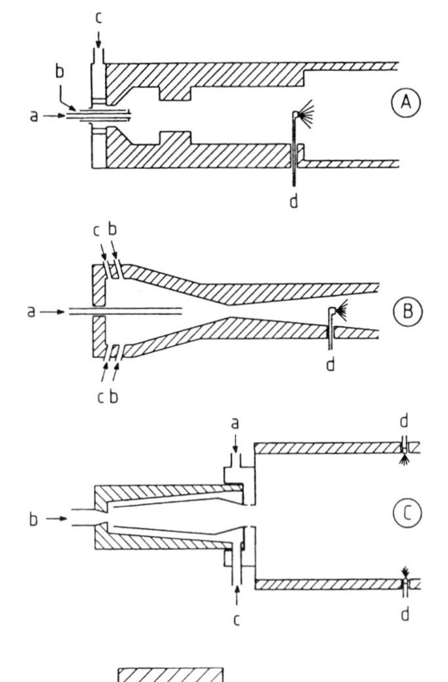

Figure 50. Furnace black reactors
A) Restrictor ring reactor; B) Venturi reactor [4.15]; C) Reactor with high-speed combustion chamber [4.16]
a) Feedstock; b) Fuel; c) Combustion air; d) Quench

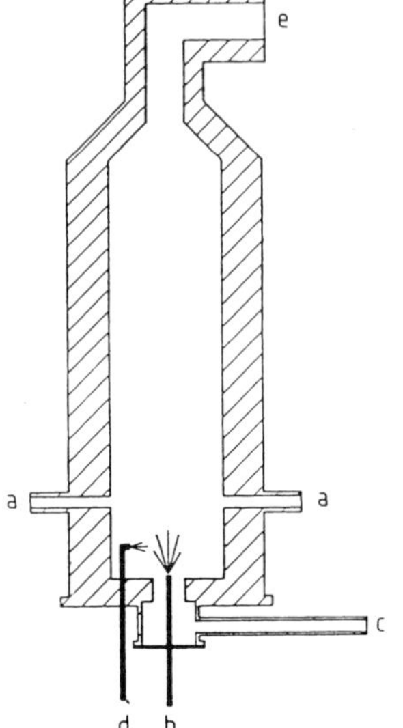

Figure 51. Vertical reactor for manufacturing semireinforcing blacks
a) Fuel inlet; b) Oil injector; c) Air conduit; d) Water spray; e) Outlet to the collecting system

the carbon black is in contact with the surrounding gases at the high reaction temperature, several reactions on the carbon surface occur (e.g., Boudouard reaction, water gas reaction), so that the chemical nature of the carbon black surface is modified with increasing residence time. When quenched to temperatures $< 900\ ^\circ C$, these reactions are stopped and a certain state of surface activity is frozen. Carbon black surface properties can also be adjusted by varying the pelletizing and drying conditions (see below).

Typical processing data for reactors with a carbon black output of 10 000 t/a (1250 kg/h) of tread black and of 14 000 t/a (1750 kg/h) of carcass black are listed in Table 28. These data show that the total mass put through the reactor amounts to 10–16 t/h. Although this is done at high streaming velocities (up to 800 m/s) and high temperatures (up to 1800 °C), modern high-performance reactors have lifetimes of 2 years and more.

The mixture of gas and carbon black leaving the reactor is cooled to temperatures of 250–350 °C in heat exchangers by counterflowing combustion air and then conducted into the collecting system. Formerly, a combination of electroflocculators and cyclones, or cyclones and filters were used [4.8], [4.17]. Currently, simpler units are preferred. Generally, the collecting system consists of only one high-performance bag filter with several chambers, which are periodically purged by counterflowing filtered gas or by pulse jets. Occasionally, an agglomeration cyclone is installed between the heat exchanger and the filter. Depending on the capacity of the production unit, the filter may contain several hundred bags with a total filter area of several thousand square meters. Usual filter loads are of the order of 0.2–0.4 $m^3\,m^{-2}\,min^{-1}$. Since the filtered gas contains 25–40 vol% water vapor, most filters operate at temperatures above 200 °C to avoid condensation. The residual carbon black content in the off-gas is less than 10 mg/m^3.

Because of the reducing atmosphere and the high temperatures in the reactor, the tail gas, which consists of 25–40 vol% water vapor, 40–50 vol% nitrogen, and 3–5 vol% carbon dioxide, also contains a certain amount of combustible gases, the amount of which depends on the feedstock and the processing conditions. These gases include 5–10 vol% carbon monoxide, 5–10 vol% hydrogen, and small amounts of methane and other hydrocarbons. The lower heating value lies between 1700 and 2100 kJ/m^3. The energy remaining in the tail gas can be calculated by using the typical overall energy balance of the furnace black process shown in Figure 52. The gas, which must be burned off for environmental reasons and its energy is used, e.g., for heating dryer drums or for the production of steam and electricity.

The fluffy carbon black coming out of the filter is pneumatically conveyed into a first storage tank. Small amounts of impurities ("grit," e.g., iron, rust, or coke particles) are either removed by magnets and classifiers or milled to an appropriate consistency.

Freshly collected carbon black has an extremely low bulk density of 20–60 g/L. To facilitate handling and further processing by the customer, it must be compacted.

Table 28. Processing data for high-performance furnace black reactors

		Semireinforcing carbon black	Reinforcing carbon black
Natural gas	m^3/h	300– 550	280– 440
Air	m^3/h	7 000–10 000	6 000– 7 500
Oil	kg/h	2 500– 3 300	2 000– 3 000
Carbon black	kg/h	1 500– 2 000	1 000– 1 500
	t/a	12 000–16 000	8 000–12 000
Yield	kg/100 kg of oil	50–65	40–60

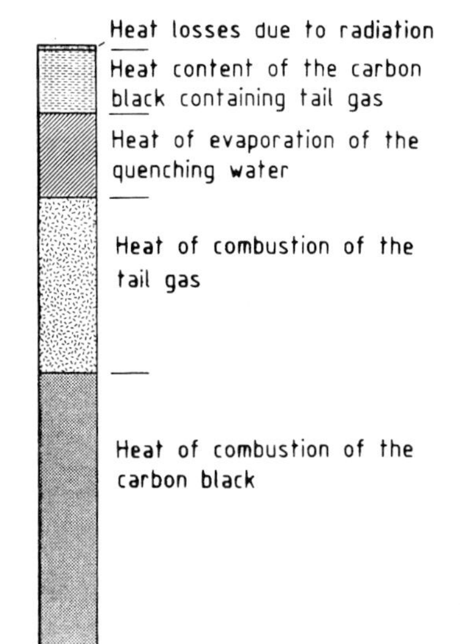

Figure 52. Typical energy balance for the manufacture of a reinforcing black

Densification by "outgassing," a process by which the carbon black is conducted over porous, evacuated drums, is the weakest form of compacting which allows the carbon black to retain its powdery state [4.18]. This form of compacting is used for certain pigment blacks which must remain very dispersible.

Other pigment blacks and the rubber blacks are compacted by granulation. Two processes are used: dry and wet pelletization. Dry pelletization is a simple, energy-saving method, but it does not work with all types of carbon black. It is mainly used for color blacks. Dry pelletization is carried out in rotating drums, where the powdery carbon black rolls to form small spheres.

The wet pelletization is used for the majority of rubber blacks. Carbon black, water, and small amounts of additives, e.g., molasses, ligninosulfonates, are mixed in special beading machines [4.19]. They usually consist of a horizontal, cylindrical trough ca. 3 m long and 0.7–1 m in diameter in the axis of which a pin shaft rotates at 300–750 rpm (Fig. 53). The pins are positioned helically around the shaft. The water containing the pelleting agents dissolved in it is injected via spray nozzles. The density is ca. 10 times that of the original carbon black. DBP absorption is also reduced during this process. The pellet strength and some application properties in rubber can be influenced by the type and amount of the pelleting agent. The size of the pellets is ca. 1–2 mm.

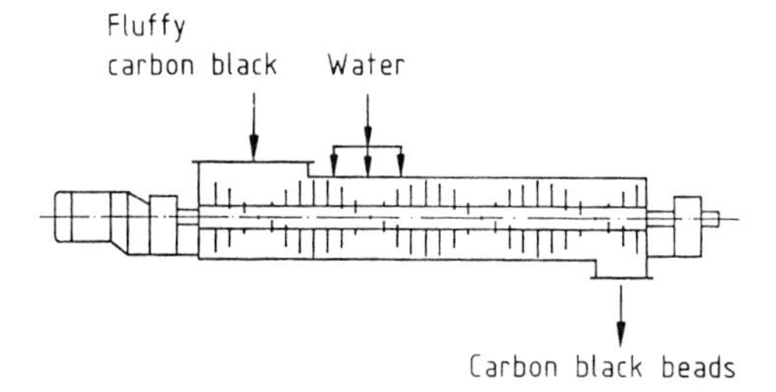

Figure 53. Pelletizer

The carbon black leaving the beading machine contains ca. 50 wt% water. It is dried in dryer drums, indirectly heated by burning tail gas. Dryer drums with a capacity of 2000 kg of carbon black per hour are 15–20 m long and 2–3 m in diameter. They are rotated at 5–15 rpm. Drying temperatures, generally between 150 and 250 °C, allow further modification of the carbon black properties.

The dried carbon black is transported via conveyor belts and elevators to the storage tank or packing station. Bulk densities of wet-pelletized carbon blacks are between 250 and 500 g/L.

A flow diagram summarizing the complete furnace black process is shown in Figure 54.

4.4.2. Gas Black and Channel Black Processes

The channel black process, used in the United States since the late 19th century, is the oldest process for producing small-particle-size carbon blacks on an industrial scale. Originally, the first reinforcing blacks were also produced by this process. In 1961, the production of channel black was about 120 000 t. The last production plant in the United States was closed in 1976, due to low profitability and environmental difficulties. Natural gas was used as the feedstock. The carbon black yield was only 3–6%.

In Germany, where natural gas was not available in sufficient amounts, the gas black process was developed in the 1930s. It is similar to the channel black process, but uses coal tar oils instead of natural gas. Yields and production rates are much higher with oil-based feedstock; this process is still used to manufacture high-quality pigment blacks with properties comparable to those of channel blacks. The gas black process has been used by Degussa on an industrial scale since 1935.

Originally, gas black was primarily used for the reinforcement of rubber. Today, almost all grades are used as color blacks in printing inks, plastics, lacquers, and coatings. High-quality oxidized gas blacks (see Section 4.4.7) are of special interest, e.g., in deep black lacquers and coatings.

In the gas black process (Fig. 55), the feed stock is partially vaporized. The residual oil is continuously withdrawn. The oil vapor is transported to the production apparatus by a combustible carrier gas (e.g., hydrogen, coke oven gas, or methane). Air may be added to the oil–gas mixture for the manufacture of very small particle size carbon black. Although this process is not as flexible as the furnace black process, various types of gas black can be made by varying the relative amounts of carrier gas, oil, and air. The carbon black properties are also dependent on the type of burners used.

A gas black apparatus consists of a burner pipe approximately 5 m long, which carries 30–50 diffusion burners. The flames burn in contact with a water-cooled drum, where about half of the carbon black formed is deposited. This black is scraped off and transported by a screw to a pneumatic conveying system. The gas black apparatus is surrounded by a steel box open at the bottom. At the top of it, fans suck the off-gas into filters, which collect the carbon black suspended in the gas.

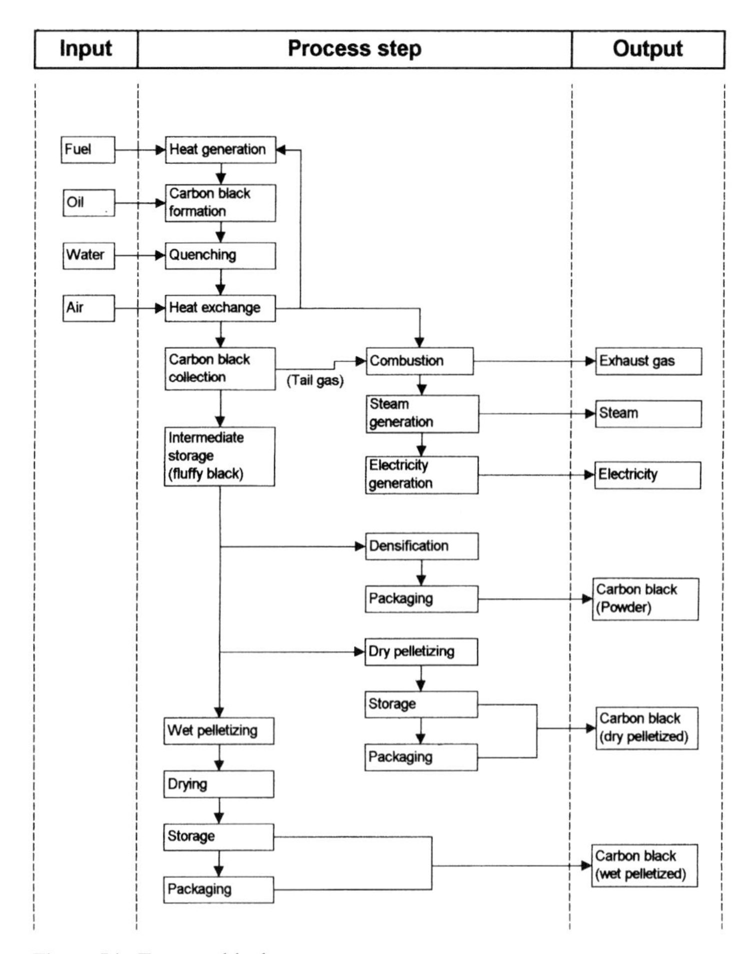

Figure 54. Furnace black process

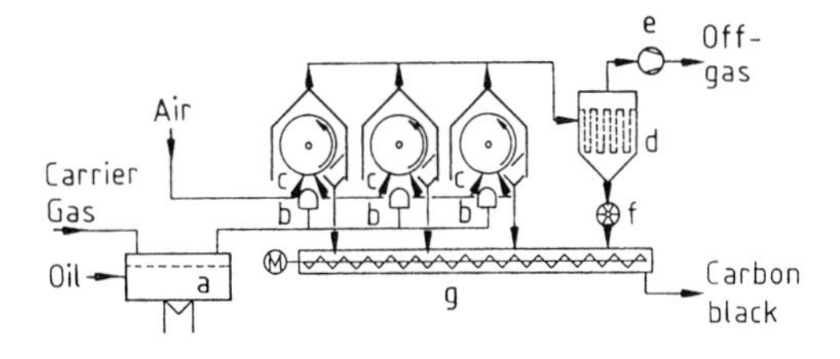

Figure 55. Degussa gas black process
a) Oil evaporator; b) Burner; c) Cooling drum; d) Bag filter; e) Blower; f) Rotary valve; g) Conveying srew

The amount of air entering the apparatus can be regulated by valves in the exhaust pipes.

Several gas black apparatus are combined to form one production unit. The whole "group" is fed by one oil vaporizer. The production rate and the yield of an apparatus depend on the type of carbon black produced. For a typical RCC black, the production rate is 7–9 kg/h and the yield is 60 %. The yield for high-quality color blacks is considerably lower (10–30 %).

To remove possible impurities, the gas black is classified and then densified, pelletized (see Section 4.4.1), or submitted to an oxidative aftertreatment (see Section 4.4.7), depending on its intended use. Since gas blacks are formed in the presence of excess air, their surface is oxidized. Acidic surface oxides are predominant.

4.4.3. Lamp Black Process

The lamp black process is the oldest industrial-scale production process [4.8], [4.17].

Currently, only a few plants still produce rather coarse blacks (mean particle diameter ≈ 100 nm) with special properties. They are used as nonreinforcing or semireinforcing blacks in rubber goods and as tinting black with a low pigment separation tendency.

The lamp black process is only partially continuous. The feedstock, oil with a high aromatic hydrocarbon content, is burned in flat steel vessels up to 1.5 m in diameter (Fig. 56). The oil is continuously introduced into the vessel to keep a constant feedstock level. The off-gas containing carbon black is sucked into a conical exhaust pipe, which is coated with a ceramic inner liner and leads to the collecting system. The properties of the carbon black can be influenced to some extent by variation of the distance between the vessel and the exhaust and the amount of air sucked into the apparatus. One lamp black apparatus can produce 100 kg/h. The production process must be interrupted at certain time intervals to remove coke-containing residues from the vessels.

4.4.4. Thermal Black Process

Some special processes for producing carbon black are based on the thermal decomposition of lower gaseous hydrocarbons in the absence of air. Natural gas and acetylene are commonly used as raw materials.

The thermal black process, which was developed in the 1930s, is still used for the production of coarse carbon blacks (nonreinforcing carbon blacks) for special applications in the rubber industry. Contrary to the above-described processes, energy generation and the pyrolysis reaction are not carried out simultaneously. Natural gas eventually blended with vaporized oil is used as both a feedstock and a fuel.

A thermal black plant consists of two furnaces, which are used in alternate heating and production periods of ca. 5-min duration (Fig. 57). Each of the cylindrical furnaces (4 m in diameter and 6 m high) contains a network of heat-resistant bricks. They are heated with natural gas and air. At a temperature of ca. 1400 °C, the air is switched off and only natural gas is introduced for pyrolysis. Since this reaction is endothermic, the temperature falls. At about 900 °C, a new heating period is necessary.

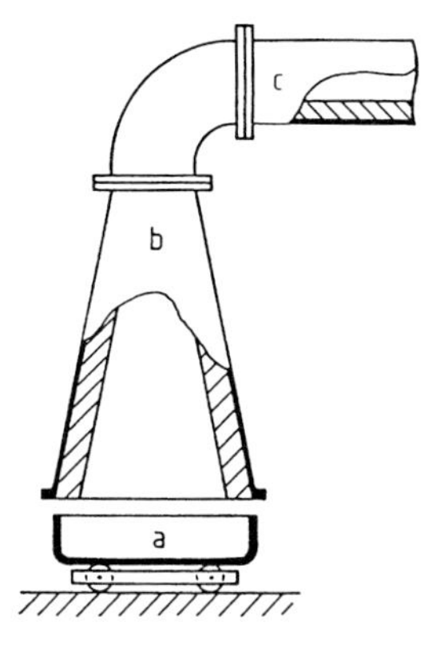

Figure 56. Lamp black process
a) Vessel filled with feedstock; b) Conical exhaust pipe; c) Pipe leading to the collecting system

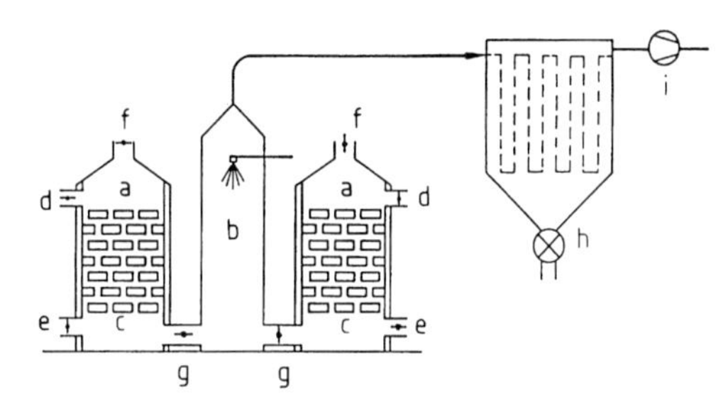

Figure 57. Thermal black process
a) Thermal black reactor; b) Cooler c) Filler bricks; d) Inlet for the feedstock; e) Inlet for the fuel; f) Outlet for the burned fuel; g) Outlet for the pyrolysis products; h) Carbon black outlet; i) Blower

The products leaving the furnace, carbon black and nearly pure hydrogen, are cooled by injecting water into an ascending channel. The carbon black is separated in the collecting system.

Carbon blacks of lower particle size can be produced by diluting natural gas with recycled hydrogen. Fine thermal blacks (FT blacks) with mean primary particle sizes of 120–200 nm are manufactured in this way. Medium thermal blacks (MT blacks) with mean particle sizes of 300–500 nm are still produced and are obtained by using undiluted feed stock. The yield of MT blacks is about 40% with respect to the amount of feedstock and fuel used.

Thermal blacks are used for mechanical rubber goods with high filler contents. Cheaper products (clays, milled coals, and cokes), however, have become increasingly important as substitutes for economic reasons. The total production of thermal blacks is, therefore, decreasing.

4.4.5. Acetylene Black Process

Acetylene and mixtures of acetylene with light hydrocarbons are the raw materials for a process that has been used since the beginning of the 20th century. Unlike other hydrocarbons, the decomposition of acetylene is highly exothermic ($\Delta H = -230$ kJ/mol).

The discontinuous explosion process is the oldest technical process. It was mainly used for the production of color blacks. Continuous processes were later developed with production rates up to 500 kg/h [4.20]. Acetylene or acetylene-containing gases are fed into a preheated, cylindrical reactor with a ceramic inner liner. Once ignited, the reaction is maintained by the decomposition heat that is evolved. The carbon black is collected in settling chambers and cyclones. Approximately 95–99 % of the theoretical yield is obtained.

The primary particles of acetylene black have different shapes than those of other carbon blacks (Fig. 58). As the increased order in the *c* direction of the crystalline regions indicates, folded sheets of carbon layers are the main structural component. Their application is limited to special uses, e.g., in dry cells, because of their relatively high price. Total worldwide production is ca. 40 000 t/a.

4.4.6. Other Production Processes

In a plasma, hydrocarbon vapors may be almost quantitatively decomposed into carbon and hydrogen [4.21]. Many producers of carbon black have done research in this field. According to numerous patent specifications, this method can be used to make small-particle carbon blacks with new properties. However, an economical plasma-based commercial process is not yet known.

The Hüls electric arc process was the only large-scale process using plasma reactions where large quantities of carbon black were produced as a byproduct in the production of acetylene. Today, this kind of carbon black is no longer used as a pigment.

Since the price of both feedstocks and fuels, and thus, the profitability of the carbon black production processes, is highly dependent on the petrochemical indus-

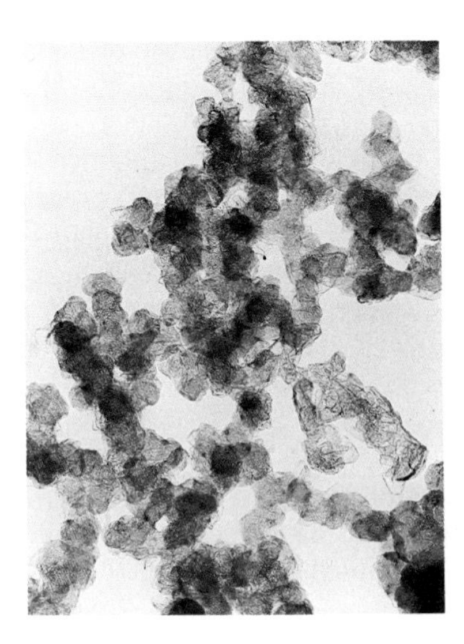

Figure 58. Electron micrograph of acetylene black

try, several attempts have been made to find new raw materials. Processes for obtaining carbon black directly from coal [4.22] or for isolating carbon black from old tires, for example, have been studied. None of them, however, have been of any commercial importance up to now. On the other hand, clay, milled coal, and coke have found limited use as substitutes for very coarse carbon blacks, primarily thermal blacks and some SRF blacks. The increasing use of precipitated silicas in tires and mechanical rubber goods, mostly in combination with organosilane coupling agents, which originally was indicative of an increasing search for new non-oil-based fillers, has led to new features of rubber properties.

4.4.7. Oxidative Aftertreatment of Carbon Black

Oxygen-containing functional groups on the surface of carbon blacks strongly influence their application properties. High contents of volatiles, i.e., high concentrations of surface oxides, decrease the vulcanization rate and improve the flow characteristics of inks. The gloss of lacquers and coatings is increased, the color tone is shifted from brownish to bluish, and jetness often increases.

Due to production conditions, only gas blacks (and channel blacks) are covered to a certain extent with acidic surface oxides. Furnace blacks contain only small amounts of oxygen in the form of basic surface oxides.

Some color blacks receive oxidative aftertreatment on a commercial scale to amend their color properties. Depending on the oxidizing agent and the reaction conditions selected, different types of surface oxides are formed in varying quantities.

The simplest method of oxidizing the carbon black surface is by aftertreating it with air at 350–700 °C. However, the degree of oxidation is limited. Higher contents of surface oxides and better process control are achieved with nitric acid [4.23], mixtures of NO_2 and air [4.24], ozone, or sodium hypochlorite solutions [4.25] as oxidizing agents. As a rule, all strongly oxidizing agents may be used, either as a gas or in solution. Most surface oxidations of carbon black are carried out at elevated temperature.

Oxidized carbon blacks may contain up to 15 wt% oxygen. They are strongly hydrophilic. Some of them form colloidal solutions spontaneously in water. In polar printing ink systems, lacquers, and coatings, a better wettability and dispersibility is achieved through surface oxidation [4.26], thus reducing binder consumption.

Surface oxidation of carbon black with nitric oxide and air can be carried out industrially in a fluidized-bed reactor [4.24]. A suitable aftertreatment unit consists of a preheating vessel, in which the carbon black is fluidized and heated, a reaction vessel to carry out the surface oxidation, and a desorption vessel, in which adsorbed nitric oxide is removed (Fig. 59). Typical reaction temperatures lie between 200 and 300 °C. Depending on the degree of oxidation, the residence time can amount to several hours. The nitric oxide acts primarily as a catalyst, the oxygen in the air being the genuine oxidizing agent.

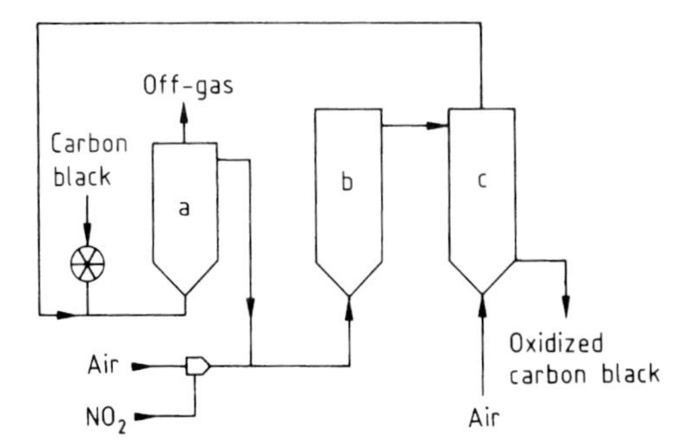

Figure 59. Equipment for the oxidative aftertreatment of carbon black in a fluidized bed
a) Fluidizing vessel; b) Reaction vessel; c) Desorption vessel

Oxidation of powdery black with ozone is also carried out on a commercial scale.
Another common method of surface oxidation is carried out during pelletization. Nitric acid, instead of water, is used as a pelletizing agent. The surface is oxidized while the wet pellets are dried at elevated temperature [4.27].

4.4.8. Environmental Problems

The furnace black process, the most economical and commercially advanced production method for carbon black, does not form toxic solid or liquid byproducts under normal operating conditions. The tail gas containing hydrogen and carbon monoxide was formerly emitted into the atmosphere, but is now burned, the energy being used partially within the process itself, e.g., for heating dryer drums in the wet-beading process and for generating steam or electricity, so that emissions of carbon black and inflammable gases are avoided.
Although their economic importance is decreasing, progress with respect to environmental problems has also been made for the older processes which are still in use.

4.5. Testing and Analysis

The chemical composition of carbon blacks (see Section 4.2), as determined by common elemental analysis methods, is of little significance for predicting their properties. Special characteristic properties are, therefore, determined for the characterization and quality control of carbon blacks. Traces of heavy metals are determined spectroscopically in the ash. Copper and manganese ions, etc., are of special interest to the rubber industry because of their interference with the aging process of rubber goods.

Electron Microscopy. Electron microscopy is one of the most important physical methods for the characterization of finely divided solids. It allows direct viewing of the shape and morphology of particles in this order of magnitude, primary particle size, particle size distribution, and aggregation.

To determine the mean primary particle size and particle size distribution, the diameters of 3000–5000 particles are measured on electron micrographs of known magnification. Spherical shape is anticipated for calculations. However, since the primary particles generally build up larger aggregates, the results may be somewhat uncertain. The specific "electron microscopic surface area" can be calculated from the primary particle size distribution. This value refers only to the outer (geometrical) surface of the particles. For porous carbon blacks the electron microscopic surface area is lower than the specific surface area according to BET (see below).

An attempt was made recently to find characteristic values for the type and degree of aggregation using electron micrographs [4.28]. However, neither visual comparisons with standard aggregates nor automatic picture analyses have led to a practical method for a quantitative characterization of the carbon black structure on a routine basis.

High-resolution phase-contrast pictures and X-ray diffraction are used to elucidate the internal structure of single primary particles.

Sorption Analysis. Specific surface areas and porosity can be calculated from the adsorption isotherm of nitrogen at − 196 °C. The method of Brunauer, Emmett, and Teller [4.29] is generally accepted for the evaluation of specific surface areas (BET surface area in square meters per gram). The two-parameter equation is applicable to carbon black. The BET surface area comprises the outer surface area as well as the surface area of the pores.

Porosity of carbon blacks can be detected by the de Boer *t*-plot method [4.30]. The total surface area and the geometrical surface area outside the pores can be determined separately from the adsorption isotherm. Special attention must be paid to the selection of a suitable master *t* curve. Due to the small diameters of most carbon black primary particles, methods for the determination of mesopores are of no importance.

Special Analytical Test Methods. Test methods which on the one hand resemble certain physicochemical methods, but on the other hand already give some indication of application properties are summarized as special analytical test methods in Table 29. They can be carried out within a short time and are used for characterization as well as production control. Since the results are influenced by the test conditions, these test methods are standardized.

Application Tests. Physicochemical and special analytical test methods allow the classification of carbon blacks and a rough estimation of their application properties. Exact data on the application properties of a carbon black in a special system, e.g., plastic material or a rubber mixture, can only be given by application tests under nearly practical conditions.

Table 29. Special analytical test methods for carbon black

Test method	Unit	Standard	Remark
Iodine adsorption	mg/g	ASTM D1510; ISO 1304	amount of iodine adsorbed from aqueous solution as a measure of the specific surface area; not applicable for oxidized carbon blacks
CTAB surface area	m^2/g	ASTM D 3765; ISO 6810	adsorption of CTAB (cetyl trimethyl ammonium bromide) from aqueous solution as a measure of the specific outer surface area
BET surface area	m^2/g	ASTM D 3037; DIN 66131; DIN 66132	total specific surface area calculated from the nitrogen adsorption isotherm by using the BET equation
External surface area	m^2/g	ASTM D 5816	STSA (Statistical Thickness Surface Area) calculated from the nitrogen adsorption isotherm
Aggregate dimension		ASTM D 3849	determination of aggregate dimensions (unit length, width, etc.) by electron microscope image analysis
DBP absorption	mL/100 g	ASTM D 2414; ISO 4656; DIN 53601	determination of the wetting point with dibutyl phthalate in a special kneeder as a measure of the carbon black structure
24M4-DBP absorption	mL/100 g	ASTM D 3493; ISO 6894	determination of DBP absorption after repeated compressing at high pressure as a measure of the permanent structure
Oil absorption	%	DIN ISO 787/5	percentage of linseed oil needed to make a barely flowable paste
Jetness, Blackness value		DIN 55979	light absorption of a carbon black paste in linseed oil, determination by visual comparison against standard blacks or by measuring the absolute light remission (DIN)
Tinting strength	%	ASTM D 3265; DIN ISO 787/16, 24; DIN 53204; DIN 53234	ability of a carbon black to darken a white pigment in a linseed oil paste; the tinting strength is the weight percentage of the standard carbon black with respect to the tested black to obtain the same gray tone, different standard white pigments and carbon black concentrations are used according to ASTM and DIN ISO
Volatiles	%	ISO 1126; DIN 53552	weight loss when calcined at 950 °C for 7 min

Table 29. (Continued)

Test method	Unit	Standard	Remark
Heating loss (moisture)	%	ASTM D 1509; DIN ISO 787/2	weight loss on drying at 125 °C for 1 h (ASTM) or 2 h at 105 °C (DIN ISO)
pH		ASTM D 1512; DIN 53200	pH of an aqueous slurry of carbon black; pH is mainly influenced by surface oxides
Extractables	%	DIN 53553	Amount of extractable material (usually by boiling toluene) in at least 8 h
Toluene discoloration		ISO 3858; ASTM D 1618	light absorption (transmission) of a toluene solution of the extracted material
Ash content	%	ASTM D 1506; DIN 53586	amount of noncombustible material after burning the carbon black at 675 °C (DIN) resp. 550 °C (ASTM, DIN)
Sulfur content		ASTM D 1619; DIN 53584	
Sieve residue		ASTM D 1514; DIN ISO 787/18	amount of coarse impurities that cannot be purged through a testing sieve by water
Pour density	g/L	ASTM D 1513; DIN 53912	measure for the densification of carbon black
Tamped density	g/L	DIN ISO 787/11	similar to bulk density; however, void volume is reduced by tamping
Pellet size distribution		ASTM D 1511	determination by means of sieve shaker
Fines content		ASTM D 1508; DIN 53583	only for pelletized blacks, percentage passing through a sieve of 125 μm mesh width
Pellet crush strength	g	ASTM D 5230; ASTM D 3313	individual pellet hardness

4.6. Transportation and Storage

The majority of the carbon black produced (up to 80 %) is shipped as bulk material; the rest is handled in bags. Generally, the large stocks of pelletized furnace blacks for the rubber industry are stored, by the producers as well as the customer, in plastic-coated steel storage tanks consisting of one or more cells with a capacity of 100 – 1000 t (300 – 3000 m^3). Carbon black is shipped in containers or tank trucks with a capacity of up to 20 t. In recent times big bags and steel bins with a capacity of up to about 1 t have reached increasing interest. Smaller amounts of rubber blacks and especially pigment blacks are stored and shipped in paper or plastic bags stacked on palettes. In some case, bags with special coating are used.

4.7. Uses

The most important types of carbon black, some typical characteristics, and the principal applications are summarized in Tables 30 and 31. A general overview of the most important application fields is given in Figure 60.

Besides their two main uses as reinforcing fillers and pigments, small amounts of carbon blacks are used by the electrical industry to manufacture dry cells, electrodes, and carbon brushes. Special blacks are used to give plastics antistatic or electrical conduction properties. Another application is the UV stabilization of polyolefins [4.31].

4.7.1. Rubber Blacks

About 90% of the carbon black produced is used by the rubber industry as a reinforcing filler in tires, tubes, conveyor belts, cables, rubber profiles, and other mechanical rubber goods. Furnace blacks are predominantly used in rubber processing. Fine-particle-size carbon blacks (reinforcing blacks) are used for the production of rubber mixtures with high abrasion resistance (e.g., tire treads). Coarser carbon blacks (semireinforcing blacks) are used in rubber mixtures requiring low heat buildup during dynamic stress (e.g., carcass mixtures). Very coarse carbon blacks (nonreinforcing blacks) are incorporated into mixtures with high elasticity and good extrusion properties.

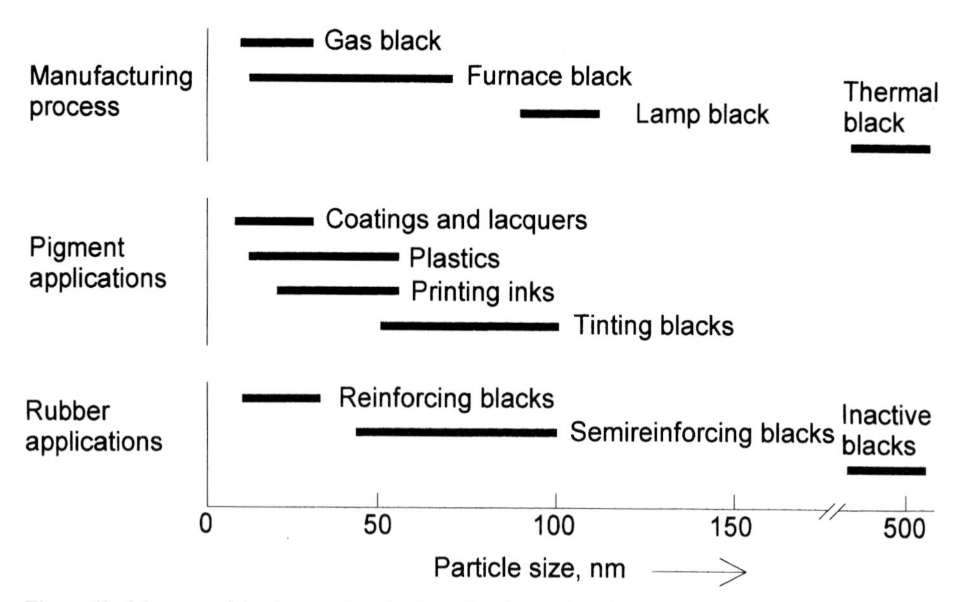

Figure 60. Mean particle sizes and typical applications of various carbon blacks

Table 30. Rubber blacks

ASTM No.	Iodine adsorption mg/g	DBP absorption ml/100 g	Use in natural and synthetic rubber
N115	160	113	Tire treads, rubber goods with high abrasion resistance (also as blend with N220 and N330 and N326)
N121	121	132	Tire treads for high performance passenger cars
N125	117	104	Tire treads for trucks, off-road tires
N220	121	114	Tire treads for trucks, tank pads, conveyor belt covers
N234	120	125	Tire treads for high performance cars, tank pads, conveyor belt covers
N326	82	72	Tire carcasses, steel cord adhesion compounds, mechanical rubber goods
N330	82	102	Tire treads, tire carcasses (blended with less active carbon blacks), tire sidewalls, mechanical rubber goods
N335	92	110	Tire treads for passenger cars
N339	90	120	Tire treads for passenger cars, abrasion resistant mechanical rubber goods
N347	90	124	Tire treads for passenger cars, abrasion resistant mechanical rubber goods
N351	86	120	Tire treads, abrasion resistant mechanical rubber goods
N356	92	154	Tire treads, cushion gum
N375	90	114	Tire treads for passenger cars and trucks, tank pads, conveyor belt covers
N539	43	111	Tire carcasses, mechanical rubber goods with good dynamic properties, extrusion compounds
N550	43	121	Tire carcasses, mechanical rubber goods with good dynamic properties, extrusion compounds
N650	36	122	Tire carcasses, extrusion compounds
N660	36	90	Tire carcasses, moulded goods
N683	35	133	Tire carcasses, tire sidewalls, mechanical rubber goods with good dynamic properties, extrusion compounds
N762	27	65	Mechanical rubber goods with excellent dynamic properties
N765	31	115	Tire carcasses, moulded goods, extrusion compounds
N772	30	65	Mechanical rubber goods with excellent dynamic properties
N774	29	72	Tire carcasses, mechanical rubber goods with excellent dynamic properties, moulded goods

Table 30 lists furnace rubber blacks which are currently used and their main application fields.

According to the ASTM classification (ASTM D 1765 and D 2516), rubber blacks are characterized by one letter and three digits. The former shows the influence of

Table 31. Pigment blacks

Type [1])	Blackness My in linseed oil	Tinting strength [2]) %	Volatiles %	BET surface area m^2/g	Mean primary particle size nm	pH	Application
Gas black							
HCC	> 270	120–122	4–6	250–350	13–15	4–5	extremely deep black
ox.	> 270	112–116	15–25	300–500	13–17	2–3	lacquers and coatings of all types, plastics, fibers
MCC	250–270	120–122	4–6	150–200	15–20	4–5	coatings, plastics, fibers
ox.	250–270	112–118	13–18	200–300	15–20	2–4	
RCC	240–250	105–115	4–6	90–110	25–30	4–5	paints, coatings, printing inks, carbon paper, carbon ribbons, plastics, fibers
ox.	240–250	100–110	10–15	100–200	25–30	3–4	
Lamp black							
	208–210	25–30	1–2	18–25	90–110	7–9	tinting black for coatings and plastics
Furnace black							
HCF	> 255	115–125	0,5–1,5	250–300	14–15	9–10	high jet coloring of plastics
MCF	250–260	120–126	0,5–1,5	150–200	16–18	9–10	high jet coloring of plastics
RCF	240–250	95–120	0,5–1,5	80–120	20–30	9–10	plastics, printing inks, carbon paper, carbon ribbons, fibers, UV stabilization of polyolefins
ox.	240–250	110–120	2–3	100–120	20–30	2,5–3,5	
LCF	215–235	60–100	0,5–1	30–65	30–60	9–10	plastics, tinting black, printing inks
ox.	215–235	65–105	1,5–2,5	30–65	30–60	2,5–4	

[1]) HCC high color channel MCC medium color channel RCC regular color channel
HCF high color furnace MCF medium color furnace RCF regular color furnace
LCF low color furnace

[2]) Reference: IRB 3 = 100%.

the carbon black on the vulcanization process (N = normal curing, and S = slow curing). The first digit of the number characterizes the primary particle size or the specific surface area (Table 32). The other two digits are free and are used to identify individual grades within the group. Within each carbon black group, considerable variation in the combination of properties is possible, e.g. by varying the carbon black structure (DBP absorption) and surface activity.

Table 32. Classification of rubber blacks according to ASTM D 1765 and ASTM D 2516

Group no. (first digit of the three-digit ASTM number)	Average particle size, nm	Specific surface area*, m^2/g
0	1– 10	> 155
1	11– 19	125–155
2	20– 25	110–140
3	26– 30	70–90
4	31– 39	43–69
5	40– 48	36–52
6	49– 60	26–42
7	61–100	17–33
8	101–200	–
9	201–500	–

* The ranges of the specific surface area may vary, depending on the structure of the carbon blacks.

4.7.2. Pigment Blacks

Quantitatively, the pigment blacks are substantially less important than the rubber blacks. They are used for the manufacture of printing inks, coloring plastics, fibers, lacquers, coatings, and paper (see Table 31). Oxidized carbon blacks are frequently used in the printing ink and coating industry. While high-color gas blacks are still predominant in lacquers, furnace blacks are becoming more and more important in plastics, coatings, and printing inks.

4.7.2.1. Pigment Properties

Due to their excellent pigment properties, carbon blacks are widely used for blackening plastics, printing inks, paints, etc., as well as for tinting in conjunction with colored and white pigments. They are insoluble in all solvents, stable against light and weathering, and insensitive to acids and alkalis, and they exhibit great depth of color and tinting strength. The depth of color (blackness, My value) is a measure of the absorption of visible light by the pigment black.

The *depth of color* is determined as follows. The carbon black is rubbed with a binder (e.g., linseed-oil) to form a paste which is applied in a thick coat on a glass plate. A colorimeter is used to measure the diffuse reflection of light through the glass. The color depth and hue of the paste can be determined from the intensity and spectral distribution of the diffusely reflected light [4.32]. The color depth can also be determined by visual comparison of the paste with pastes of known blackness under very bright illumination.

The *tinting strength* is a measure of the ability of a carbon black to darken lighter-colored pigments. In accordance with DIN 53 204 and 53 234 (see Table 29), it is determined by blending the black and a white pigment with linseed oil to form a homogeneous paste, then comparing the paste with standard pastes containing a reference black.

The *mean primary particle sizes* of pigment blacks lie in the range 10–100 nm; *specific surface areas* are between 20 and 1000 m^2/g. The specific surface area, determined by N_2 adsorption and evaluation by the BET method [4.29], is often cited as a measure of the fineness of a black. Blacks with specific surface areas $> 150\ m^2/g$ are generally porous. The BET total specific surface area is larger than the geometric surface area measured in the electron microscope, the difference being due to the pore area resp. the pore volume.

Another important property is the *structure* (see Section 4.1) which characterizes the coalescence of primary particles into aggregates resembling chains or bunches of grapes. The dibutyl phthalate (DBP) absorption is commonly accepted as a measure for the carbon black structure. Due to absorption phenomena the DBP number increases with increasing specific surface area. Oil absorption (or, in general, the vehicle demand) is as well an indicator for the structure, but it also depends on the wettability of the black surface. Since linseed oil is a polar system, oil absorption declines as the concentration of surface oxides rises.

The *chemical composition of the black surface* is important for the wettability and dispersion qualities [4.33]. For polar systems (e.g., paint binders, many printing ink systems, and liquid plastic resin systems), the presence of oxygen-containing functional groups is particularly advantageous. Acidic surface oxides (carboxyl, lactole, phenol, and carbonyl oxygen) and basic oxides (pyrone-like structures) can be detected by chemical methods [4.34]. As a result of the manufacturing process, furnace blacks always contain small amounts of basic surface oxides, whereas on gas blacks acidic oxides are by far predominatic. Controlled oxidation can greatly increase the quantity of acid surface oxides in gas blacks. Acid oxides can also be introduced onto furnace blacks. Table 33 indicates typical surface oxides contents.

The *content of volatiles* is often used as a measure of the degree of oxidation of carbon blacks. This is defined as the weight loss on heating at 950 °C (DIN 53552). Typical values for the weight fraction of volatiles are 1–3% for furnace blacks, 4–8% for gas blacks, and up to 25% for oxidized gas blacks.

The three parameters, mean primary particle size (or specific surface area), structure (or aggregate size), and surface chemistry (e.g. surface oxides), largely determine the application characteristics of carbon blacks. A summary of how these parameters affect color and performance appears in Table 31.

Table 33. Functional groups at the surface of selected typical gas and furnace blacks

Grade	Type	Acidic oxides, µg-equiv/g	Basic oxides, µg-equiv/g
FW 1	gas	910	unknown**
FW 2	gas*	1950	unknown**
FW 200	gas*	2250	unknown**
Printex G	furnace	50	40
Printex 60	furnace	70	100

* After oxidation. ** Gas blacks also contain small amounts of basic oxides, but their quantitative determination is difficult.

4.7.2.2. Blacks for Printing Inks [4.35]

With a consumption of about 150 000 t/a carbon blacks are the most important pigments used in the printing-ink industry. The required properties of the pigment black vary rather widely, depending on the printing process and the type of binder.

Printing inks are classified as physically or chemically drying. The most important forms of *physical drying* are the absorption of low-volatility solvents into the paper (principally in newspaper printing) and the evaporation of volatile solvents (flexographic and rotogravure printing). With "heat-set" inks, the printed sheet is heated; "moisture-set" inks contain water-insoluble vehicles dissolved in glycol, which are precipitated by steam treatment.

The *chemical drying* of printing inks involves oxidation (where unsaturated oils are cross-linked by atmospheric oxygen) or polymerization of binder constituents by UV light. The first setting is often by striking in, even with chemically-dried printing inks.

Blacks for *newspaper printing* usually have moderate fineness and adequate color depth but are still relatively easy to disperse. In order to attain high printing speeds, inks must have adequate flowability. Blacks of moderate and higher structure are used because particles of very low structure can migrate into the absorbent newspaper along with the vehicle. While originally the letterpress process was the main printing process for newspapers, today the offset and the flexo printing have become predominant. Gas blacks of grade RCC (Table 31) are especially suitable for polar systems; they are highly wettable and easily dispersible due to their volatiles content, exhibit relatively low oil absorption, and give inks with very high hiding power. For largely nonpolar systems based on mineral oil, furnace blacks are better suited. For economic reasons of cost, furnace blacks of moderate and high structure are being increasingly used.

Letterpress printing call for high-viscosity inks (5–50 Pa · s) that do not run over the edges of the types.

Offset printing needs also high viscosity inks with high pigment concentrations. Only a relatively small amount of ink is applied to the paper due to the repeated division of the ink film. Pigments with a high hiding power are therefore needed. These requirements are satisfied by the classical RCC and LFI gas blacks (Table 31), and furnace blacks with very low structure and mean primary particle sizes of ca. 20–40 nm. Coarse blacks are more easily dispersed and exhibit a blue hue. Blue-tinted blacks are generally preferred because they are visually perceived as deeper in color than brown-tinted ones. The hue of black printing inks is often corrected by the addition of colored pigments (chiefly blue pigments such as iron blue, see Section 3.6). Despite the high concentration of carbon black (up to ca. 20 %), a lustrous print with a brilliant appearance can be obtained due to the low binder consumption of gas blacks and low-structured furnace blacks.

Untreated or oxidized blacks are used, depending on the polarity of the binder system and the desired coloristic effects. UV-drying printing inks combine relatively high diffuse reflectance with good hiding power and relatively low reactivity toward free radicals; coarse oxidized furnace blacks are particularly suitable. In oxidatively drying binders, oxidized blacks are again used on account of their good rheological and coloristic properties. In the dosage of drying agents (siccatives) it has to be taken into account that some heavy-metal salts are selectively adsorbed by the acidic surface oxides.

Offset printing imposes special requirements on the wetting behavior of carbon black and binder. The printing plate is wetted partly by water and partly by the printing ink. The carbon black and the binder must not be wettable by water in order to prevent bleeding of the ink while on the printing plate. This requirement is optimally satisfied by untreated furnace blacks. As in letterpress, high-viscosity inks with high carbon black contents are used. Since the ink films are very thin, these blacks must exhibit high hiding power. Therefore, usually low structure blacks which allow a higher degree of filling are preferred.

In contrast to the letterpress and offset processes, *gravure (intaglio) printing* calls for low-viscosity inks (0.1 Pa · s). The ink dries physically by solvent evaporation. The carbon black concentration is relatively low (6–12%). Medium and coarse furnace blacks (primary particle size 30–25 nm) of low structure are used, these permit relatively high pigment contents at low viscosity and high gloss of the printed matter. Fine-grained blacks (particle sizes 20–50 nm) of higher structure are only used for printing on absorbent papers. In order to prevent damage to the printing cylinder by the doctor blade, the blacks must be absolutely free of hard, coarse particles (grit).

Flexographic printing inks are similar to those used in gravure processes but have a somewhat higher carbon black concentration (10–15%).

4.7.2.3. Blacks for Paints

The black coloring of paints continues to be a major application of gas blacks. In most paint systems, they offer performance advantages over furnace blacks: easier dispersibility, better stability against flocculation, higher gloss, and better weather resistance. Further improvements in rheological properties, dispersibility, and coloristic qualities are achieved by oxidative aftertreatment of gas blacks. The primary effect of this operation is to change the volatiles content, resulting in a deepening of the color (enhancement of blackness) and a shift of the hue toward the blue (see Fig. 61), giving a visual impression of deeper blackness.

Gas blacks are used in various fineness grades (particle size 10–27 nm), depending on the desired color depth of the paint film. Figure 62 illustrates the relationship between particle size and blackness. Oxidized furnace blacks with blackness extending into the lower HCC range (Table 31) have come on the market, but do not offer the same performance as gas blacks.

A highly concentrated dispersion of carbon black is first prepared with a portion of the binder and solvent. The viscosity of this concentrate is a function of the particle size, structure, and surface chemistry of the black, the type of binder and its interaction with the pigment black, and the proportions of black, binder, and solvent. The final paint is made from the concentrate by adding more binder and solvent, its carbon black concentration is 3–8% referred to the solids content. Wetting agents are sometimes added to improve dispersibility and prevent flocculation. A number of concentrates for paint manufacture e.g., carbon black–nitrocellulose chips or carbon black–alkyd resin pastes, can be obtained from paint producers.

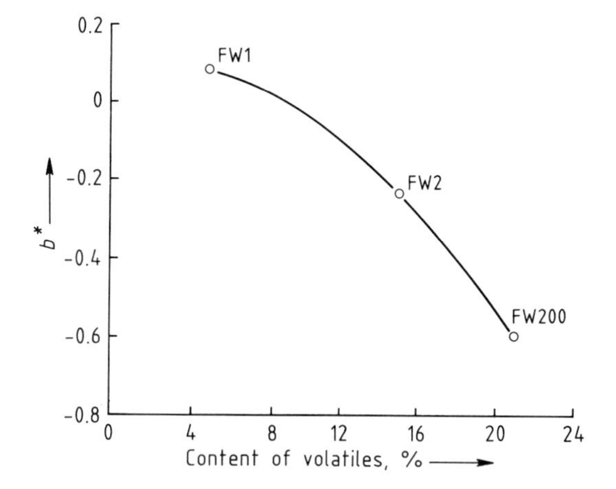

Figure 61. Shift in hue of gas blacks from FW1 to FW2 and FW200 (Degussa)
b^* = ordinate in the color coordinate system [4.32]. Negative b^* values denote a high content of blue

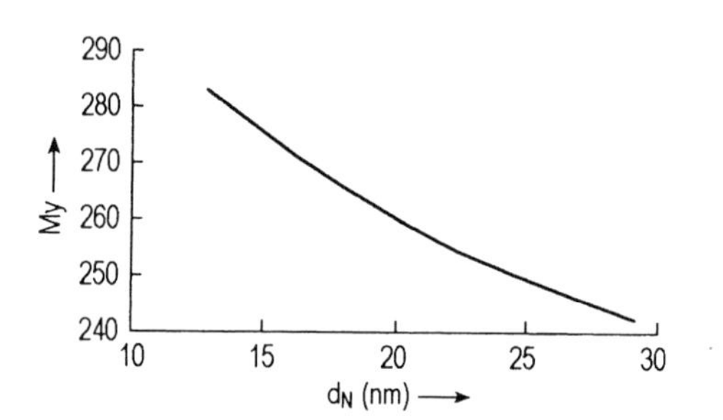

Figure 62. Blackness value My of gas blacks in linseed oil as a function of mean primary particle size (d_N)

The blacks used for tinting colored and white paints are chiefly coarse types such as lamp black and low surface area furnace blacks. Because of their relatively large primary particle size, these are readily dispersible and show little tendency to float when mixed with other pigments. In gray tints, fine blacks give a brown undertone while coarse ones give a blue undertone. Coarse blacks, e.g. lamp black, have lower tinting strengths than blacks with smaller particle size.

4.7.2.4. Blacks for Plastics

Unlike other pigments, carbon black used in plastics processing not only yields coloristic effects but also modifies the electrical properties, provides heat and UV resistance, and may act as a filler to modify mechanical properties.

Tinting of Plastics. Carbon black is used in large amounts for the black and gray tinting of plastics, including polyethylene, polypropylene, poly(vinyl chloride)

(PVC), polystyrene, ABS polymers, and polyurethanes. In tonnage terms, the largest application is in polyolefins. Pelletized black is usually employed.

Black concentrations of 0.5–2% are required for full-tone tinting. For crystal-clear plastics, the addition of up to 1% carbon black is generally sufficient. Plastics with marked intrinsic color (e.g., ABS) are tinted with blacks of higher tinting strength at concentrations of 1–2%. Transparent tints have black contents of 0.02–0.2%.

The color depth (blackness) increases as the primary particle size and the degree of aggregation decline. The blackness achieved by a given carbon black, however, also depends on the polymer (Fig. 63). Optically critical deep-black tintings are obtained with HCC and HCF gas and furnace blacks (Table 31). For general black tinting, where optimal brilliance and color depth are not essential, lower-priced blacks of moderate fineness belonging to classes MCC, RCC, and MCF (Table 31) are employed. Coarse furnace blacks and flame black are suitable for tinting.

Most of the carbon black used in the plastics industry is produced by the furnace black process (see Section 4.4.1). Fine furnace blacks often appear more bluish than coarse ones in incident light (full-tone tinting) but more brownish in transmitted light (transparent tinting) and in gray tints. By controlling the conditions of black manufacture, however, substantial shifts in hue can be achieved. Bluish carbon blacks are commonly preferred because of the impression of greater color depth.

Blacks for the tinting of products used for food packaging and storage must comply with relevant legal provisions.

Dispersion is crucial for tinting quality. In order to develop its full coloristic properties, carbon black requires much greater shear forces for dispersion than other pigments. Optimal dispersion i.e., complete disintegration of agglomerates into discrete particles (carbon black aggregates) and coating with the polymer is more difficult the finer the black and the lower its degree of aggregation (black structure). Unsatisfactory distribution of the black has a negative effect on the tinting

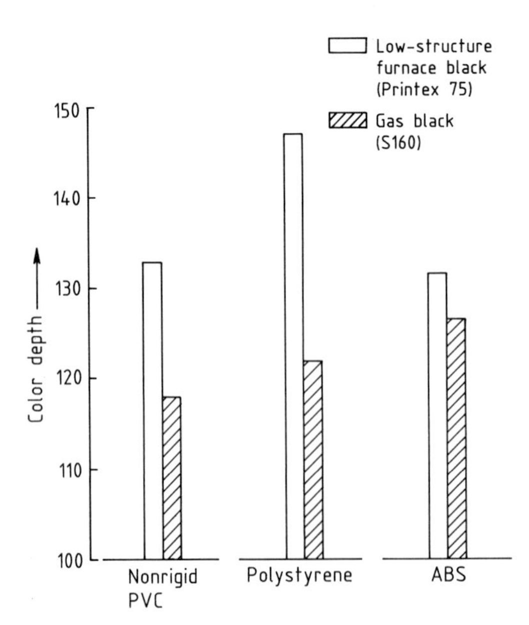

Figure 63. Color depth of two typical blacks in plastics (plotted to a visual scale).

strength and becomes a problem especially in gray tints and in the black tinting of opaque plastics. The depth of color in black-tinted transparent plastics, on the other hand, is only slightly affected by a small proportion of poorly dispersed black. Poor dispersion can also cause surface defects (specks) in plastic parts and mechanical defects, especially in films and fibers.

Good distribution of the black in the plastic is achieved by incorporating it in two stages. In the first step a carbon black–plastic concentrate (master batch) is prepared, for example in kneaders. This contains ca. 20–50% carbon black, depending on the binder absorption of the black. In a second operation, the concentrate is diluted with the appropriate polymer to give the final black content. Dispersion of the black takes place during the preparation of the concentrate. The high black concentration increases the viscosity and enables the medium to transmit higher shear forces than would be possible if 1–2% black were incorporated directly.

The carbon black plastic concentrates are often not prepared by the plastics processor. They may be purchased, from the pigment-black manufacturer in the form of chips, pellets, or powders. Black–plasticizer pastes (e.g., black/DOP pastes for tinting plastisols) and water- and solvent-based dispersions (e.g., for fiber manufacture) are also commercially available.

Conductive Fillers. The incorporation of sufficiently large amounts of blacks can impart antistatic (resistivity 10^6–10^9 $\Omega \cdot$ cm) or conductive ($\leq 10^6$ $\Omega \cdot$ cm) properties to plastics.

The conductivity of a plastic with carbon black filler increases with decreasing particle size and increasing degree of aggregation of the black, and increasing carbon black concentration. Surface oxides and hydrogen-containing surface groups of the black diminish the conductivity. The distribution of the black in the plastic is also crucial for the conductivity. As the black is incorporated into the plastic, the conductivity initially increases, passes through a maximum, and, if the black is subject to excessive shear force decreases. Conditions are optimal when the black is uniformly distributed in the polymer but is not so widely dispersed that the black particles are completely surrounded by the medium. Only then can bridges form between the particles in loose contact, thus promoting the flow of current. Orientation of the black particles (e.g., in extrusion or injection molding) permits anisotropic conductivity. Conductivity also depends on the polymer system due to variations in wettability or local concentration by partial crystallization of the polymer. For example, in order to obtain the same conductivity nonrigid PVC requires roughly double the black concentration than does polypropylene.

Blacks used to produce conductive and antistatic plastics are chiefly high-structure furnace blacks with relatively fine particles and low contents of volatile constituents. Black concentrations are around 10–40% for conductive systems. Antistatic plastics (e.g., cable sheathing and floor coverings) contain 4–15% carbon black.

Relatively coarse-grained oxidized gas blacks with high volatiles contents are especially suitable for plastics with good electrical insulation (e.g., cable sheathing compounds, plastics for high-frequency welding, toners).

UV Stabilization. Many polymers, above all polyethylene, are degraded by UV radiation under atmospheric conditions. Blacks can confer long-term stability to such polymers, firstly by absorbing the UV radiation and secondly by acting as free-radical acceptors to inactivate the active intermediate species formed in the degradation process. The stabilizing action increases with decreasing particle size and with increasing black concentration up to about 2–3%. Blacks with primary particle sizes of 20–25 nm are commonly used.

The radical-accepting ability of blacks used as stabilizers can interfere with the free-radical cross-linking of polymers.

4.8. Economic Aspects

In 1995, approximately 6 to 6.5×10^6 t of carbon black were produced. Production capacities were estimated at the same time to be 8×10^6 t/a (Table 34). The annual growth rate, on the average 7.9% per year between 1965 and 1975 in the United States, has decreased substantially, primarily in respect of rubber blacks due to longer tire life and the fact that the car market is reaching a saturation. Therefore, the future growth rate of the carbon black market is expected to be rather limited and will not exceed 1 to 2% per year.

Nearly 40% of the total worldwide production capacity of carbon black is concentrated in the United States and Western Europe. A detailed survey of the capacities in Western Europe is given in Table 35.

More than 90% of the total amount of carbon blacks produced are used as reinforcing fillers in rubber, of which 65–70% go into the tire industry and an additional 25–30% are needed for the production of mechanical rubber goods. Less than 10% of all carbon blacks produced are used for nonrubber purposes.

4.9. Toxicology and Health Aspects

As far as toxicology is concerned, one must distinguish between soot, which is formed by the uncontrolled combustion of coal and oil, and carbon black, which is industrially produced under precisely defined conditions. Commercial carbon blacks are characterized by an atomic ratio H:C of < 0.1, low ash content, and high adsorption capacity. The soluble organic fraction (extractable materials) is less than 0.5 wt%.

The "chimney-sweeper cancer" described by P. Pott in England as early as 1775, which is basically related to soot but not to carbon black, was the starting point of intensive research on analytical test methods for polycyclic aromatic hydrocarbons (PAH) in carbon blacks. Increasingly refined test methods have been developed (e.g., column, thin-layer, and paper chromatography, gas chromatography, UV and fluorescence spectrophotometry, and mass spectroscopy) to detect traces of such substances and to investigate the ability of carbon black surfaces to adsorb and to desorb PAHs.

Table 34. Carbon black production capacity (1996)

Region	Capacity, 10^3 t
North America	1815
Western Europe	1310
Eastern Europe	1545
Asia	2630
South America	480
Africa, Australia	185
Total	7965

Table 35. Carbon black production capacity in Europe (1995/96)

Country	Capacity, 10^3 t
United Kingdom	180
Federal Republic of Germany	277
France	280
Italy	200
Netherlands	160
Spain	90
Sweden	40
Others	83
Total	1310

Many years of experience in the carbon black producing and processing industry have clearly shown that there is no health hazard attributable to this product. Allegations occasionally being made to the effect that carbon blacks impair human health, could be refuted. In most cases it was possible to attribute these claims to a confusion between carbon blacks and soot or coal dust. Carbon black differs markedly from soot, the unwanted, uncontrolled by-product of combustion found in chimneys and the ambient air [4.36], [4.37].

Commercial carbon blacks (not aftertreated) consist of more than 95% of carbon, and small amounts of hydrogen, oxygen and sulphur. They have an atomic ratio H:C of < 0.1 and a low ash content. In most of the carbon blacks the total extractable matter (with toluene) is less than 0.15%, only parts of it are polycyclic aromatic hydrocarbons (PAHs). Due to this low content and the highly adsorbent properties of carbon black, the extractable matter is bound very tightly on the surface and can only be removed with powerful solvents [4.38]. – In contrast, soots vary widely in composition and properties. Besides carbon they contain variable amounts of inorganic and organic by-products. Because of their uncontrolled genesis the content of extractable matter is often very high ($> 25\%$) and consists among others of a wide range of PAHs. The high content and loose binding of the PAHs is to a great extent responsible for the mutagenicity and carcinogenicity of soots.

Detailed studies of workers in the carbon black industry and among carbon black consumers in the USA and Europe have shown that carbon black causes neither structural damages of the lungs nor lung tumours. No association between cumulative carbon black exposure and the incidence of respiratory disease could be detected [4.39]–[4.41].

In small animal tests no irritation of the mucous membranes or eyes was observed. Also no changes could be detected when the skin was exposed to whole carbon black. Conversely, animals exposed to the benzene extract of the same carbon black developed a significant number of malignant tumours [4.42]. If adsorbed on the carbon black surface, the PNAs are biologically inactive. Tests in five different systems for genetic activity of a furnace black containing 294 ppm toluene extractable PNAs indicated that no mutagenic activity could be attributed to whole carbon black [4.43]. The results of several studies provided no indications of pathological effects in the gastrointestinal tracts when carbon blacks were ingested in rodents [4.44].

Dangerous materials other than PAHs, such as polychlorinated bi- and terphenols, polychlorinated dioxins and polychlorinated hydrofurans were not found in carbon blacks. Nitrosamines could not be detected in carbon black, but they may be formed in rubber compounds, if rubber chemicals containing secondary amines are used. The total amine content of carbon black is less than 0.01% and the aromatic amine content is therefore less than this. No heavy metal is present in amounts above 0.002%. Carbon blacks therefore conform to all known regulations that limit these impurities.

Recently conducted lifelong inhalation studies on various species of animals that involved titanium dioxide, iron oxide, talc, diesel soot as well as carbon black have shown that these fine dusts considered hitherto as inert are lungtoxic [4.45], [4.46]. Of the various species used in the studies including rats, mice and hamsters only the rat species developed identifiable lung tumours. In all cases the rats were exposed to high dust levels for prolonged periods of time, to the point that the animals experienced a phenomenon termed "lung overload". The effect is not restricted to carbon black. It appears from the results of long-term inhalation studies that any insoluble, low toxicity particle will cause lung tumour in rats if deposited chronically at high enough doses [4.47]. Extrapolating the incidence of lung tumours in rats inhaling inert insoluble particles, such as carbon black, to human responses must be seriously questioned. The tests were carried out under such extreme overload conditions, never experienced in the workplace today, that the lungs of the rats were unable to clear themselves during non-exposure periods.

Such conditions with high carbon black levels would not occur under today's environment of carbon black production plants. Most of the carbon black producing countries have adopted the TLV (Threshold Limit Value) of USA of 3.5 mg/m^3 [4.48]. Extensive sampling in a number of carbon black plants in Europe [4.49] provided the results that the occupational exposure standard (OES) for carbon black is 3.5 mg/m^3 for an 8-hr TWA (time weighted average) and 7 mg/m^3 for the short-term exposure limit (STEL).

In a former exhaustive review in 1984 carbon black was classified by the International Agency for Research on Cancer (IARC) [4.50], an agency of the WHO, into category 3: *Not classifiable as to its carcinogenicity to humans*. This classification was confirmed in 1987. However, based on the results of the inhalation studies with rats, IARC [4.51] re-classified carbon black in 1995 in group 2B: *The agent is possibly carcinogenic to humans*. The criteria applied are: *Inadequate evidence in humans and sufficient evidence in experimental animals.*

Only if strong evidence can be provided that the mechanism of carcinogenicity in rats does not operate in humans, a new re-classification of carbon black by IARC in group 3 might be possible.

Food Contact Regulations. Carbon blacks, which according to their application properties may be used for food products, cosmetics, drinking water pipes, food packaging materials, and toys, must comply with local regulations. The testing methods for the approval of carbon blacks for such applications are different for different countries. The general aim, however, is to give limitations for the content of PAHs and of heavy metals. In general, the amount of heavy metals is far below the limitations. With respect to PAHs, either the total amount of extractable material or the content of special species is limited. Since industrial carbon blacks contain only trace amounts of strongly adsorbed PAHs, the majority of commercial grades comply with these regulations.

5. Specialty Pigments

This chapter deals with pigments that produce special effects: magnetic pigments (Section 5.1), anticorrosive pigments (Section 5.2), luster pigments (Section 5.3), transparent pigments (Section 5.4), and luminescent pigments (Section 5.5).

5.1. Magnetic Pigments

5.1.1. Iron Oxide Pigments

Ferrimagnetic iron oxide pigments are used in magnetic information storage systems such as audio and video cassettes, floppy disks, hard disks, and computer tapes Cobalt-free iron(III) oxide and nonstoichiometric mixed-phase pigments have been used since the early days of magnetic tape technology. Currently, γ-Fe_2O_3 [*1309-37-1*] and Fe_3O_4 [*1317-61-9*] (the latter in small amounts) are mainly used in the production of low-bias audio cassettes [iron oxide operating point IEC I standard (International Electrotechnical Commission)], and studio, broadcasting, and computer tapes.

Production. The shape of the pigment particle is extremely important for ensuring good magnetic properties. Isometric iron oxide pigments produced by direct precipitation are seldom used. Since 1947, needle-shaped γ-Fe_2O_3 pigments have been prepared with a length to width ratio of ca. 5:1 to 20:1 and a crystal length of 0.1–1 μm [5.1].

Anisometric forms of Fe_3O_4 with the spinel structure or γ-Fe_2O_3 with a tetragonal superlattice structure do not crystallize directly. They are obtained from iron compounds that form needle-shaped crystals (usually α- and γ-FeOOH, see Section 3.1.1) [5.2]–[5.4]. The oxyhydroxides are converted to Fe_3O_4 by dehydration and reduction. Reducing agents may be gases (hydrogen, carbon monoxide) or organic compounds (e.g., fatty acids). The particle geometry is retained during this process.

Since the pigments are subjected to considerable thermal stress during this conversion, the FeOOH particles are stabilized with a protective coating of silicates [5.5], phosphates [5.6], chromates [5.7], or organic compounds such as fatty acids [5.8].

Finely divided stoichiometric Fe_3O_4 pigments are not stable to atmospheric oxidation. They are therefore stabilized by partial oxidation or by complete oxidation to γ-Fe_2O_3 below 500 °C.

In an alternative process, the starting material consists of needle-shaped particles of α-Fe_2O_3 instead of FeOOH pigments [5.9], [5.10]. The synthesis is carried out in a hydrothermal reactor, starting from a suspension of $Fe(OH)_3$, and crystal growth is controlled by means of organic modifiers.

Properties. Magnetic pigments with very different morphological and magnetic properties that depend on the field of application and quality of the recording medium are used. The largest particles (length ca 0.6 μm) are used in computer tapes. The noise level of the magnetic tape decreases with decreasing particle size. Fine pigments are therefore being used increasingly for better quality compact cassettes.

The magnetic properties may be determined by measurement of hysteresis curves on the powder or magnetic tape.

Table 36 shows some quality requirements for the most important applications of magnetic pigments. Column 4 gives the coercive field strength (H_c) required for information storage materials. The coercive field is the magnetic field required to demagnetize the sample.

The saturation magnetization M_s is a specific constant for the material and for magnetic iron oxides is principally determined by the Fe^{2+} ion content. The ratio of remanent magnetization to saturation magnetization (M_r/M_s) for the tape depends mainly on the orientation of the pigment needles with respect to the longitudinal direction of the tape, and should approach the theoretical maximum value of unity as closely as possible.

Apart from the morphological and magnetic properties, usual pigment properties such as pH value, tap density, soluble salt content, oil absorption, dispersibility, and chemical stability are of great importance for the manufacture of magnetic recording materials.

Producers of magnetic iron oxides include BASF and Bayer (Germany); Ishihara, Sakai, Titan K., and Toda K (Japan); 3 M, Mitsui, Magnox, and ISK Magnetics (USA); Saehan Media (Korea); and Herdilla (India).

World production of cobalt-free magnetic iron oxides in 1995 was ca. 13000 t, of which ca. 96% was used in compact cassettes and audio tapes, and ca. 4% in computer tapes.

5.1.2. Cobalt-Containing Iron Oxide Pigments

Cobalt-containing iron oxides form the largest proportion (ca. 75%) of magnetic pigments produced today. Due to their high coercivity they can be used as an alternative to chromium dioxide for the production of video tapes, high-bias audio tapes (CrO_2 operating point), and high-density floppy disks.

Production. The iron oxide pigments described in Section 5.1.1 are either doped or coated with cobalt:

Table 36. Some quality requirements for iron oxide and metallic iron magnetic pigments

Field of application	Pigment type	Approximate particle length, µm	Specific surface area, m^2/g	Coercive field strength H_c, kA/m	Saturation magnetization, M_s, $\mu T \cdot m^3/kg$	M_r/M_s
Computer tapes	γ-Fe_2O_3	0.60	13–17	23– 25	86– 90	0.80–0.85
Studio radio tapes	γ-Fe_2O_3	0.40	17–20	23– 27	85– 92	0.80–0.85
IEC I compact cassettes						
standard (iron oxide operating point)	γ-Fe_2O_3	0.35	20–25	27– 30	87– 92	0.80–0.90
high grade	Co–γ-Fe_2O_3	0.30	25–37	29– 32	92– 98	0.80–0.90
IEC II compact cassettes (CrO_2 operating point)	Co–γ-Fe_2O_3,	0.30	30–40	52– 57	94– 98	0.85–0.92
	Co–Fe_3O_4	0.30	30–40	52– 57	98–105	0.85–0.92
IEC IV compact cassettes (metal operating point)	metallic iron	0.35	35–40	88– 95	130–160	0.85–0.90
Digital audio (R-DAT)*	metallic iron	0.25	50–60	115–127	130–160	0.85–0.90
½″ Video	Co–γ-Fe_2O_3	0.30	25–40	52– 57	94– 98	0.80–0.90
	Co–Fe_3O_4	0.30	30–40	52– 57	98–105	0.80–0.90
Super-VHS video	Co–γ-Fe_2O_3	0.20	45–50	64– 72	94– 96	0.80–0.85
8-mm video	metallic iron	0.25	50–60	115–127	130–160	0.85–0.90

* R-DAT: rotary digital audio tape.

1) Body-doped pigments contain 1–5% cobalt that is uniformly distributed throughout the bulk of the pigment particles. It is either incorporated during production of the FeOOH precursor or precipitated as the hydroxide onto one of the intermediate products [5.11] using cobalt(II) salts as the cobalt source.
2) Cobalt-coated pigment particles (2–4 wt% Co) consist of a core of γ-Fe_2O_3 or nonstoichiometric iron oxide phase, and a 1–2 nm coating of cobalt ferrite with a high coercivity [5.12]. The coating can be produced by adsorption of cobalt hydroxide, or epitaxial precipitation of cobalt ferrite in a strongly alkaline medium [5.13], [5.14]. Surface-coated pigments show better magnetic stability than doped pigments.

Properties. Pigments with a coercive field strength of 50–56 kA/m are used in video cassettes, high-bias audio cassettes (chromium dioxide operating point IEC II), and high-density floppy disks. Depending on the quality of the tape, the particle size varies between 0.2 and 0.4 μm (see Section 5.1.1, Table 36).

Pigments with a higher coercive field strength (ca. 70 kA/m) and smaller particle size (particle length ca. 0.15–0.2 μm) are used for super VHS cassettes.

Pigments treated with only small amounts of cobalt (0.5–1 wt% Co, coercive field strength ca. 31 kA/m) are used as an alternative to cobalt-free γ-Fe_2O_3 pigments for high-quality, low-bias audio cassettes. The other parameters described in Section 5.1.1 are also of importance for cobalt-treated pigments.

Cobalt-containing pigments are mainly produced by the magnetic iron oxide producers listed in Section 5.1.1. World production for 1995 was 40 000 t, of which the highest proportion (ca. 85%) was used for video tapes.

5.1.3. Chromium Dioxide

In the course of the development of pigments for magnetic information storage, CrO_2 was the first pigment material that gave a higher recording density than γ-Fe_2O_3. In the field of audio recording this led to the IEC II standard or "chrome position".

Production and Chemical Properties. Of the several chemical reactions for the formation of CrO_2 [5.15], only the conversion of an intimate mixture of Cr(III) and Cr(VI) compounds under hydrothermal conditions has been developed into an industrial process [5.16]. The synthesis is carried out in autoclaves at ca. 350 °C and 300 bar. The process comprises two consecutive partial reactions as indicated (Fig. 64) which allows a clear separation of nucleation and crystal growth, thus leading to a narrow particle size distribution. The crystal structure of CrO_2 allows the direct formation of smooth, needle-shaped particles which is essential for the magnetic properties of the pigment. Pure CrO_2 slowly disproportionates in the presence of water. The CrO_2 crystal surface of commercial pigments is therefore topotactically converted to β-CrOOH which serves as a protection layer (Fig. 65) [5.17]. The conditions for this reaction have a great influence on the chemical properties of the

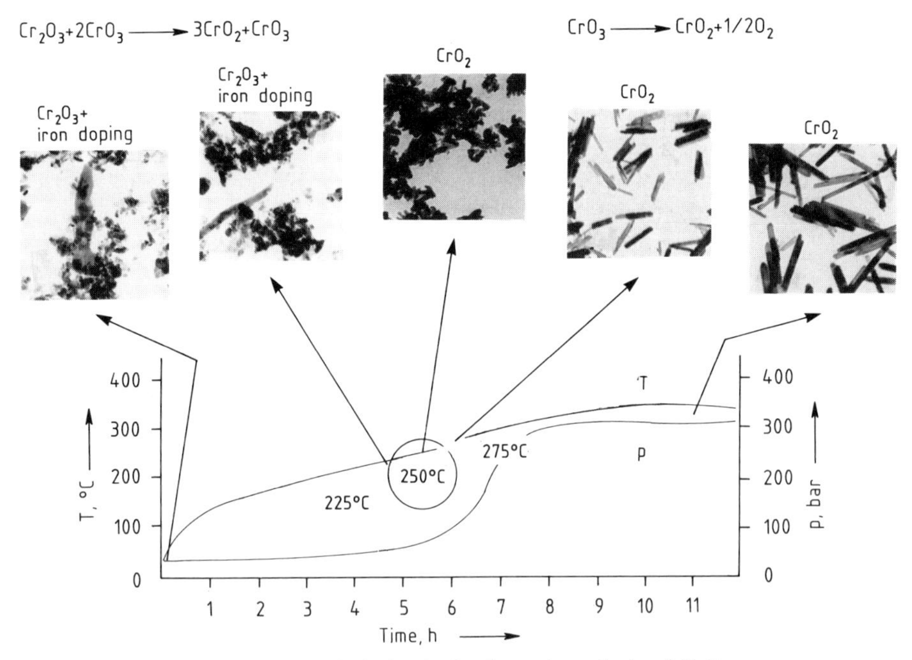

Figure 64. Nucleation and growth during hydrothermal synthesis of CrO_2

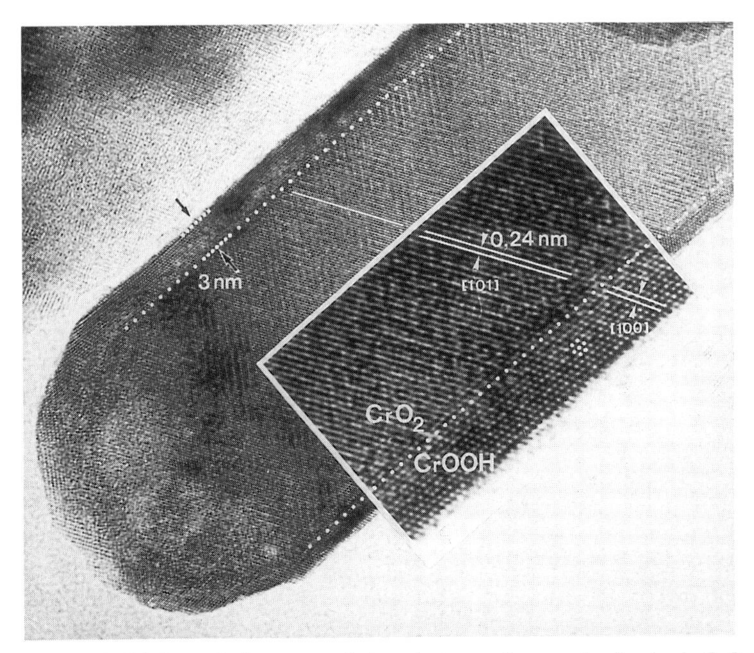

Figure 65. High-resolution transmission electron micrograph of a single CrO_2 particle showing the topotactically grown crystalline core–shell structure

pigment surface [5.18], [5.19]. Thus, the isoelectric point in water can vary between pH 3 and 7 [5.20], which affects dispersibility. In the absence of moisture, CrO_2 is stable up to ca. 400 °C; above this temperature it decomposes to form Cr_2O_3 and oxygen.

Physical Properties. Chromium dioxide [*12018-01-8*], chromium(IV) oxide, CrO_2, is a ferromagnetic material with a specific saturation magnetization M_s/ϱ of 132 A m²/kg at 0 K corresponding to the spin of two unpaired electrons per Cr^{4+} ion. The M_s/ϱ value of CrO_2 at room temperature is ca. 100 A m²/kg [5.21]; CrO_2 magnetic pigments reach values of 77–92 A m²/kg. The material crystallizes with a tetragonal rutile lattice in the form of small needles which have the desired magnetic shape anisotropy. The morphology of the particles can be varied with several dopants, particularly antimony and tellurium [5.22]. The coercive field strength (in addition to shape) can be controlled by doping with transition metal ions which modify the magnetocrystalline anisotropy of the material; the Fe^{3+} ion being industrially important [5.23]. Depending on the iron content up to ca. 3 mol%, H_c values of CrO_2 may vary between 30 and 75 kA/m [5.24], [5.25]. Iron doping also increases the Curie temperature from 115 to ca. 170 °C. Because the Fe^{3+} ions are coupled antiferromagnetically with Cr^{4+} ions in the crystal lattice, the saturation magnetization (M_s/ϱ) of CrO_2 decreases with increasing dopant levels [5.26]. Properties of CrO_2 pigments are listed in Table 37.

Other important properties of CrO_2 when used as a magnetic pigment are its black color, electrical conductivity (2.5–400 Ω^{-1} cm^{-1} [5.27]) and relatively high crystal hardness (Mohs hardness 8–9 [5.20]). Therefore, coating formulations based on CrO_2 require less or even no additives such as carbon black (good conductivity, black color) or refractory oxides such as alumina.

Uses and Economic Aspects. Chromium dioxide is used exclusively for magnetic recording media, e.g., tapes for audio, video, and computer applications. An application of particular interest depends on its relatively low Curie temperature. This allows thermomagnetic duplication at temperatures low enough for the base polymer of magnetic tapes, and is exploited in a commercial, high-speed copying process [5.28]. The TMD duplication system uses a mirror master tape made with metal particles and a CrO_2 copy tape. The two tapes pass through the system with intimate

Table 37. Typical properties of CrO_2 pigments

Field of application	Mean particle length, nm	Specific surface area, m²/g	Coercivity, kA/m	Specific saturation magnetization, Am²/kg
Magnetic stripes	280	17	28–30	78–80
IEC-2 compact cassettes	250	25–27	37–45	78–80
Digital compact cassette	230–250	28–32	50–65	72–75
1/2″ video	240	28–30	48–51	75–77
Super-VHS video	200–230	30–43	55–65	72–75
IBM data cartridge	250	25–27	40–42	72–78

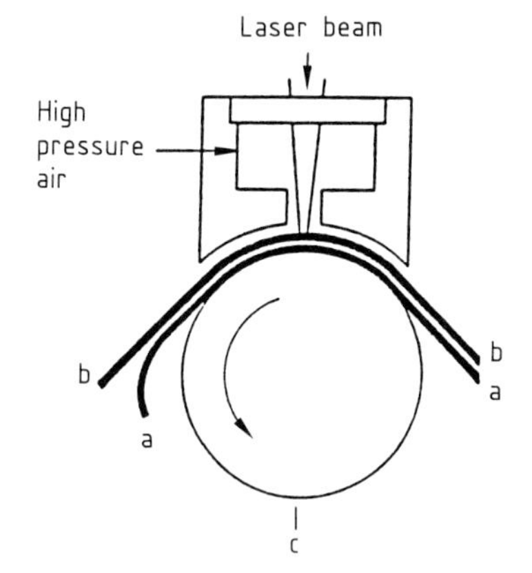

Figure 66. Thermal magnetic duplication (TMD process [5.29])
a) Master tape; b) Copy tape; c) Print wheel

coating–coating contact and are heated locally to about 30 °C above the Curie temperature of the copy tape but well below the Curie temperature of the iron particles (Fig. 66). Above the Curie temparature, the coercivity and residual magnetization of the magnetic particles in the copy tape become zero. If the magnetic particles in the copy tape are cooled below their Curie temperature while still in contact with the master tape they attain the magnetization pattern of the mirror master tape and finally yield a copy tape with read-out signals comparable to those obtained with real-time recording. The TMD system currently operates at 135 times the real-time speed in direct recording.

Chromium dioxide may also be used in combination with cobalt-modified iron oxides (see Section 5.1.2) in the production of magnetic recording media. The world production of CrO_2 in 1995 amounted to 8000 t, ca. 11 % of the total consumption of magnetic pigments. Producers are BASF and Du Pont.

5.1.4. Metallic Iron Pigments

The magnetization of iron is more than three times higher than that of iron oxides. Metallic iron pigments can have a coercive field strength as high as 150 kA/m, depending on particle size. These properties are highly suitable for high-density recording media. Oxidation-resistant products based on metallic pigments first became available in the late 1970s.

Production. Metallic iron pigments are commercially produced by the reduction of acicular (needle-shaped) iron compounds [5.30]. As in the production of magnetic iron oxide pigments, the starting materials are iron oxide hydroxides (see Section 3.1.1) or iron oxalates, which are reduced to iron in a stream of hydrogen either directly or via oxidic intermediates.

Due to their high specific surface area, metallic pigments are pyrophoric, so that passivation is necessary. This can be achieved by slow, controlled oxidation of the particle surface [5.31].

Properties. The coercive field strength of metallic iron pigments is primarily determined by their particle shape and size, and can be varied between 30 and 150 kA/m. Pigments for analog music cassettes ($H_c \approx 90$ kA/m) usually have a particle length of 0.35 μm (see Section 5.1.1, Table 36). The length to width ratio of the pigment needles is ca. 10:1. Finely divided pigments (particle length ca. 0.25 μm) with a coercive field strength of 120 kA/m are used for 8-mm video and digital audio cassettes (R-DAT), tapes used by television organizations (ED Beta, Betacam SP MP, M II, Digital Video D2), and for master video cassettes (mirror master tapes). In the field of data storage, a small quantity is used in micro floppy disks.

Metallic pigments have a higher specific surface area (up to 60 m^2/g) and a higher saturation magnetization than oxidic magnetic pigments. Their capacity for particle alignment corresponds to that of the oxides (see Section 5.1.1, Table 36).

Economic Aspects. The largest producers of metallic iron pigments are Dowa Mining, Kanto Denka K., and Mitsui Toatsu.

World consumption in 1995 was ca. 1800 t, of which ca. 75% was used in the manufacture of video tapes ca. 14% for audio and 11% for data storage applications. Consumption is expected to increase.

5.1.5. Barium Ferrite Pigments

Barium ferrite pigments have been considered for several years for high-density digital storage media [5.32], [5.33]. They are very suitable for preparing unoriented (e.g., floppy disks), longitudinally oriented (conventional tapes), and perpendicularly oriented media. In the latter the magnetization is oriented perpendicular to the coating surface. They are required for perpendicular recording systems which promise extremely high data densities, especially on floppy disks. Barium and strontium ferrites are also used to prevent forgery of magnetic stripes, e.g., in cheque and identity cards.

Properties. Hexagonal ferrites have a wide range of structures distinguished by different stacking arrangements of three basic elements known as M, S, and Y blocks [5.34]. For magnetic pigments, the M-type structure (barium hexaferrite [*12047-11-9*] $BaFe_{12}O_{19}$) is the most important. The magnetic properties of M-ferrite can be controlled over a fairly wide range by partial substitution of the Fe^{3+} ions, usually with combinations of di- and tetravalent ions such as Co and Ti. Barium ferrite crystallizes in the form of small hexagonal platelets. The preferred direction of magnetization is parallel to the *c*-axis and is therefore perpendicular to the surface of the platelet. The specific saturation magnetization of the undoped material is ca. 72 A m^2/kg and is therefore somewhat lower than that of other magnetic oxide pigments. In barium ferrite the coercive field strength is primarily determined by the magnetocrystalline anisotropy and only to a limited extent by particle morphology. This is the reason why barium ferrite can be obtained with extremely uniform magnetic properties. Barium ferrite pigments have a brown color and chemical properties similar to those of the iron oxides.

Production. There are three important methods for manufacturing barium ferrite on an industrial scale: the ceramic, hydrothermal, and glass crystallization methods. The main producers are Toshiba and Toda.

Ceramic Method. Mixtures of barium carbonate and iron oxide are reacted at 1200–1350 °C to produce crystalline agglomerates which are ground to a particle size of ca. 1 μm. This method is only suitable for the high-coercivity pigments required for magnetic strips [5.35].

Hydrothermal Method. Iron [Fe(III)], barium, and the dopants are precipitated as their hydroxides and reacted with an excess of sodium hydroxide solution (up to 6 mol/L) at 250–350 °C in an autoclave. This is generally followed by an annealing treatment at 750–800 °C to obtain products with the desired magnetic properties. Many variations of the process have been described [5.36]–[5.40], the earliest report being from 1969 [5.41]. In later processes, hydrothermal synthesis is followed by coating with cubic ferrites, a process resembling the cobalt modification of iron oxides (see Section 5.1.2). The object is to increase the saturation magnetization of the material [5.42]–[5.44].

Glass Crystallization Method. This process was developed by Toshiba [5.45]. The starting materials for barium ferrite production are dissolved in a borate glass melt. The molten material at ca. 1200 °C is quenched by pouring it onto rotating cold copper wheels to produce glass flakes. The flakes are then annealed to crystallize the ferrite in the glass matrix. In the final stage the glass matrix is dissolved in acid. In a variation of this process, the glass matrix is produced by spray drying [5.46].

Magnetic Recording Properties. Typical values of physical properties of barium ferrite pigments used in magnetic recording are given in Table 38.

Barium ferrite is highly suitable for high-density digital recording mainly because of its very small particle size and its very narrow switching field distribution. It also has a high anhysteretic susceptibility and is difficult to overwrite [5.47]. This is partly explained by positive interaction fields between particles in the coating layer [5.48]. The high anhysteretic susceptibility makes barium ferrite media particularly suitable for the anhysteretic (bias field) duplicating process [5.49]. Besides the TMD process

Table 38. Typical properties of barium ferrite pigments

Application	Specific surface area, m^2/g	Platelet diameter, nm	Platelet thickness, nm	H_c, kA/m	M_s/ϱ, A m^2/kg
Unoriented (floppy disk)	25–40	40– 70	15– 30	50– 65	50–65
Oriented	25–60	40–120	10– 30	55–100	50–65
Magnetic strips	12–15	100–300	50–100	220–440	60–70

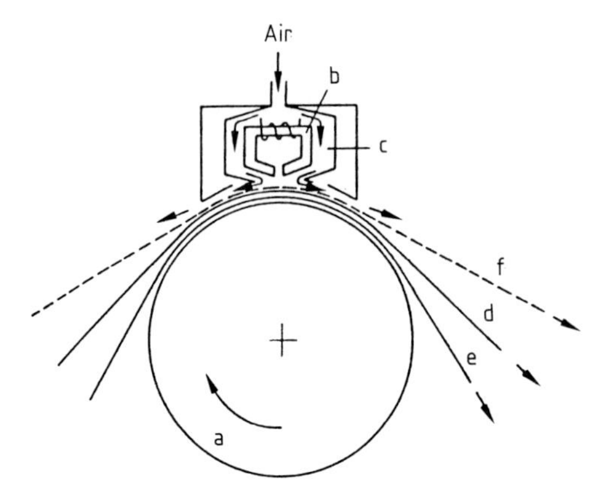

Figure 67. Anhysteretic duplication process [5.50]
a) Transfer drum; b) Transfer magnetic field head; c) Air chamber; d) Mirror master tape; e) Copy tape; f) Drive tape

(Section 5.1.3.) this is the second faster-than-real-time video tape duplicating process. The signal is transferred from the mirror master tape to the copy tape by applying a high-frequency bias field while the two tape coatings are in close contact with each other. The amplitude of the bias field is equal to the maximum coercivity of the copy tape, but should not exceed the minimum coercivity of the mirror master tape. Under the influence of both the bias field and the field of the master tape, the magnetic particles in the copy tape are magnetized in exactly the opposite pattern of the mirror master tape. The current copy speed is approximately 100 times standard playing speed for VHS vido tapes. Figure 67 shows the principle of the method.

Unlike many other magnetic materials used in high-density recording, barium ferrite, being an oxide, is not affected by corrosion [5.51]. Processing of the pigment can be problematic, e.g., applying orienting fields can easily lead to unwanted stacking of the particles which has adverse effects on the noise level and the coercive field strength of the magnetic tape. A marked temperature dependence of the magnetic properties was a problem in the early days, but this can be overcome by appropriate doping [5.52].

5.2. Anticorrosive Pigments

5.2.1. Principles

This chapter deals with the protection of metal surfaces against corrosion by means of coatings that contain anticorrosive pigments. The degree of corrosion protection depends not only on the pigment, but also on the binder, and these must complement each other chemically. The uses of anticorrosive pigments in paint binders are summarized in Table 39. For standards, see Table 1 ("Corrosion testing").

Table 39. Possible uses of anticorrosive pigments in various primers

Anticorrosive pigment	Air-drying alkyd primers	Alkyd–melamine primers	2-Component epoxy primers	2-Component acrylic isocyanate primers	Chlorinated rubber primer	Aqueous polymer dispersions	Poly(vinyl-butyral) primers	Electro-deposition coatings	Mirror coatings**	References
Zinc phosphate	+	+	+	+	+	+	o	o	–	[5.67]–[5.71]
Chromium phosphate	+	o	–	–	–	–	–	o	–	[5.72]
Aluminum phosphate	+	+	+	–	+	+	+	–	–	[5.68], [5.70], [5.73], [5.74]
Zinc phosphomolybdate	+	+	+	–	+	+/o	+	o	–	[5.68], [5.71], [5.75]
Zinc calcium phosphomolybdate	+	–	+	–	–	+/o	–	+/o	–	[5.75]
Phosphate multiphase pigments (see Table 45)										
Product 1	+	+	o	+	+	+	o	o	–	[5.67]
Product 2	+	+	+	+	+	+	o	o	–	[5.67]
Product 3	+	+	+	–	+	+	+	o	–	[5.68]
Zinc hydroxyphosphite	+	–	+	–	+	+	–	–	–	[5.76]
Lead cyanamide	–	+	–	–	–	–	–	+	+	[5.67]
Zinc cyanamide	–	+	–	–	–	–	–	o	+	[5.67], [5.77]
Barium metaborate	+	+	–	–	–	+	–	–	–	[5.78], [5.79]
Zinc borophosphate	o	–	o	–	–	+	+	–	–	[5.67]
Zinc potassium chromate	+	+	+	+	+	+	+	–	–	[5.70]
Zinc tetrahydroxychromate	+	+	–	–	–	+	+	–	–	[5.70]
Strontium chromate*	+	+	+	–	–	–	+	+	–	[5.70]
Red lead	+	+	+	–	+	–	+	–	o	
Zinc salt of 5-nitroisophthalic acid	+	+	+	+	+	o	+	o	–	[5.67]

\+ Commonly used (good anticorrosive properties); o Anticorrosive properties depend on the formulation; – Seldom used
* Mainly on aluminum. ** Mirror coatings are used to protect the thin metal surface (silver, copper) of metal-coated glass against oxidation or reaction with H_2S or SO_2.

Anticorrosive pigments may be divided into three types [5.53]:

1) *Pigments with a physical protective action* are chemically inert and are termed *inactive* or *passive*. An example is micaceous iron oxide [5.54], [5.55]. These lamellar pigments are packed in layers; they lengthen the pathways and obstruct the penetration of ions. They improve adhesion between the substrate and the coating, absorb UV radiation, and protect the underlying binder (Fig. 68), [5.56]–[5.60].
2) *Pigments with chemical protective action* contain soluble components and can maintain a constant pH value in the coating. They are termed *active* and their action depends on reactions in the interfacial areas between the pigment and substrate, pigment and binder, or between pigment and ions that penetrate into the coating [5.61], [5.62]. An example is red lead. Redox reactions can occur to form protective compounds (oxides or oxide hydrates that may contain pigment cations). Saponification of the binder or neutralization of the acidic decomposition products enable a definite pH value to be maintained in the coating.
3) *Pigments with an electrochemical protective action* passivate the metallic surface [5.63], [5.64]. Those that prevent corrosion of the iron by forming a protective coating (e.g., phosphate pigments) are regarded as being active in the anodic region of the metal surface (anodic protection) [5.65]. Pigments that prevent rust formation due to their high oxidation potential (e.g., chromates) are said to be active in the cathodic region (cathodic protection) [5.66].

Corrosion of iron is explained by the position of iron in the electrochemical series of the elements (Fe/Fe^{2+}: -0.44 V). In steel, local anode and cathode areas are found due to the presence of phases containing, for example, carbon, carbides, and oxides. These latent local cells are activated by moisture, oxygen, and current-carrying electrolytes and the following reactions occur between the anode areas consisting of iron, and the cathode areas containing carbides or oxides.

Anode: $Fe \rightarrow Fe^{2+} + 2e^- \; 2\,Fe^{2+} + O_2 + 2\,H_2O + 2\,OH^- \rightarrow 2\,Fe(OH)_3 \rightarrow \underset{\text{Rust}}{2\,FeO(OH)} + 2\,H_2O$

Cathode: $O_2 + 2\,H_2O + 4e^- \rightarrow 4\,OH^-$

The rust can promote further corrosion. The OH^- ions formed at the cathode produce local high alkalinity and cause hydrolysis of, for example, esterified binders, resulting in detachment of the primer from the substrate [5.57].

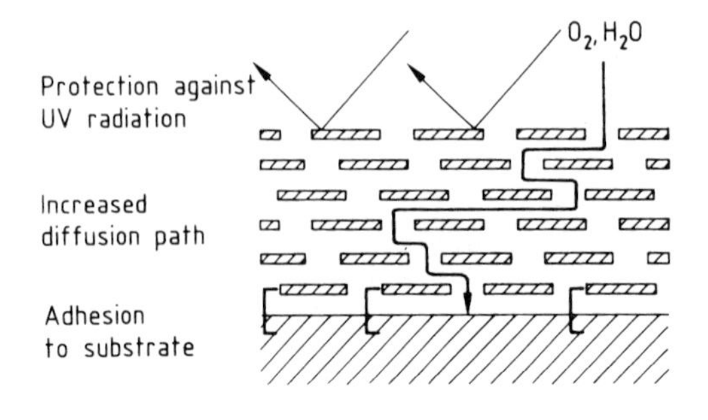

Figure 68. Schematic representation of the physical anticorrosion effects of flake pigments

Active anticorrosive pigments inhibit one or both of the two electrochemical partial reactions. The protective action is located at the interface between the substrate and the primer. Water that has diffused into the binder dissolves soluble anticorrosive components (e.g., phosphate, borate, or organic anions) out of the pigments and transports them to the metal surface where they react and stop corrosion. The oxide film already present on the iron is thereby strengthened and sometimes chemically modified. Any damaged areas are repaired with the aid of the active substance. Inhibition by formation of a protective film is the most important mode of action of the commoner anticorrosive pigments.

5.2.2. Phosphate Pigments

The most important phosphate-containing pigments are zinc phosphate [5.67]–[5.69], [5.71]; basic zinc phosphate [5.70], [5.71]; chromium phosphate [5.72]; aluminum triphosphate [5.73]; barium phosphate [5.53]; aluminum zinc phosphate [5.68]; and zinc/iron phosphate [5.71]. Phosphate ions form protective coatings of basic iron(III) phosphate on an iron surface. Their composition is described in [5.64].

The reactions of phosphate pigments in binders and on metal surfaces were described by G. MEYER [5.80], and J. RUF [5.81]. They postulated that basic complexes are formed from phosphate pigments containing water of crystallization and they react with inorganic ions or with carboxyl groups of the binder. Reaction of the pigment with the binder results in physical changes such as deterioration of adhesion or gelling. In paints that contain hydroxy- and carboxy-functional binders, gelling during storage can be prevented by using an excess of alcohol as the solvent.

Toxicological data for phosphate pigments are described in Section 5.2.14.

5.2.2.1. Zinc Phosphate [5.67]–[5.71]

The most important phosphate-containing pigment is zinc phosphate [*7779-90-0*], $Zn_3(PO_4)_2 \cdot 4\ H_2O$, M_r 458.1. It can be used with a large number of binders (see Table 39) and has a very wide range of uses [5.67], [5.68], [5.70], [5.71], [5.82].

Zinc phosphate is usually produced on an industrial scale from zinc oxide and phosphoric acid, or from zinc salts and phosphates [5.83]. Composition and properties are given in Table 40.

The mechanism of the action of zinc phosphate is shown in Figure 69. Zinc phosphate dihydrate pigment is hydrated to the tetrahydrate in an alkyd resin binder [5.84]. The tetrahydrate is then hydrolyzed to form zinc hydroxide and secondary phosphate ions which form a protective film of basic iron(III) phosphate on the iron surface [5.80]. The anticorrosive action of zinc phosphate depends on its particle size distribution. Micronization improves the anticorrosive properties [5.85]–[5.87].

The effect of corrosion-promoting ions on the anticorrosive properties of zinc phosphate is described in [5.88], [5.89].

Table 40. Composition and properties of zinc and chromium phosphate pigments

Property (standard)	Commercial products		
	Zinc phosphate $Zn_3(PO_4)_2 \cdot 2–4\ H_2O$	Micronized zinc phosphate $Zn_3(PO_4)_2 \cdot 2–4\ H_2O$	Chromium phosphate $CrPO_4 \cdot 3\ H_2O$
Metal content, wt% (ISO 787, part 2)	51	51	ca. 27
Phosphate content, wt% (ISO 787, part 2)	49	< 49	ca. 47
Chloride content, wt% (ISO 787, part 13)	0.01	0.05	< 0.1
Sulfate content, wt% (ISO 787, part 13)	< 0.1	0.05	0.04
Water-soluble content, wt% (ISO 787, part 3)	< 0.1	0.1	0.4
Sieve residue, wt% (ISO 787, part 18)	0.01	0.01	max. 0.3
Density, g/cm^3 (ISO 787, part 10)	3.2	3.4	2.5
Specific surface area, m^2/g (DIN 66131/66132)	1.0	4.3	22
Loss on ignition, wt% (ISO 787, part 2)	10–11	13	26
Oil absorption value, g/100 g (ISO 787, part 5)	20	24	65
Conductivity, μS (ISO 787, part 14)	100	300	ca. 500
pH (ISO 787, part 9)	6.2–7	7	6.5
Color	white-beige	white-beige	green

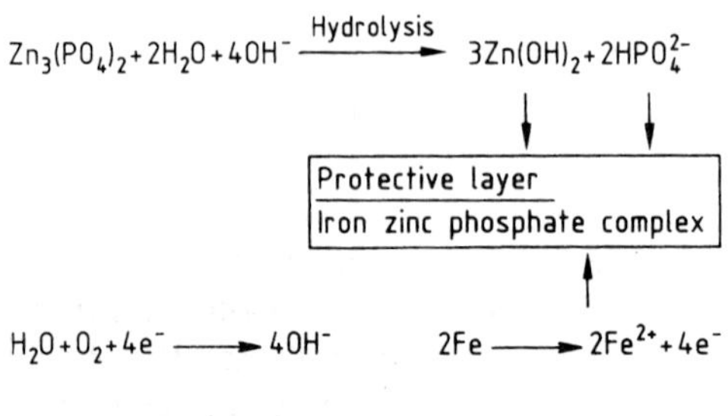

Figure 69. Passivation of iron by zinc phosphate [5.80]

Trade names are as follows:

Zinc phosphate: Sicor ZNP/M, ZNP/S (BASF, Germany); Heucophos ZP10 (Dr. H. Heubach, Germany); Halox Zinc Phosphate (Halox Pigments, USA); Phosphinox PZ20 (SNCZ, France); Hispafos N2 (Colores Hispania, Spain).

Basic zinc phosphate: Heucophos ZPO (Dr. H. Heubach, Germany); Hispafos SP (Colores Hispania, Spain).

5.2.2.2. Aluminum Phosphate

Commercial aluminum phosphate anticorrosive pigments consist of aluminum zinc phosphate hydrates [*66905-65-5*], [5.68], [5.70], or zinc-containing aluminum triphosphate [*13939-25-8*], [5.73]. Their compositions and properties are listed in Table 41. Their applications in commercial binders are listed in Table 39. Aluminum zinc phosphate hydrate pigments are produced by reacting acid solutions of aluminum hydrogen phosphate with zinc oxide and alkali aluminate. The precipitated pigment is filtered off from the mother liquor, washed, dried, and ground [5.90]. Commercial aluminum triphosphate pigments contain ions of trimeric phosphoric acid which form stable aluminum-containing iron phosphate complexes [5.73].

The aluminum phosphate pigments give good adhesion of the paint film to the metallic substrate.

Trade names are as follows:

Zinc aluminum phosphate: K-White 105 (Teikoku Kako, Japan); Heucophos ZPA (Dr. H. Heubach, Germany); Phosphinal PZ04 (SNCZ, France).

Aluminum triphosphate: K-White 82, -84 (Teikoku Kako, Japan).

Table 41. Composition and properties of aluminum phosphate anticorrosive pigments

Property	Aluminum triphosphate (commercial product 1; K-White 82) [5.73]	Aluminum triphosphate (commercial product 2; K-White 84) [5.73]	Aluminum zinc phosphate hydrate (commercial product 3; Heucophos ZPA, Phosphinal PZ04) [5.68], [5.70]
Al content, wt%	5.5–7.7	4.7–6.9	4–5
Zn content, wt%	11.6–14.9	21.3–24.5	35–39
P_2O_5 content, wt%	42.0–46.0	36.0–40.0	30/37
SiO_2 content, wt%	13.0–17.0	11.0–15.0	
Loss on ignition (600 °C), wt%	ca. 10	ca. 8	10–16
Density, g/cm^3	3.0	3.1	3.2
Oil absorption value, g/100 g	35	30	30/35
Particle size, μm	0.5–10	0.5–10	0.5–10
pH value	6–7	6–7	6.5
Water-soluble content, wt%	max. 1.0	max. 1.0	0.1
Color	white	white	white

5.2.2.3. Chromium Phosphate [5.53], [5.72]

Chromium phosphate [*16453-74-0*], $CrPO_4 \cdot 3\ H_2O$, M_r 201.0, is produced from chromium(III) salts and alkali phosphates. The physical and chemical properties are listed in Table 40; important areas of use are given in Table 39, [5.72].

Chromium phosphate has a low solubility. It is therefore nearly always used in combination with other anticorrosive pigments. It is an extremely good long-term inhibitor, but is less effective during the initial phase of corrosion protection.

5.2.2.4. New Pigments Based on Metal Phosphates

Developments in the field of phosphate pigments are aimed at correcting the deficiencies of phosphate (such as the moderate corrosion protection against accelerated weathering in the salt spray test) during the initial phase of corrosion protection. Improvements include crystallization to form new structures, replacement of cations (e.g., of zinc by calcium), modification of anions (phosphovanadates and phosphocarbonates), optimization of the particle size distribution, providing the zinc ions and phosphate ions in separate compounds, and use of synergistic effects. New pigments are listed in Table 42. Some of them are commercially available, e.g., the zinc molybdatophosphates and zinc calcium molybdatophosphates. Trade names include Heucophos ZMP [5.68], Moly-White ZNP and MZA [5.75], and Actirox 102 and 106 [5.71], [5.102].

5.2.2.5. Multiphase Phosphate Pigments

Multiphase anticorrosive phosphate pigments are based on appropriate combinations of inorganic phosphates with sparingly soluble, electrochemically active organic corrosion inhibitors [5.67]. Examples include Sicor SPO (zinc phosphate with zinc salt of 5-nitroisophthalic acid, BASF) [5.67]; Sicor NOP (zinc phosphate with calcium salt of 5-nitroisophthalic acid, BASF) [5.67]; and Heucophos ZPZ (basic zinc phosphate with 0.3% organic components, Heubach) [5.68]. The multiphase pigments are produced in one combined synthesis. A slurry of precipitated zinc phosphate is mixed intensively with the zinc or calcium salt of 5-nitroisophthalic acid (Section 5.2.13), filtered, washed, and ground until the specified conductivity is reached. They are more effective than physical mixtures of pigments produced in separate processes. Multiphase phosphate pigments are used in place of zinc chromate if the anticorrosive effect of pure zinc phosphate is insufficient in the initial weathering phase (e.g., in the salt spray test). The very good anticorrosive effect of multiphase pigments in air-drying alkyd resins or in alkyd–melamine stoving enamels is comparable to that of zinc chromate, even in open air weathering. Commercial multiphase pigments differ mainly in the solubility of the organic corrosion inhibitor. Systems with more soluble inhibitors are used in the binders with a low swelling capacity (e.g., epoxy resins or chlorinated rubber).

Table 42. New anticorrosive metal phosphate pigments

Composition	Binder	Metal substrate	Reference
$CaZn_2(PO_4)_2 \cdot 2\ H_2O$	alkyd resin chlorinated rubber	iron	[5.91]
$CaHPO_4 \cdot 2\ H_2O/ZnO$	polyvinyl butyral phenolic resin	aluminum	[5.92]
$CaHPO_4 \cdot 2\ H_2O/ZnO$	alkyd resin	iron	[5.93]
$MgHPO_4 \cdot 3\ H_2O/ZnO$	polyvinyl butyral phenolic resin	aluminum	[5.92]
$MgHPO_4 \cdot 3\ H_2O/ZnO$	alkyd resin	iron	[5.93]
$CaHPO_4 \cdot 2\ H_2O/MgHPO_4 \cdot 3\ H_2O$	alkyd resin	iron	[5.94]
$[0.5\ MgHPO_4 \cdot 0.5\ MgCO_3] \cdot x\ H_2O$	epoxy polyamide	aluminum	[5.95]
$[0.4\ SrHPO_4 \cdot 0.6\ SrCO_3] \cdot x\ H_2O + 0.1 - 3\%$ fluoride	epoxy polyamide	aluminum	[5.95]
$TiO_2/ZrO_2/SiO_2 \cdot x\ P_2O_5$	alkyd–melamine	iron	[5.96]
$(Fe_x,Cr_y,Na_z)PO_4 \cdot x\ H_2O$	epoxy resin	iron	[5.97]
$Al(H_2PO_4)_3 \cdot x\ H_2O/ZnO$	alkyd resin		[5.98]
Phosphate glasses containing			
ZnO/Al_2O_3	alkyd resin		[5.99]
MgO, SiO_2, Al_2O_3, CaO, B_2O_3	alkyd resin		[5.100]
Mg/Ca phosphovanadate	epoxy, alkyd– melamine	iron	[5.101]
$[Zn_3(PO_4)_2 \cdot 2{-}4\ H_2O] \cdot [x\ ZnMoO_4]$	alkyd	iron	[5.68]
$(Zn,Ca)_3(PO_4,MoO_4)_2 \cdot x\ H_2O$	alkyd, epoxy	iron	[5.75]

The compositions and properties of some commercial multiphase anticorrosive pigments are listed in Table 43. Their anticorrosive properties in various binders are shown in Table 39.

Trade names include Sicor SPO, -NOP (BASF, Germany); Heucophos ZPZ (Dr. H. Heubach, Germany); and Novinox PZ02 (SNCZ, France).

5.2.3. Other Phosphorus-Containing Pigments

Iron Phosphide. Commercial iron phosphide anticorrosive pigments usually consist of Fe_2P [*12751-22-3*], M_r 142.7, with traces of FeP and SiO_2. The pigment is a powder with a metallic gray color and contains 70 % Fe, 24 % P, 2.5 % Si, and 3.0 % Mn. The density is 6.53 g/cm^3 and the mean particle size is ca. 3–5 µm.

Table 43. Multiphase anticorrosive phosphate pigments (commercial products)

Property	Commercial product 1 (Sicor SPO, BASF) [5.67]	Commercial product 2 (Sicor NOP, BASF) [5.67]	Commercial product 3 (Heucophos ZPZ, Heubach) [5.68]
Zinc content, wt%	63	62	58
Phosphate content, wt%	28	30	39
Organic inhibitor content, wt%	2.5	0.5	0.3
Loss on ignition, wt%	9	9	10–12
Water-soluble content, wt%	0.9	0.5	0.8
Chloride/sulfate content, wt%	0.02	0.02	<0.05
pH	7	7	7
Conductivity, μS	260	240	650
Density, g/cm^3	3.7	3.8	3.7
Specific surface area, m^2/g	7	5	
Oil absorption value, g/100 g	ca. 20	ca. 20	16
Color	white	white	white
CAS registry no.	[*92308-56-0*]	[*92308-55-9*]	[*34807-26-6*]

Iron phosphide anticorrosive pigments are recommended by manufacturers as replacement materials for zinc dust to reduce the price of zinc-rich paints [5.103]. For toxicological data, see Section 5.2.14.

A *trade name* for iron phosphide is Ferrophos (Hooker Chemicals & Plastics Corp., USA).

Zinc Hydroxyphosphite. Commercial zinc hydroxyphosphite [*64539-51-1*] is a white, nontoxic pigment with basic character.

The pigment has the following physical properties [5.76]:

Oil absorption value, g/100 g	15–20
Density, g/cm^3	3.96
pH, 2% suspension	6.5–7.5
Mean particle size, μm	2–3
Water-soluble content, $g/100\ cm^3$	< 0.01
Specific conductivity, Ω/cm	9700

Its anticorrosive properties in various binders are given in Table 39.

A *trade name* for zinc hydroxyphosphite is Nalcin 2 (National Lead Chemicals, USA; Kronos Titan, Germany).

5.2.4. Borosilicate Pigments

Borosilicate pigments [*59794-15-19*] usually contain calcium or zinc ions in a matrix of silicon dioxide and boron trioxide, $x\ (Ca,Zn) \cdot y\ SiO_4 \cdot z\ BO_3$. Aqueous slurries of this white pigment are alkaline (pH > 9). Borosilicate pigments are recommended as nontoxic alternatives to basic lead silicate. Their main field of application is in waterborne binders and electrodeposition coatings [5.69], [5.104], [5.105].

A *trade name* for calcium borosilicate is Halox CW 291, -2230 (Halox Pigments, USA).

5.2.5. Borate Pigments

Barium metaborate [*13701-59-2*], $BaO \cdot B_2O_3 \cdot H_2O$, [5.78] and zinc borophosphate [*84012-98-6*], $ZnO \cdot x\ B_2O_3 \cdot y\ P_2O_5 \cdot 2\ H_2O$, are colorless pigments. Their properties are listed in Table 44; their uses in various binders are listed in Table 39. They both have a relatively high water solubility. Barium metaborate is coated with silica to reduce this.

The anticorrosive effects of these pigments depend mainly on their ability to maintain high pH values in the coating. They are most effective in the initial phase of corrosion protection. The borate ion neutralizes acidic foreign ions and binder decomposition products, but is also thought to act as an anodic passivator, forming a protective film. The effectiveness of metal borate pigments in aqueous air-dried anticorrosive coatings is described in [5.106].

The metal borate pigments are classified as having a relatively low toxicity [5.53], [5.107]. For toxicological data, see Section 5.2.14.

Examples of *trade names* are Butrol (Buckmann Laboratories, USA) for barium metaborate, and Sicor BZN (BASF, Germany) for zinc borophosphate.

5.2.6. Chromate Pigments

Anticorrosive chromate pigments are summarized in Table 45. Their possible combinations with various binders are listed in Table 39. Standards for specifications of chromate pigments are listed in Table 1.

The anticorrosive action of the chromate pigments is based both on chemical and electrochemical reactions [5.66], [5.108]–[5.113]. Electrochemical passivation and chemical reaction are illustrated in Figure 70 [5.114], [5.115]. Passivation is based on electrochemical processes in the cathodic region. In addition, a protective film is also

Table 44. Properties of metal borate pigments

Property	Barium metaborate	Zinc borophosphate
Density, g/cm^3	ca. 3.3	2.8
pH	9–10	8
Water-soluble content, wt%	0.4	1–1.5
Use	primer, topcoat	primer

Table 45. Metallic anticorrosive chromate pigments

Name	Formula	CAS registry no.	Synonym	Appearance	Solubility in water, g/L
Zinc chromate	$ZnCrO_4$	[*13530-65-9*]	chromic acid–zinc salt (1:1) zinc yellow C.I. Pigment Yellow 36	lemon yellow powder	insoluble
Zinc tetraoxychromate	$ZnCrO_4 \cdot 4\ Zn(OH)_2$	[*15930-94-6*]	basic zinc chromate ZTO chromate zinc tetrahydroxychromate C.I. Pigment Yellow 36	yellow powder	ca. 0.04
Basic zinc potassium chromate	$4\ ZnO \cdot K_2O \cdot 4\ CrO_3 \cdot 3\ H_2O$	[*37300-23-5*]	basic zinc chromate zinc chromate pigment lemon yellow C.I. Pigment Yellow 36	lemon yellow triclinic flakes	2.5–5
Zinc potassium chromate	$KZn_2(CrO_4)_2OH$	[*11103-86-9*]	chromic acid–potassium zinc salt (2:2:1) zinc yellow C.I. Pigment Yellow 36	yellow powder	sparingly soluble
Strontium chromate	$SrCrO_4$	[*7789-06-2*]	strontium chromate A strontium yellow C.I. Pigment Yellow 32	yellow powder	ca. 2

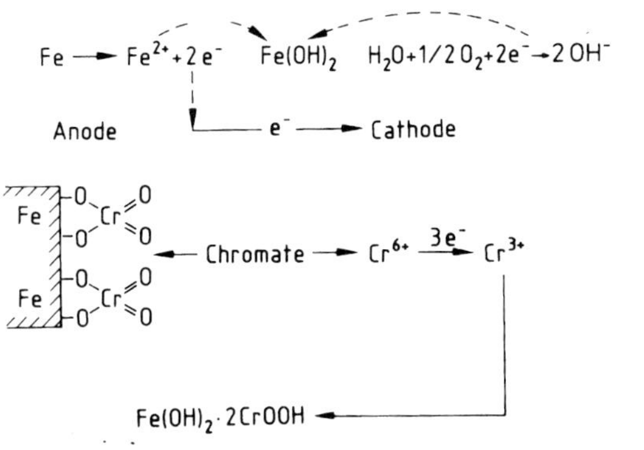

Figure 70. Passivation of iron by chromate pigments

formed by reaction of chromate ions with metal ions at the surface of the substrate to form metal oxide hydrates.

The anticorrosive properties of this class of pigments depend on:

1) The content of water-soluble chromate ions
2) The ratio of water-soluble chromate ions to water-soluble corrosion-promoting ions (chloride and sulfate ions)
3) The active pigment surface in the coating (i.e., particle size distribution and dispersibility)

Chromate-containing pigments are classified as toxic; their use is therefore very limited and they must be appropriately labeled [5.116], [5.117].

Zinc-Containing Chromate Pigments. Zinc chromate is produced by reacting an aqueous slurry of zinc oxide or hydroxide with dissolved chromate ions followed by neutralization, or by precipitation of dissolved zinc salts with dissolved chromate salts. Zinc tetraoxychromate is produced from zinc oxide and chromic acid in an aqueous medium. Basic zinc potassium chromate is obtained by reacting an aqueous slurry of zinc oxide with potassium dichromate and sulfuric acid. The pigments are washed, filtered, dried, and ground.

Trade names include Zinkchromat CZ 20–CZ 40 (SNCZ, France); Zinkchromat 1W (BASF, Germany); and Zinktetraoxichromat TC 20 – TC 40, LOW DUST (SNCZ, France).

Strontium Chromate. Strontium chromate is precipitated from solutions of sodium dichromate and strontium chloride, followed by filtration, washing, drying, and grinding. A primer composition based on strontium and calcium chromates is described in [5.70], [5.118].

A *trade name* for strontium chromate pigments is Strontiumchromat L 203 S–L 203 E (SNCZ, France).

Lead Silicochromate [5.119]. Lead silicochromate [*11097-70-4*], 4 $(PbCrO_4 \cdot PbO) + 3 (SiO_2 \cdot 4\ PbO)$, is an orange powder. This pigment is a core pigment, in which the active pigment substance ($PbCrO_4$) is precipitated on to an inert core (SiO_2).

In the production process used by National Lead Industries, a solution of chromic acid is added to an aqueous slurry of lead oxide, finely ground silica, and a small amount of basic lead acetate:

$$2\ PbO + CrO_3 \longrightarrow PbCrO_4 \cdot PbO$$

The product is heated in a rotary kiln causing the excess lead oxide to react with the basic lead chromate, forming tetrabasic lead chromate:

$$3\ PbO + PbCrO_4 \cdot PbO \longrightarrow PbCrO_4 \cdot 4\ PbO$$

On raising the temperature the following reaction takes place on the surface of the silica:

$$4\ (PbCrO_4 \cdot 4\ PbO) + 3\ SiO_2 \longrightarrow 4\ (PbCrO_4 \cdot PbO) + 3\ (SiO_2 \cdot 4\ PbO)$$

The pigment contains 47.0 wt% PbO, 5.4 wt% CrO_3, and 47.6 wt% SiO_2. Properties are as follows: density 5.2 g/cm^3, oil number 15 g/100 g, tamped volume 163 cm^3/100 g, sieve residue (> 42 μm) 0.004 wt%, moisture content 0.1 wt%, and pH ca. 7.

A *trade name* for lead silicochromate is Onor (National Lead Chemicals, USA) [5.76].

5.2.7. Molybdate Pigments

Molybdenum-based anticorrosive pigments offer a nontoxic alternative to the zinc chromate pigments [5.120]. They all have a neutral color (white) but the pure compounds are very expensive. To produce economically competitive pigments molybdate and phosphate pigments are combined, or molybdate compounds are applied to inorganic fillers (e.g., zinc oxide, alkaline-earth carbonates, or talc) [5.75], [5.121]–[5.123].

Commercially important pigments include basic zinc molybdates ($ZnMoO_4$ [*13767-32-3*], ZnO [*1314-13-2*], sodium zinc molybdates ($Na_2Zn(MoO_4)_2$ [*11139-48-3*]), basic calcium carbonate zinc molybdate, and basic calcium carbonate zinc phosphate molybdate pigments ($CaMoO_4$ [*7789-82-4*], $CaCO_3$ [*471-34-1*], ZnO [*1314-13-2*], $Zn_3(PO_4)_2$ [*7779-90-0*]). Properties are given in Table 46.

Phosphate-containing molybdate pigments are especially suitable for water-thinnable or latex-based binders, because they improve adhesion to iron substrates. The other molybdate pigments are mainly used in solventborne binder systems.

Unlike chromate ions in chromate pigments, the MoO_4^{2-} ions in molybdate pigments are not chemically reduced in most coatings. Hence, they are ineffective for cathodic protection. Their protective action is assumed to be due to activity in the anodic region, similar to that of phosphate ions. As with the protective phosphate films, molybdate films are very resistant to chloride and sulfate [5.124]. The duration of maximum activity depends on the metal ions used in the pigment and is probably due to differences in solubility.

Trade names are as follows:

Molybdenum-based: Moly-White 101, 212, 331 (The Sherwin-Williams Company, USA).

Zinc molybdenum phosphate: Actirox 102, -106 (Colores Hispania, Spain); Moly-White ZNP, -MZA (The Sherwin-Williams Company, USA); Heucophos ZMP (Dr. H. Heubach, Germany).

5.2.8. Lead and Zinc Cyanamides

Lead cyanamide [*20837-86-9*], $PbCN_2$, M_r 247.23, is a lemon yellow powder. Zinc cyanamide [*20654-08-4*], $ZnCN_2$, M_r 105.41, is a white to beige powder. Their properties are listed in Table 47 [5.125].

The cyanamides are active anticorrosive pigments which have a passivating action under alkaline conditions. The action of lead in anticorrosive pigments is discussed in Section 5.2.10.1. Heavy metal oxides, especially of iron and manganese, catalyze the conversion of cyanamide to urea even below 20 °C [5.126]:

$$H_2N-C\equiv N+H_2O \longrightarrow H_2N-CO-NH_2$$

Table 46. Properties of molybdenum-containing pigments

Property	Zinc molybdate pigment	Calcium zinc molybdate pigment	Basic sodium zinc molybdate pigment	Basic calcium zinc phosphomolybdate pigment
Density, g/cm^3	5.06	3.00	4.00	3.00
Oil absorption value, g/100 g	14	18	14	16
Mean particle size, µm (Fisher subsieve sizer)	0.65	1.88	1.02	2.2
pH	6.5	8.5		8

Table 47. Properties of lead and zinc cyanamides

Property	Lead cyanamide	Zinc cyanamide
Density, g/cm^3	ca. 6.8	ca. 3.1
Apparent density, g/cm^3	550–750	ca. 250
Metal content, wt%	ca. 83	ca. 60
CN_2 content, wt%	ca. 16	ca. 34
Specific surface area, m^2/g	2	ca. 50
pH	9–11	8.5
Conductivity, µS	<300	<1000

At alkaline pH, however, the dimerization of cyanamide to dicyanodiamide is favored. Dicyanodiamide acts as a corrosion inhibitor:

$$2\,H_2N-C\equiv N \longrightarrow (H_2N)_2C=N-C\equiv N$$

In strongly alkaline solution, hydrolysis to ammonia takes place:

$$H_2N-C\equiv N + H_2O + 2\,OH^- \longrightarrow 2\,NH_3 + CO_3^{2-}$$

The main fields of application are in mirror coatings [5.77], [5.125], electrodeposition coatings, and primers (see Table 39).

Cyanamide pigments are produced from industrial-grade calcium cyanamide which is first dissolved. Sulfide and phosphide impurities are precipitated as iron or lead salts [5.127]–[5.132] or oxidized [5.133]–[5.135] and filtered off together with graphite impurities. The pure calcium cyanamide is reacted in an aqueous medium with soluble lead or zinc salts or with a slurry of lead oxide or zinc oxide [5.127], [5.129], [5.133], [5.136]–[5.138]. The pigments are filtered, washed, dried, and ground. Zinc cyanamide [5.139] and pure lead cyanamide are not explosive. An explosion reported during the production of lead cyanamide was caused by contamination with small amounts of acid or nitrates [5.140].

Zinc cyanamide is nontoxic, but the toxicological classification of lead cyanamide has to take its lead content into account. For toxicological data, see Section 5.2.14.

Trade names include Bleicyanamid LY 80, Zinkcyanamid (BASF, Germany).

5.2.9. Ion-Exchange Pigments

Ion-exchange pigments were developed as nontoxic alternatives to the chromate pigments. They consist of a silicate carrier (zeolite [5.141] or amorphous silica gel [5.142]) to which calcium ions are bound. Commercial ion-exchange pigments have the following properties [5.143]–[5.145]:

pH	ca. 9
Conductivity, μS	ca. 300
Ca content, wt%	ca. 5
Water-soluble fraction, %	ca. 0.9
Oil absorption value, g/100 g	40–60
Density, g/cm^3	1.8
Mean particle size, μm	3–5
Color	white
Apparent density, g/L	500–600

Ion-exchange pigments should be classified as active anticorrosive pigments. They act by exchanging their calcium ions for hydrogen ions in the paint film. In this way acidic chemical substances are neutralized. The calcium ions then bind to the metal oxide surface. The pigments have good anticorrosive properties owing to their high pH value, but the danger of pinholing (depending on the binder) should be pointed out.

A *trade name* for ion-exchange pigments is Shieldex AC3, AC5 (W. R. Grace, UK).

5.2.10. Metal Oxide Pigments

5.2.10.1. Red Lead

Red lead [*1314-41-6*], Pb_3O_4, M_r 685.57, crystallizes in the tetragonal system and is a red powder with a density of 9.1 g/cm^3. It decomposes at ca. 500 °C at atmospheric pressure. For standard specifications, see Table 1.

Red lead should be regarded as the lead salt of orthoplumbic acid, H_4PbO_4, i.e., it is lead(II) orthoplumbate, Pb_2PbO_4, in which PbO_6 octahedra are linked by Pb(II) ions [5.146].

Red lead is produced industrially by oxidizing lead monoxide (PbO) at ca. 460–480 °C with agitation in a stream of air for 15–24 h. Most red lead is used in the glass, ceramic, and accumulator industries where an apparent density of < 2 g/mL is adequate. For the paint industry, however, highly dispersed red lead is normally necessary, with a sieve residue of < 0.1% on a 0.063 mm sieve (ISO 787, part 18) and an apparent density of 1.3–2.0 g/mL (ISO 510, DIN 55516).

The electrochemical action of red lead results from the fact that lead has valencies of 2 and 4 in lead orthoplumbate: Pb(IV) compounds are reduced to Pb(II) in the cathodic region [5.147]. The chemical anticorrosive effect is a result of lead soaps that are formed when fatty acids in the binder react with the red lead. The lead soaps permeate the paint film as lamellae, and give good mechanical strength, water resistance, and adhesion to the steel surface. Furthermore, the corrosion-promoting chloride and sulfate ions are precipitated by lead(II) ions [5.148].

Possible combinations of red lead with various binder systems are listed in Table 39 [5.114]. Red lead is still used for heavy-duty anticorrosion applications, especially for surfaces bearing residual traces of rust. In waterborne paints, red lead has no advantages over zinc phosphate [5.149].

A *trade name* for red lead is Bleimennige, hochdispers (Heubach & Lindgens, Germany).

5.2.10.2. Calcium Plumbate

Calcium plumbate [*12013-69-3*], Ca_2PbO_4, density 5.7 g/cm^3, is a beige powder formed from lead monoxide and calcium oxide at ca. 750 °C in a stream of air [5.150].

The anticorrosive properties of calcium plumbate are inferior to those of red lead [5.114]. Calcium hydroxide is formed as a hydrolysis product when water penetrates through a primer that contains calcium plumbate. The pH at the metal surface then increases to ca. 11–12 which inhibits corrosion.

The most important use is in primers for zinc-coated substrates. The pH change occurring on hydrolysis of the calcium plumbate etches the zinc surface which improves adhesion of primers, especially on hot-dip galvanized steel [5.151].

5.2.10.3. Zinc and Calcium Ferrites

Many paint formulations contain iron oxide as an extender. It is a physically protective anticorrosive pigment (only to a small extent). In order to obtain a chemically protective anticorrosive pigment with active constituents the iron oxide is heated with oxides or carbonates of alkaline earths (CaO, $CaCO_3$) or zinc (ZnO) to form pigments of the ferrite type [5.152], [5.153]. The following systems have been reported for alkyd resin primers: 2 $CaO \cdot Fe_2O_3$, $CaO \cdot Fe_2O_3$ [*11119-52-9*] and Zn(-Mg)$O \cdot Fe_2O_3$ [*12063-19-9*]. In the coating these pigments are hydrolyzed with water to form alkaline-earth hydroxides or zinc hydroxide which prevent corrosion by increasing the pH. Alkaline-earth soaps are also formed in certain binder media [5.154]. However, the pigment volume concentration must be high to ensure good results [5.155].

Only one zinc ferrite pigment has attained economic significance. Its properties are as follows [5.156]:

Water-soluble salts, wt%	max. 0.6
Oil absorption value, g/100 g	ca. 22
pH	9–11
Density, g/cm^3	5.0

A *trade name* for zinc ferrite is Anticor 70 (Bayer, Germany; Mobay Chemical Corporation, USA).

5.2.10.4. Zinc Oxide

Zinc oxide [*1314-13-2*], ZnO, M_r 81.37, is a white powder that is usually used in combination with active anticorrosive pigments. For a detailed description, see Section 2.3. It has the following physical properties:

Oil absorption value, g/100 g	20–24
Density, g/cm^3	5.6
Apparent density, cm^3/100 g	100
Mean particle size, μm	0.11–0.22
BET surface area, m^2/g	3–10

The inhibiting action of zinc oxide is based on its ability to react with corrosive substances and to maintain an alkaline pH in the coating. It also reacts with acidic components of the binder to form soaps and absorbs UV light.

The lead content of commercial zinc oxide depends on the manufacturer and is in the range 0.002–1.5%. For a zinc oxide coating to be considered lead-free, the lead content must be less than 1.5%.

5.2.11. Powdered Metal Pigments

5.2.11.1. Zinc Dust

Zinc dust [*7440-66-6*], Zn, A_r 65.37, *mp* 419.4 °C, density 7.14 g/cm^3, is a free-flowing blue-gray powder composed of spheroidal particles. It is produced by melting zinc in a crucible, vaporizing it at ca. 900–950 °C, and condensing and sifting the product. Alternatively, molten zinc is atomized with a nozzle to produce dust, which is then sifted. Properties of commercial zinc dust pigments are listed in Table 48 [5.157], [5.158]. For standard specifications, see Table 1 ("Zinc dust pigments").

The action of zinc dust in primers with organic binders is based on sealing effects and electrochemical processes. The zinc reacts with water and atmospheric oxygen that diffuse into the binder, forming zinc hydroxide which is then neutralized by sulfuric acid (from SO_2 in the air) and hydrochloric acid (from Cl-containing substances in the air, e.g., NH_4Cl). This causes an increase in volume and decreases permeability. The corrosion products of zinc also have an anticorrosive action [5.159]. Cathodic protection takes place when the zinc and iron come into contact; the zinc content in the primer must be at least 94–96% [5.160]–[5.163]. Zinc dust coatings are used in large quantities for structural steel, including underwater steel construction and shipbuilding. Zinc dust is also used in inorganic binder systems (alkali silicates or alkyl silicates) in the form of two-component systems [5.164].

A *trade name* is Zinc Dust Ultra 25 and 35 (Lindgens & Söhne, Germany).

Table 48. Properties of commercial zinc dust pigments

Property	Zinc dust Ultra 25	Zinc dust Ultra 35
Total zinc content, wt%	> 99.0	> 99.0
Metallic zinc content[a], wt%	94–96	94–96
Lead content, wt%	< 0.005	< 0.005
Cadmium content, wt%	< 0.005	< 0.005
Iron content, wt%	0.003	0.003
Copper content, wt%	0.001	0.001
Acid-insoluble material, wt%	< 0.1	< 0.1
Sieve residue[b], wt%	< 0.01	< 0.01
Mean particle size[c], μm	2.8–3.2	3.3–3.8

[a] $KMnO_4$ method. [b] Sieve residue on 16900 sieve (DIN 4188, aperture: 45 μm). [c] Air permeability method.

5.2.11.2. Lead Powder

Lead powder [*7439-92-1*], Pb, A_r 207.19, *mp* 327.5 °C, density 11 g/cm^3, is a dark gray powder containing 99% metallic lead and ca. 0.5% lead(II) oxide. It is produced by spraying molten lead at ca. 0.5 MPa and sifting. Owing to its high surface area (particle size 1–15 μm), it is liable to oxidize, and is therefore supplied in airtight packages or as a paste.

Lead powder can be combined with many binders [5.165], [5.166]. It does not affect the stability or viscosity of the paint. Binders that absorb only small amounts of water are particularly suitable (e.g., epoxy resins, chlorinated rubber). When formulating paints based on lead powder, care must be taken not to dilute it with other pigments and extenders by more than 5 vol%.

Lead powder coatings are mainly used for protecting against aggressive chemicals. They have a high UV reflection and extremely good elasticity. Lead powder pigments and pastes are also used in radiological protection.

5.2.12. Flake Pigments

Anticorrosive flake pigments are summarized in Table 49. They increase the barrier resistance of a coating towards water and aggressive gases by increasing the length of the diffusion path in the coating [5.57]–[5.59] (see Section 5.2.1). The interaction between pigment and binder should be as water-resistant as possible to prevent the diffusion of water from the surface of the flakes through the coating.

In principle, all lamellar minerals may be used as barrier pigments, e.g., micaceous iron oxide [5.167]–[5.169], layer silicates (mica), linear polymeric silicates (wollastonite), and talc [5.170]. However, untreated mica and talc are not very suitable because they are highly permeable to water [5.57]. The surface can be modified with, for example, silanes or titanates, to reduce water permeability and improve adhesion [5.171]. If the chemistry of the treated pigment surface is adapted to the functional groups of the binder, the pigment can be used in aqueous anticorrosive dispersions [5.172]–[5.174].

Table 49. Anticorrosive flake pigments

Pigment	CAS registry no.	Formula	Density, g/cm^3	Color
Micaceous iron oxide	[*1317-60-8*]	Fe_2O_3 (specular hematite)	4.6–4.8	dark gray
Talc	[*14807-96-6*]	$Mg_3Si_4O_{10}(OH)_2$	2.6–2.8	white, light green to brown
Muscovite	[*114755-46-3*]	$KAl_2Si_3AlO_{10}(OH,F)_2$	2.8	gray-white
Phlogopite	[*114733-61-8*]	$KMg_3AlSi_3O_{10}(OH,F)_2$	2.8–2.9	brown
Flake aluminum	[*7429-90-5*]	Al	2.7	silver
Flake zinc	[*7440-66-6*]	Zn	7.1	gray

Flake aluminum pigments with varying platelet thicknesses and shapes are used for corrosion protection [5.175]. For standard specifications, see Table 1 ("Aluminum pastes and pigments"). They are coated with a water-repellent, fatty film and are therefore particularly suitable for conventional solventborne coating systems. They have outstandingly good weather resistance [5.175]–[5.177].

Flake zinc pigments have a barrier effect and also act by a cathodic anticorrosive mechanism. Compared with zinc dust coatings, flake zinc pigments are formulated with lower pigment volume concentrations [5.178].

5.2.13. Organic Pigments

Unlike soluble inhibitors, organic anticorrosive pigments are sparingly soluble organic compounds or metallic salts of organic acids and are used in binders instead of, or in addition to, inorganic anticorrosive pigments. Inhibitors are not viable substitutes for active anticorrosive pigments in coatings [5.179]. Organic anticorrossive pigments were developed to replace toxic, chromate-based anticorrosive pigments.

Trade names include Sicorin RZ (BASF, Germany); Alkophor 827 (Henkel, Germany); and Irgakor 252 (Ciba-Geigy, Switzerland).

The zinc salt of 5-nitroisophthalic acid [*60580-61-2*] [5.49] is produced as an easily dispersible pigment by a wet chemical process from 5-nitroisophthalic acid and zinc oxide [5.180]. It has the following properties:

Color	colorless
Density, g/cm^3	2.5
Oil absorption value, g/100 g	40
Bulkiness, cm^3/100 g	650
Specific surface area, m^2/g	16
Heat resistance, °C	ca. 300
Zinc content, wt%	ca. 44
Organic content, wt%	ca. 49
Chloride content, wt%	ca. 0.02
Sulfate content, wt%	ca. 0.03
Nitrate content, wt%	ca. 0.01
Volatile content, wt%	ca. 5
Water-soluble content, %	ca. 1.6
pH	7.2
Conductivity, μS	ca. 500

A less expensive replacement for zinc chromate is obtained by combining the zinc salt of 5-nitroisophthalic acid with zinc phosphate pigments (Section 5.2.2.5) [5.65], [5.67], [5.181]. A concentration of 0.5–2.0% based on the liquid coating is recommended.

The electrochemical anticorrosive action of the zinc salt of 5-nitroisophthalic acid is comparable with that of zinc potassium chromate [5.182].

Table 39 gives possible combinations with binders. The compound is nontoxic (see Section 5.2.14).

Other Organic Anticorrosive Pigments. The following organic anticorrosive pigments are either commercial products or have been reported in the literature:

1) (2-Benzothiazolylthio)succinic acid [5.183], [5.184]
2) Zinc mercaptobenzothiazole [5.185], [5.186]
3) Basic zinc salt of *N*-benzosulfonylanthranilic acid [5.67], [5.187]

They are used in the binder at concentrations of 0.5–2.0 %, if possible in combination with zinc phosphate or other suitable anticorrosive pigments. For toxicological data, see Section 5.2.14.

5.2.14. Toxicology

The change in the toxicological classification of lead- and chromate-based anticorrosive pigments has resulted in reappraisal of the formulation of anticorrosive coatings.

Acute oral toxicities (LD_{50}) of anticorrosive pigments are given in Table 50.

Table 50. Acute oral toxicity of anticorrosive pigments

Pigment	LD_{50} (rat), mg/kg	Reference
Zinc phosphate	> 5000	[5.67]
Aluminum triphosphate	>11000	[5.73]
Multiphase phosphate pigment	> 5000	[5.67]
Iron phosphide	>21500	[5.99]
Barium metaborate	850	[5.78]
Zinc potassium chromate*	> 640	[5.67]
Zinc tetraoxychromate*	> 5000	[5.67]
Calcium, zinc carbonate/molybdate	> 5000	[5.75]
Zinc molybdate	11800	[5.75]
Lead cyanamide	3500	[5.67]
Zinc cyanamide	< 1800	[5.67]
Red lead	>15000	[5.188]
Zinc oxide	>10000	[5.189]
Zinc salt of 5-nitroisophthalic acid	>10000	[5.67]
	>	
Basic zinc salt of *N*-benzosulfonylanthranilic acid	4000	[5.67]
Zinc mercaptobenzothiazole	> 5000	[5.185]
2-Benzothiazolylthiosuccinic acid	5000	[5.184]

*Classified as a carcinogenic substance in Section III A1 of the list of MAK values.

5.3. Luster Pigments

As shown in Figure 71 A, *conventional pigments* interact with light by absorption and/or diffuse scattering (see Section 1.3). Luster pigments comprise nacreous pigments (Section 5.3.1) and metal effect pigments (Section 5.3.2).

Nacreous pigments simulate the nacreous luster of natural pearls (Fig. 71 C), which consist of alternating, transparent layers of materials with a high ($CaCO_3$) and a low (protein) refractive index. Small platelets of a high-refractive-index material (i.e., the nacreous pigment) are oriented in parallel alignment in a matrix of lower refractive index, e.g., a paint binder or plastic (Fig. 71 D). Synthetic nacreous pigments are transparent or light-absorbing platelet crystals. They may also have a multilayer structure, the layers having different refractive indices and light absorption properties.

Metal effect pigments (Fig. 71 B) consist of small metallic flakes (mainly aluminum or Cu/Zn bronze) which act as small mirrors that reflect almost all of the incident light.

5.3.1. Nacreous and Interference Pigments

Synthetic or natural pigments used to achieve lustrous, brilliant, or iridescent color effects by interference on thin optical films are called nacreous or pearlescent pigments [5.190]–[5.194]. They were originally used to simulate the appearance of

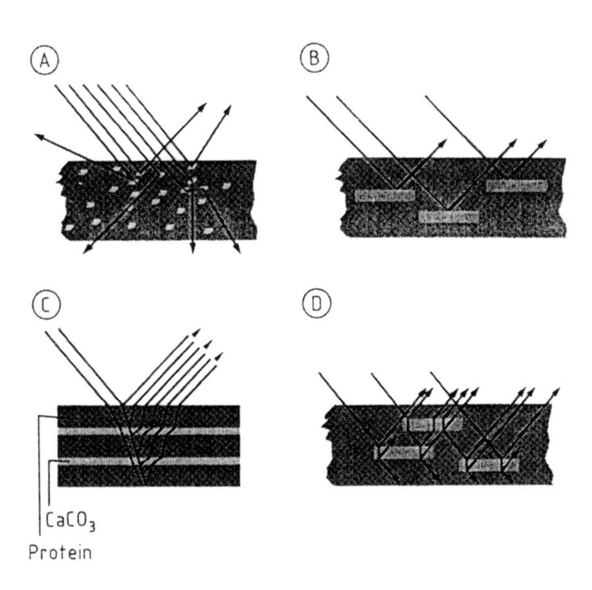

Figure 71. Optical principles of conventional and luster pigments
A) Conventional pigment that absorbs and scatters light; B) Metal effect pigment with complete regular reflection; C) Natural pearl composed of alternating layers of protein and $CaCO_3$; D) Nacreous pigment: the pearl is simulated by parallel orientation of the pigment platelets

natural pearls. The visual effects produced by reflection and transmittance of light by thin multilayer films are not restricted to pearls and shells but are widespread in nature [5.195]. Extending GREENSTEIN's definition [5.191], nacreous (pearlescent) pigments are "Thin platelets of high refractive index which partially reflect and partially transmit light", or contain layers of this kind. DIN 55943 (April 1990) proposes the comprehensive term "interference pigments" for nacreous pigments consisting of materials with a high refractive index and additional absorption characteristics (e.g., Fe_2O_3). However, in this article the term "nacreous pigments" is used.

The use of pearls and nacreous shells for decorative purposes goes back to ancient times (e.g., in Chinese wood intarsia). The history of pearl pigments dates back to 1656, when the French rosary maker JAQUIN isolated a silky lustrous suspension from fish scales (pearl essence) and applied this to small beads to create artificial pearls [5.196]. It took more than 250 years to isolate the pearl essence material (guanine platelets) and understand the pearl effects [5.197]. Attempts were made to create synthetic pearl colors as organic or inorganic, transparent, highly refractive coatings and pearl pigments as crystalline platelets [5.198]. From 1920 onwards hydroxides, halides, phosphates, carbonates, and arsenates of zinc, calcium, barium, mercury, bismuth, lead, and other cations were produced for this purpose. Only the traditional natural pearl essence (Section 5.3.1.2), basic lead carbonate (Section 5.3.1.3), and bismuth oxychloride (Section 5.3.1.4) are still of importance.

The strong demand for pearl effects came from the growing coatings and plastics industries who wanted to improve the acceptance and popularity of their products. Furthermore, nacreous pigments also allowed artists and designers to create new visual effects similar to those found in nature [5.199], [5.200]. The breakthrough for nacreous pigments came with the invention of mica coated with metal oxides (Section 5.3.1.5) [5.201]. Mica-based nacreous pigments now account for $>$ 85% of the world market.

Important nacreous pigments and producers are listed below:

Natural pearl
- Engelhard Corp./Mearl, N. J., USA

Basic lead carbonate
- J. Jaeger GmbH, Viechtach, Germany
- Poliperl, S. A., Argentina
- Semo Ltd., Seoul, South Korea

Bismuth oxychloride
- Engelhard Corp./Mearl, N. J., USA
- Rona, N. Y., USA (subsidiary of Merck KGaA)

Metal oxide-mica
- BASF, Ludwigshafen, Germany
- Kemira OY, Pori, Finland
- Engelhard Corp./Mearl, N. J., USA
- Merck KGaA, Darmstadt, Germany (and overseas subsidiaries: E. M. Industries, N. Y., USA; Merck Japan Ltd., Onahama, Japan)

Nacreous pigments are used to obtain pearl, iridescent (rainbow), or metallic effects, and in transparent color formulations to obtain brilliance or two-tone color and luster flops (change with viewing angle). The most important applications are plastics, coatings, printing inks, cosmetics, and automotive paints. The two major nacreous pigment producers, Merck KGaA and Engelhard Corp./Mearl, do not publish production data.

Table 51 shows an overview of pigments with luster effects. Effect pigments can be classified as metal platelets, oxide-coated metal platelets, oxide-coated mica platelets, platelet-like mono-crystals and comminuted PVD-films (Physical Vapor Deposition). Aims of new developments are new effects, colors, improvement of hiding power, increase of the interference color, increase of light and weather stability and improved dispersibility characteristics. Of special interest are pigments which are toxicologically safe and which can be produced by ecologically acceptable processes.

5.3.1.1. Optical Principles

Nacreous and Interference Pigments. The optical principles of nacreous (interference) pigments are shown in Figure 72 for a simplified case of nearly normal incidence without multiple reflection and absorption. At the interface P_1 between two materials with refractive indices n_1 and n_2, part of the beam of light L_1 is reflected (L'_1) and partially transmitted (i.e., refracted) (L_2). The intensity ratios depend on n_1 and n_2. In a multilayer arrangement, as found in pearl or pearlescent and iridescent materials (Fig. 71 C), each interface produces partial reflection. After penetration through several layers, depending on the size and difference between n_1 and n_2,

Table 51. Overview of effect pigments [2.202]

Pigment type	Examples
metallic platelets	Al, Zn/Cu, Cu, Ni, Au, Ag, Fe (steel), C (graphite)
oxide coated metallic platelets	surface oxidized Cu-, Zn/Cu-platelets, Fe_2O_3 coated Al-platelets
coated micra platelets	non absorbing coating: TiO_2 (rutile), TiO_2 (anatase), ZrO_2, SnO_2, SiO_2
	selectively absorbing coating: FeOOH, Fe_2O_3, Cr_2O_3, TiO_{2-x}, TiO_xN_y, $CrPO_4$, $KFe[Fe(CN)_6]$, colorants
	totally absorbing coating: Fe_3O_4, TiO, TiN, $FeTiO_3$, C, Ag, Au, Fe, Mo, Cr, W
platelet-like monocrystals	BiOCl, $Pb(OH)_2 \cdot 2PbCO_3$, α-Fe_2O_3, α-$Fe_2O_3 \cdot nSiO_2$, $Al_xFe_{2-x}O_3$, $Mn_yFe_{2-y}O_3$, $Al_xMn_yFe_{2-x-y}O_3$, Fe_3O_4, reduced mixed phases, Cu-phthalocyanine
comminuted thin PVD-films	Al, Al (semitransparent)/SiO_2/Al/SiO_2/Al (semitransparent)

virtually complete reflection is obtained, provided that the materials are sufficiently transparent.

In pigments that simulate natural pearl effects, the simplest case is a platelet-shaped particle with two phase boundaries P_1 and P_2 at the upper and lower surfaces of the particles, i.e., a single, thin, transparent layer of a material with a higher refractive index than its surroundings. For small flakes with a thickness of approx. 100 nm, the physical laws of thin, solid, optical films apply [5.203].

Multiple reflection of light on a thin solid film with a high refractive index (Fig. 72) causes interference effects in the reflected light and in the complementary transmitted light. For the simple case of nearly perpendicular incidence, the intensity of the reflectance (I) depends on the refractive indices (n_1, n_2), the layer thickness (d), and the wavelength (λ) [5.204], [5.205]:

$$I = \frac{A^2 + B^2 + 2AB\cos\theta}{1 + A^2B^2 + 2AB\cos\theta} \qquad A = \frac{n_1 - n_2}{n_2 + n_1} \qquad B = \frac{n_2 - n_1}{n_2 + n_1} \qquad \theta = 4\pi\frac{n_2 d}{\lambda}$$

With the given n_1 and n_2 values, the maximum and minimum intensities of the reflected light – seen as interference colors – can be calculated and agree well with experimental results [5.206]. Refractive indices of materials commonly used in nacreous pigments follow:

Vacuum/air	1.0	$Pb(OH)_2 \cdot 2PbCO_3$	2.0
Water	1.33	BiOCl	2.15
Proteins	1.4	Carbon (diamond)	2.4
Plastics	1.4–1.7	Fe_3O_4	2.4
Mica	1.5	TiO_2 (anatase)	2.5
$CaCO_3$ (aragonite)	1.68	TiO_2 (rutile)	2.7
Natural pearl (guanine, hypoxanthine)	1.85	Fe_2O_3 (hematite)	2.9

In practice, platelet crystals are synthesized with a layer thickness d calculated to produce the desired interference colors (iridescence) [5.206], [5.207]. Most nacreous pigments now consist of at least three layers of two materials with different refractive indices (Fig. 73). Thin flakes (thickness ca. 500 nm) of a material with a low refractive index (mica) are coated with a highly refractive metal oxide (e.g., TiO_2, layer thickness ca. 50–150 nm). This results in particles with four interfaces that constitute a more complicated but still predictable thin film system. The behavior of more

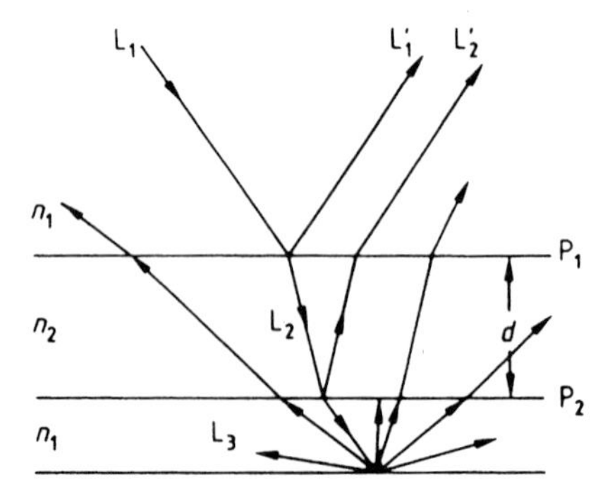

Figure 72. Simplified diagramm showing nearly normal incidence of a beam of light (L_1) from an optical medium with refractive index n_1 through a thin solid film of thickness d with refractive index n_2

L'_1 and L'_2 are regular reflections from phase boundaries P_1 and P_2. L_3 represents diffuse scattered reflection from the transmitted light

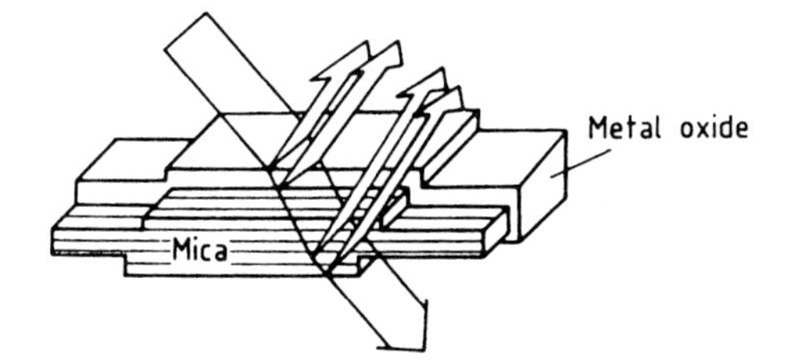

Figure 73. Simplified scheme of light reflection at the phase boundaries of a metal oxide-mica pigment

complex multilayer pigments containing additional thin, light-absorbing films can also be calculated if appropriate optical parameters are known [5.206].

Color effects depend on the viewing angle [5.207]–[5.211]. Nacreous pigment platelets split white light into tow complementary colors that depend on the platelet thickness. The reflected (interference) color dominates under regular (maximum) reflection, i.e., when the object is observed at the angle of regular reflection (Fig. 74 B). The transmitted part dominates at other viewing angles under diffuse viewing conditions provided there is a nonabsorbing (white) or reflecting background (Fig. 74 A). Variation of the viewing angle, therefore, produces a sharp gloss (reflectance) peak and the color changes between two extreme complementary colors [5.207], [5.210]. The resulting complex interplay of luster and color is measured goniophotometrically in reflection and at different angles [5.206]. No general standard measurement geometries have yet been specified. However, colorimetric analysis always includes measurements close to regular conditions and under diffuse conditions. This can be done, for example, by tilting a standard pigmented film on a drawdown card (Fig. 74). The colorimetric data are interpreted according to CIE $L^*a^*b^*$ data. A nacreous pigment is characterized by a minimum of three $L^*a^*b^*$

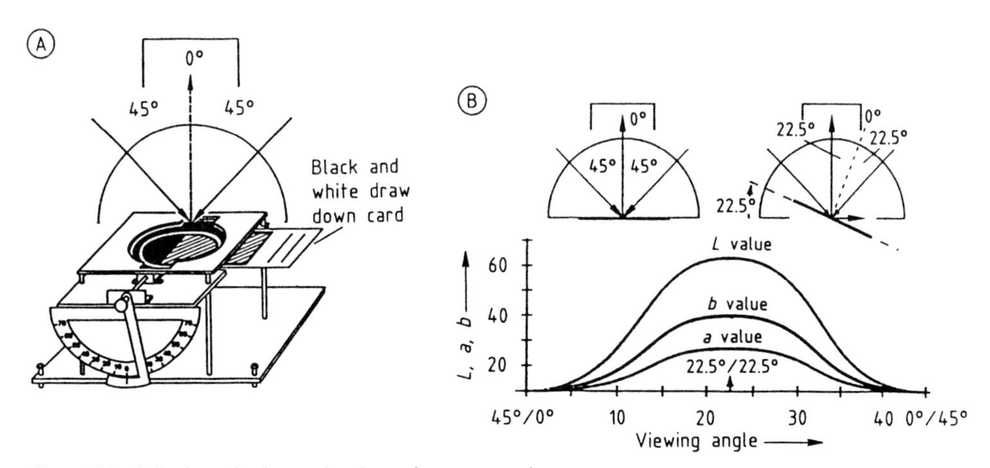

Figure 74. Colorimetric determination of nacreous pigments
The platelet particles are incorporated in a lacquer of defined thickness on a standard black and white drawdown card. The card is mounted on a rotatable sample holder and measured under (A) 0°/45° (diffuse reflection) and (B) 22.5°/22.5° (regular reflection, luster)

data sets measured under different conditions (e.g., 0°/45° black background, 22.5°/22.5° black background, 0°/45° white background). An analysis of these data specifies a pigment on the basis of its hiding power, luster, and hue [5.209]. The influence of other measurement geometries, especially of different commercially available instruments, is discussed in [5.206], [5.207].

Against a black background or in a blend with carbon black, the transmitted light is absorbed and the reflected interference color is seen as the mass tone (i.e., overall color) of the material. In blends of nacreous pigments with absorbing colorants, the particle size of the latter must be well below the scattering limit, i.e., they must be transparent (see Section 5.4). The nacreous effect or iridescent reflection is otherwise quenched by the hiding pigments. This also applies to blends with strongly reflecting metal effect pigments (e.g., aluminum). Blends of pigments with different interference colors obey an additive mixing law, (e.g., blue + yellow = white) instead of the subtractive color mixing of pure absorption pigments (e.g., blue + yellow = green) [5.207].

The behavior of combinations of interference pigments with absorption and/or metal effect pigments is too complicated to predict and needs thorough goniophotometric analysis. The nacreous pigment effects generally dominate under regular (gloss) viewing conditions and the absorbing colorant under diffuse viewing conditions [5.211].

5.3.1.2. Natural Pearl Essence [5.194], [5.196], [5.212]–[5.216]

Natural pearl essence (Essence d'Oriente, Fish Silver) is a pigment suspension derived from fish scales, skin, or bladder. It is the oldest commercial pearl luster pigment, consisting mainly of a mixture of the purines guanine [*73-40-5*] (75–97%) and hypoxanthine [*68-94-0*] (25–3%).

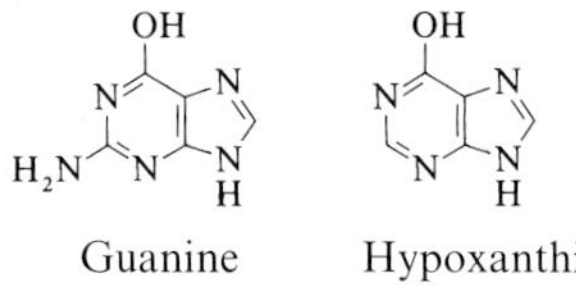

Guanine Hypoxanthine

The ratio of these two purines depends on the fish species (mainly herrings, sardines, and other white fish) and their geographical origin (Japan, Norway, the northeast U.S., and Canadian coastal border).

The pigments are formed in the fish scales as platelet-shaped crystals (0.05 μm × 1–10 μm × 20–50 μm). A commercial synthetic process for producing purines with this crystal shape has not been found. An aqueous suspension of fish scales is, therefore, extracted with organic solvents to dissolve and remove the proteins. The remaining dispersion contains purine crystals and scale which are separated from one another by a complicated washing and phase-transfer process [5.216].

Due to its tendency to agglomerate in dry form, natural pearl essence is handled as a 22–25% dispersion in various media (e.g., nitro-cellulose lacquer for nail enamel, aqueous or alcoholic media for lotions or shampoos). It is used almost exclusively in cosmetic applications [5.216]. Despite the high price, natural pearl essence has several advantages over synthetic pearlescent pigments: It is less fragile, has a low density (1.6 g/cm^3) which reduces settling in liquid formulations, and has a very high, but soft luster (n = 1.79–1.91, the highest value being along the shortest axis). The world market for natural pearl essence in 1995 was estimated to be under 50 t.

5.3.1.3. Basic Lead Carbonate [5.190]–[5.194], [5.217], [5.218]

The first commercially successfull synthetic nacreous pigments were hexagonal platelet crystals of lead salts: Thiosulfate, hydrogen phosphate, hydrogen arsenate and most important nowadays, basic carbonate. Basic lead carbonate [*1319-46-6*] $Pb(OH_2) \cdot 2PbCO_3$, M_r 775.7, is precipitated from aqueous lead acetate or lead propionate solutions with carbon dioxide:

$$3\,Pb(OCOCH_3)_2 + 2\,CO_2 + 4\,H_2O \rightarrow Pb(OH)_2 \cdot 2\,PbCO_3 + 6\,CH_3COOH$$

Under appropriate reaction conditions, regular hexagonal platelets (ca. 50 nm thick and 20 µm in diameter) can be obtained. The high refractive index (n = 2.0), high aspect ratio (> 200), and the extremely even surface of basic lead carbonate make it an optical match to natural pearl essence.

The platelet thickness can be adjusted to produce interference colors by modifying reaction conditions. When aligned with its plane orthogonal to the incident light, the platelet crystal behaves as a thin, solid, optical film (see Fig. 72) with two phase-shifted reflections from the upper and lower crystal planes (the phase boundaries).

The crystals are mechanically sensitive and their high density (6.4 g/cm^3) results in fast sedimentation. In view of their agglomeration tendency and occupational health (toxicity) risks, they are not produced in powder form, but are flushed from the aqueous phase into suitable organic solvents or resins and handled as stabilized dispersions.

Currently, use of basic lead carbonate is limited to artificial pearls, buttons, and bijouterie. Due to the low chemical stability of this pigment and toxicity problems, it is being increasingly replaced by bismuth oxychloride and mica-based pigments. Worldwide production of basic lead carbonate pigment in 1995 was ca. 1000 t.

5.3.1.4. Bismuth Oxychloride [5.190]–[5.194], [5.219]

Powders containing bismuth compounds have long been used for decorative purposes to generate a shiny luster or lustrous colors (e.g., facial cosmetic powder in ancient Egypt, imitation pearls made by coating glass and ceramic beads). Bismuth oxychloride [*7787-59-9*], BiOCl, M_r 260.4, was the first synthetic nontoxic nacreous pigment. It is produced by hydrolysis of acidic bismuth solutions in the presence of chloride ions. Precipitation conditions may be varied (concentration, temperature, pH, pressure) or surfactants added to obtain the desired crystal quality. The virtually tetragonal bipyramidal structure is thereby "squashed" into a flat platelet.

Pure BiOCl is available in three grades with different nacreous effects that depend on the aspect ratios and crystal size:

1) Low- or medium-luster powder (aspect ratio 1:10 to 1:15), mainly used as a highly compressible, white, lustrous filler with excellent skin feel
2) Dispersion of high luster quality (aspect ratio 1:20 to 1:40) consisting of square or octagonal platelets in nitro-cellulose lacquers (nail polish) or castor oil (lipsticks)
3) Dispersion of very high luster quality (aspect ratio > 1:50) consisting of lens-shaped platelets in nitro-cellulose lacquer, castor oil, or butyl acetate

Pigments consisting of BiOCl-coated mica or talc and blends of BiOCl with other organic or inorganic colorants are also available.

The dominant market for BiOCl is still the cosmetic industry. Its low light stability (it turns from silver white to metallic grey in sunlight), relatively high price, fast settling (high density, 7.73 g/cm^3), and mechanical sensitivity limit its use in technical applications. Although the darkening reaction is not yet understood, low-luster grades with improved light stability are available. Some manufacturers promote the combination with UV stabilizers for technical purposes. Uses are in the button industry, bijouterie, printing, and for X-ray contrast in catheters. The current world market is ca. 500 t/a.

5.3.1.5. Metal Oxide – Mica Pigments

The dominant class of nacreous pigments is based on platelets of natural mica coated with thin films of transparent metal oxides (see Fig. 73). Mica minerals are sheet silicates (laminar). Nacreous pigments are usually based on transparent muscovite [*9941-63-5*] but some are based on natural or synthetic phlogopite [*110710-26-4*]. Although muscovite occurs worldwide, few deposits are suitable for pigments; it is biologically inert and approved for use as a filler and colorant [5.220], [5.221].

Selection and pre-processing of the mica substrate are two of the key factors which determine the quality and appearance of nacreous pigments. The aspect ratio of the final pigment depends on the particle size distribution of the mica platelets which have a thickness of 300–600 nm and various diameter ranges (e.g., 5–25, 10–50, 30–110 µm). Since light is regularly reflected from the planes of the metal-oxide-coated mica and partially scattered from the edges, brilliance and hiding power are inversely related to each other.

A mica pigment coated with a metal oxide has three layers with different refractive indices and four phase boundaries P_1–P_4: (P_1) TiO_2 (P_2) mica (P_3) TiO_2 (P_4) (Fig. 75). Interference of light is generated by reflections of all six combinations of phase boundaries, some of which are equal: $P_1P_2 = P_3P_4$, $P_1P_3 = P_2P_4$, P_1P_4, and P_2P_3. The thickness of the mica platelets varies in accordance with a statistical distribution. Consequently, interference effects involving the phase boundaries between the mica substrate and the oxide coating add together to give a white background reflectance. The interference color of a large number of particles, therefore, depends only on the thickness of the upper and lower metal oxide coating layers [5.191], [5.204], [5.205].

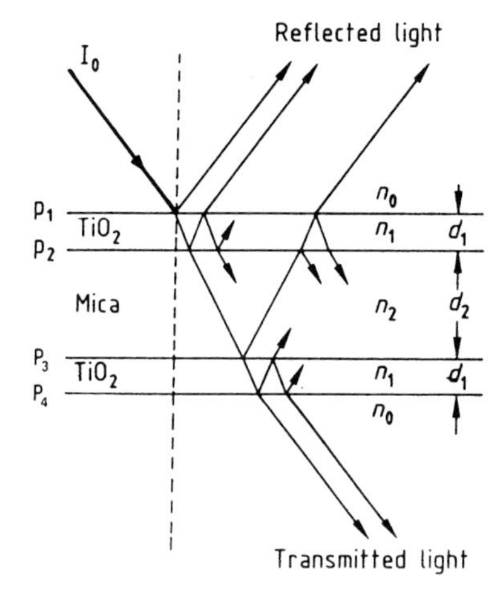

Figure 75. Multilayer, thin-film pigment consisting of a TiO_2 coating (high n_1) surrounding a mica platelet (low n_2)
The four phase boundaries (P_1–P_4) are indicated

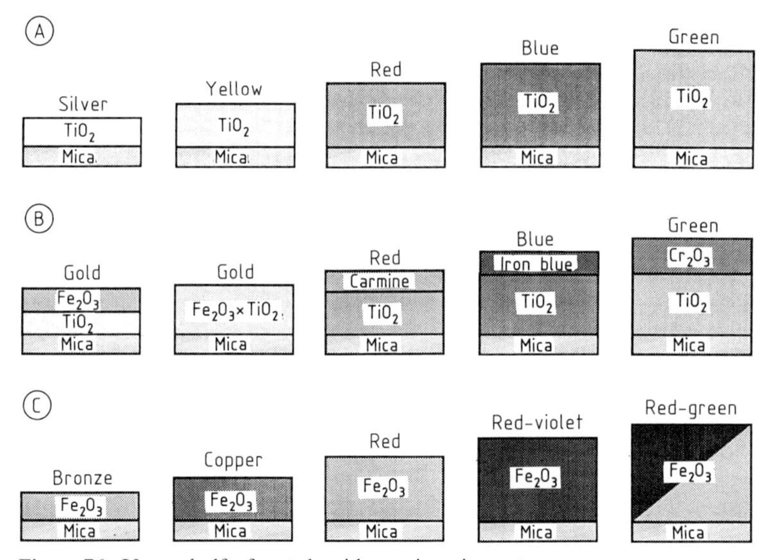

Figure 76. Upper half of metal oxide – mica pigments
Increasing layer thickness of metal oxide causes different interference colors in reflection. Combination with absorption colorants (e.g., Fe_2O_3) produces metallic effects.
A) Interference colors; B) Combination pigments; C) Metallic colors

The development of the mica-based pigments started with pearlescent colors (Fig. 76 A, TiO_2 – mica). It was followed by brilliant, mass-tone-colored combination pigments (i.e., mica, TiO_2, and another metal oxide) with one color (interference color same as mass tone) or two colors (interference and mass tone different) that depend on composition and viewing angle (Fig. 76 B). In the 1980s further development was made by coating mica particles with transparent layers of iron(III) oxide (Fig. 76 C) [5.222].

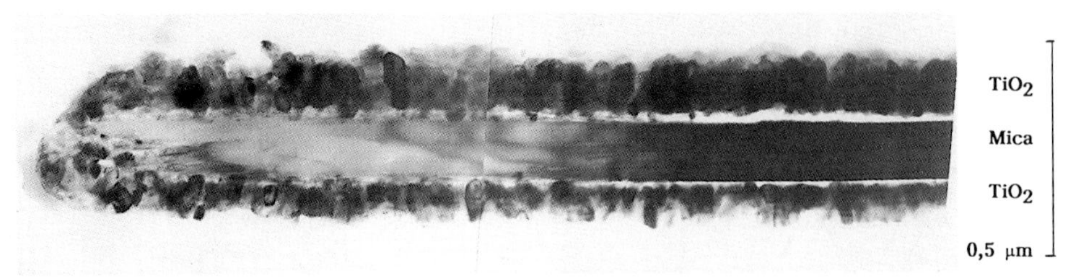

Figure 77. Cross section of a TiO_2-mica pigment (transmission electron micrograph)

Titanium Dioxide – Mica (Fig. 77). The first multilayer pigments were marketed in the 1960s as TiO_2-coated muscovite micas [5.193], [5.201]. Two different processes are used for coating mica in aqueous suspension on a commercial basis:

1) Homogeneous hydrolysis

$$TiOSO_4 + Mica + H_2O \xrightarrow{100\,^\circ C} TiO_2 - Mica + H_2SO_4$$

2) Titration

$$TiOCl_2 + 2\,NaOH + Mica \longrightarrow TiO_2 - Mica + 2\,NaCl + H_2O$$

The pigments are then dried and calcined at 700–900 °C. The titration (chloride) process is preferred for interference pigments with TiO_2 layers because it is easier to control.

Chemical vapor desposition in a fluidized bed has also been proposed [5.223]:

$$TiCl_4 + 2\,H_2O + Mica \xrightarrow{>100\,^\circ C} TiO_2 - Mica + 4\,HCl$$

Only the TiO_2 anatase crystal modification is formed on the mica surface. Small amounts of SnO_2 are, therefore, used to catalyze conversion to the rutile structure with its higher refractive index, brilliance, color intensity, and superior weather and light resistance [5.192]–[5.194].

The sequence of interference colors obtained with increasing TiO_2 layer thickness agrees with theoretical calculations in the color space [5.206], [5.224], [5.225]. An experimental development of $L^*a^*b^*$ values is given in Figure 78.

TiO_2-mica pigments are used in all color formulations of conventional pigments where brilliance and luster are required in addition to color, i.e., in plastics, coatings, printing, and cosmetics. A major market for silver white pigments (pearl pigments, "white metallic") is the plastics industry.

Table 52 contains a comparative overview of TiO_2-mica, basic lead carbonate, bismuth oxychloride, and natural fish silver pigments. Some further physical data are summarized in Table 53.

Iron Oxide-Mica. Iron(III) oxide is suitable, like titanium dioxide, for coating of mica platelets. It combines a high refractive index (metallic luster) with good hiding power and weather resistance.

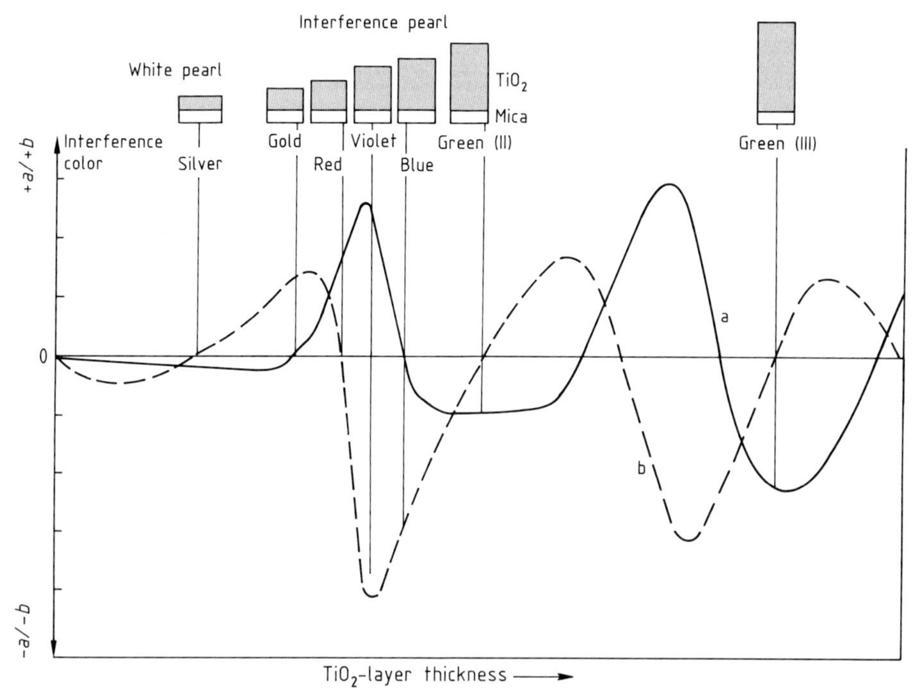

Figure 78. Experimental dependence of interference colors on the TiO_2 layer thickness on mica expressed in the Hunter $L^*a^*b^*$ scale (a, b only)

Fe_2O_3-mica pigments are produced by precipitation of iron(II)- or iron(III)-ions in aqueous mica suspensions and calcination of the so obtained coated particles at 700–900 °C [5.190], [5.192]–[5.194]:

$$2\ FeCl_3 + Mica + 3\ H_2O \longrightarrow Fe_2O_3/Mica + 6\ HCl$$

It is also possible to produce iron oxide – mica pigments by a direct CVD fluidized bed process where iron pentacarbonyl is oxidized and Fe_2O_3 is deposited on the mica surface [5.227]:

$$2\ Fe(CO)_5 + 1.5\ O_2 + Mica \longrightarrow Fe_2O_3/Mica + 10\ CO$$

Iron(III) oxide crystallizes independently of the synthesis route in the α-modification (hematite) after calcination. Brilliant, intense colors are obtained with 50–250 nm layers of Fe_2O_3. Absorption and interference colors are formed simultaneously and vary with layer thickness of iron oxide. Especially, the red shades are extremely intensive because interference and absorption enhance each other (Fig. 79). It is possible to produce an intense green-red flop with different viewing angles at a layer thickness similar to a green interference [5.228].

Table 52. Properties and applications of pearlescent pigments [5.226]

Pearlescent pigment	Advantages	Disadvantages	Main application field
natural fish silver	very low density high luster nontoxic light stable	high price low hiding power limited availability	nail lacquers
basic lead carbonate	very high luster good hiding power low price light stable	high density chemically and thermally of limited stability; toxic	buttons bijouterie
bismuth oxychloride	very high luster good hiding power nontoxic	limited light stability high density	decorative cosmetics buttons bijouterie
titanium dioxide-mica	high luster good hiding power (depending on the particle size) highest thermal, chemical, and mechanical stability nontoxic low price low density	inferior luster in comparison with top qualities of basic lead carbonate and bismuth oxychloride	plastics lacquers cosmetics printing inks ceramic products

Table 53. Technical data of pearlescent pigments

Pearlescent pigment	Shape	Particle size (μm)	Thickness (nm)	Density (g/cm^3)
natural fish silver	needles, longish platelets	10–40	40–50	1.6
basic lead carbonate	hexagonal crystals	4–20	40–70	6.4
bismuth oxychloride	flat tetragonal bipyramidal crystals	5–30	100–700	7.7
titanium dioxide-mica	platelets	1–200	200–500	3

Combination Pigments on the Basis of Mica. Simple blending of transparent absorption pigments with pearlescent pigments is only one way to attain new coloristic effects. On the other hand, it is possible to produce nacreous pigments coated with a layer of transparent absorption colorants to realize more pronounced brilliant colors with a sharper color flop, which means the change of color with the viewing angle. An additional advantage of such pigments is the elimination of dispersion problems associated with transparent absorption pigments due to their small particle size and high surface area.

One possibility for attractive combination pigments is the coating of TiO_2-mica pigments with an additional layer of an inorganic or organic colorant. The thickness of the TiO_2 layer is decisive for the brilliance or interference effect under regular viewing conditions, whereas, the transparent colorant dominates at all other viewing

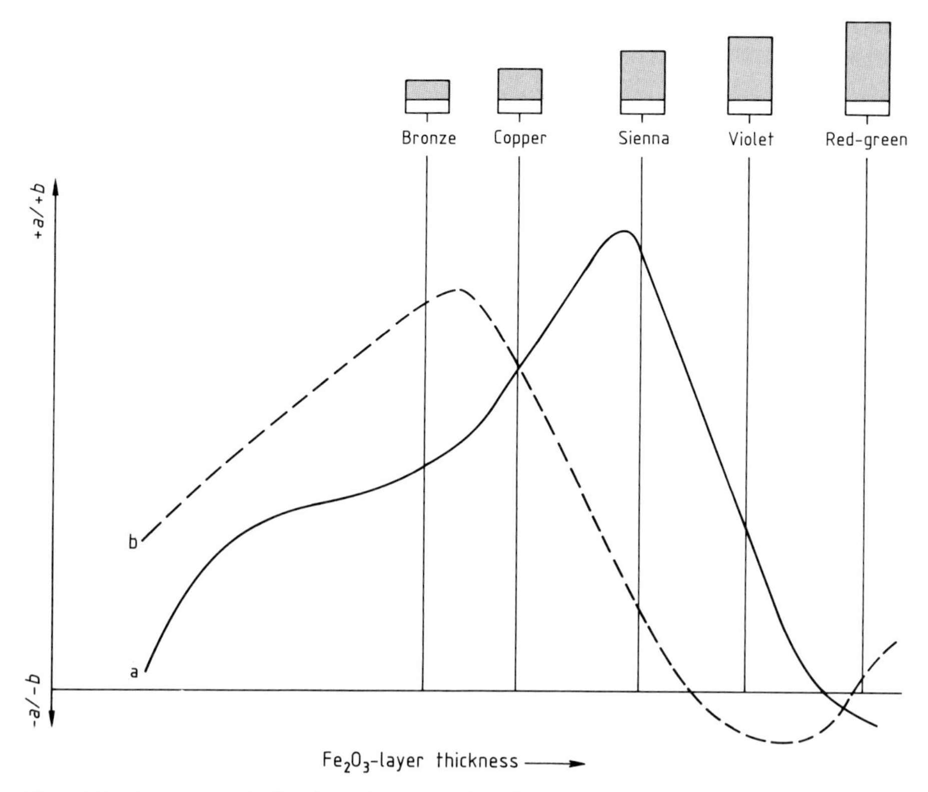

Figure 79. Experimental *a/b* values (Hunter scale) of Fe_2O_3-mica with increasing layer thickness of Fe_2O_3

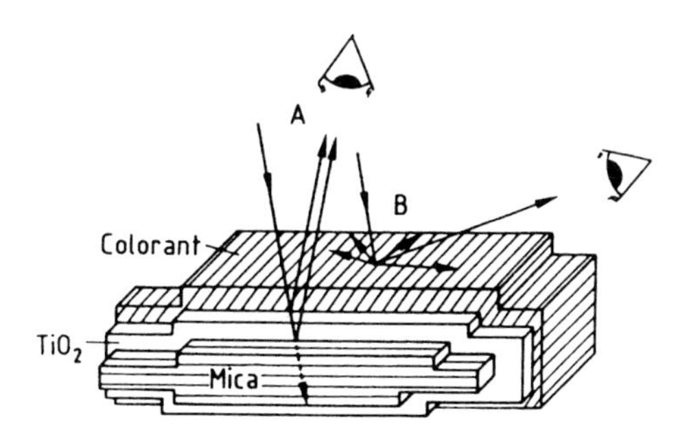

Figure 80. Combination pigment consisting of TiO_2-mica coated with an additional layer of an absorption colorant
Under regular conditions (A), a brilliant color effect dominated by thin film reflection is visible. All other viewing angles (B) show the color of the transparent absorption colorant (Section 4.4, "Transparent Pigments")

angles (Fig. 80). A deep, rich color with a luster flop at all angles is attained for the case, whereby the colorant and interference color are matched. If the interference color and the mass-tone of the colorant are different, a color flop (two-tone pigments) is seen in addition to the luster flop.

Iron(III) oxide is the most important metal oxide for combination with titanium dioxide on mica flakes. Brilliant golden pigments result which can be applied for several purposes. Two routes are used to synthesize these pigments, and different structures are formed [5.190], [5.192]–[5.194]. In the first case, a thin layer of Fe_2O_3

is coated on the surface of a TiO_2-mica pigment. The overall interference color is the result of both metal oxide layers. The mass-tone is determined by the Fe_2O_3 layer, and interesting gold pigments (e.g., reddish gold) are possible. In the second case, co-precipitation of iron and titanium oxide hydroxides on mica particles and calcination leads to greenish gold pigments. Interference and mass-tone can be explained as above. The mass-tone in this case is, however, further modified due to the additional formation of the highly refractive yellowish iron titanate phase Fe_2TiO_5 (pseudobrookite) [5.229].

Other inorganic colorants used, instead of iron oxide for combination pigments, are Cr_2O_3 (green), iron blue, cobalt blue, Fe_3O_4 (black) and carbon black. It is interesting, that in the case of the two black colorants, the interference color is seen as the mass-tone. There is an analogy to blend with black pigments in a color formulation, where the transmitted part of the light is absorbed.

Coating of TiO_2-mica pigments with an organic colorant for a mass-tone or two-tone pigment is done by precipitation or deposition on the pigment surface in aqueous suspension, assisted by complexing agents or surfactants. A second route is fixing the colorants as a mechanically stable layer using suitable additives.

5.3.1.6. New Developments for Pearlescent Pigments and Flakes

Coloristic Variations on the Basis of Mica. Mica platelets can be coated with a variety of compounds to produce novel pigments. Solid state reactions and CVD-process enlarge the possibilities for the synthesis of mica pigments. In addition, the calcination of the materials in the presence of inert (e.g. N_2, Ar) or reactive gases (e.g. NH_3, H_2, hydrocarbons) allows the formation of phases which are not producible by working in air. Table 54 contains a summary of nacreous mica pigments with special coloristic properties.

Reduction of suitable aqueous metal salts by electroless plating yields metal-coated mica pigments. They are less expensive than noble metal flakes (Au, Ag) and their brilliant metallic appearance is comparable with that of metallic flakes [5.192], [5.193].

Functional Pigments. At first metal oxide – mica pigments were developed purely for their excellent coloristic property. Meanwhile, they are also interesting for functional uses. In coatings with a high content of platelet fillers, an advantageous overlapping roof-tile type arrangement is possible that provides close inter-particle contact, barrier effects, and dense covering. The composition of the oxide layer on the mica surface and its thickness are always responsible for the physical properties like electrical conductivity, magnetism, IR-reflectivity, laser markability.

Table 55 contains data of the functional properties of some metal oxide – mica pigments.

New Developments Based on Non-Micaceous Systems. The class of single crystal lustrous pigments is not limited to the non-absorbing types like bismuth oxychloride and basic lead carbonate. Recent developments are absorbing pigments such as platelet-like graphite, laminar phthalocyanines and flaky iron oxides. These flakes

Table 54. Examples of mica-based pearlescent pigments with special coloristic properties

Pigment composition	Preparation	Remarks	References
TiO_{2-x}/TiO_2/Mica	TiO_2/Mica + H_2 (Ti, Si) T > 900 °C (solid state reaction)	grey, blue-grey	[5.230]
TiO_xN_y/TiO_2/Mica	TiO_2/Mica + NH_3 T > 900 °C (solid state reaction)	grey, blue-grey	[5.227]
$FeTiO_3/TiO_2$/Mica	Fe_2O_3/Mica + H_2 T > 600 °C (solid state reaction)	grey (ilmenite pigments)	[5.231]
Fe_3O_4 Mica	Fe_2O_3/Mica + H_2 T ≈ 400 °C (solid state reaction)	black	[5.232]
	Fe^{2+} + O_2 + Mica (precipitation)	black	[5.232]
	$Fe(CO)_5$ + O_2 + Mica (CVD-process)	black	[5.227]
TiN/Mica	$TiCl_4$ + NH_3 + Mica (CVD-process)	gold	[5.202]
TiO_2/C/Mica	$TiOCl_2$ + C + Mica (precipitation) calcination under N_2	silver-grey, interference colors (Carbon inclusion pigments, Fig. 81)	[5.233]
$BaSO_4/TiO_2$/Mica	Ba^{2+} + SO_4^{2-} + $TiOCl_2$ + Mica (precipitation)	Low luster pigments	[5.234]
Fe_3O_4/Mica (mica surface only partially coated)	Fe^{2+} + O_2 + Mica (precipitation)	Transparent colors	[5.209]

consist of pure iron oxide or mixed phase pigments, e.g. $Al_xFe_{2-x}O_3$, $Mn_yFe_{2-y}O_3$, or $Al_xMn_yFe_{2-x-y}O_3$. Hexagonal platelet crystals with a diameter of 5–50 µm are grown under hydro-thermal or flux conditions [5.237], [5.238]. They can be reduced to the corresponding isomorphous magnetites or used as substrates for additional coating with metal oxides [5.239].

The hematite platelets show a predominantly metallic effect. Very thin particles with a thickness of 50 to 400 nm display a pale copper gloss, which is indicative of interference. The shade can be varied and the properties of the platelets can be controlled by doping. Al or Mn are incorporated by substitution of Fe in the hematite lattice, and Si is incorporated interstitially [5.240]. Laminar iron oxide pigments are interesting because of their excellent fastness to light, outdoor exposure and their good mechanical stability. Main applications up to now are automotive lacquers and cosmetics.

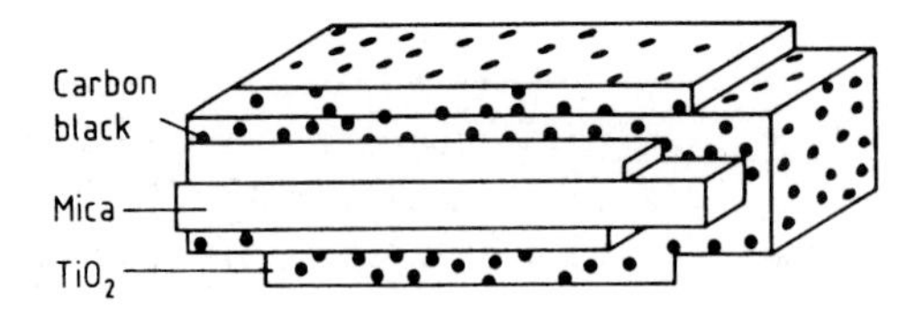

Figure 81. Carbon black particles are finely distributed within the TiO_2 layer to form a silver-grey metallic-like pigment

Table 55. Functional metal oxide – mica pigments

Pigment composition	Property	Application	References
$(Sn, Sb)O_2$/Mica $Sn(O,F)_2$/Mica	electrically conductive	conductive flooring, antistatic packaging materials, light colored primed plastic surfaces which can be electrostatically painted in further coating process, light colored conducting surfaces in clean room conditions for dust reduction	[5.235]
Fe_3O_4/Mica	magnetic	magnetic surfaces	[5.194] [5.232]
TiO_2/Mica	IR-reflective	IR-reflecting plastic sheets, e.g. for domed and continuous roof lights	[5.236]
TiO_2/Mica $(Sn, Sb)O_2$/Mica	laser sensitive	laser marking of plastics, coatings	[5.233]

Novel developments in nacreous and special effect pigments include the group of comminuted thin PVD films (Physical Vapor Deposition). This group includes ground single layer or multilayer film particles. The production of the film is the first step of the synthesis and is produced by physical methods. In a second step, the film is ground to small platelets. Examples are pigments derived from amorphous or nano-crystalline aluminum or pigments consisting of a sequence of layers. These layers act as transmission filters towards each side, a typical sequence is as follows: A specular metal layer (e.g. Al, Cr), a dielectric layer with low refractive index (e.g. MgF_2), and a semi-transparent metal layer. Such compositions are known as Fabry Perot structures [5.241]. The optical effect of these pigments is characterized by a very high gloss with extremely brilliant colors. The pigments show an extreme color flop when viewed from different observation angles. PVD film pigment can be used in the graphics industry, particularly as thin film security device pigments in security printing [5.242].

Several developments are concentrated on the search for substrate containing systems, where mica is replaced by other platelet-like materials. So it is possible to replace mica by kaoline or talc to produce bright conductive pigments by coating with $(Sn,Sb)O_2$ [5.243].

On the other hand, new pigments based on transparent silica flakes show extremely strong optical effects, which are different to mica pigments [5.244]. Angle dependent colors and other effects, achieved by the combination of these SiO_2-flakes with thin titania and/or iron oxide coating layers, have led to a new generation of pearl luster pigments. They add a whole new dimension to the existing possibilities of color stylings with luster pigments.

Other platelet crystals with a high aspect ratio are reported for phtalocyanine [5.245], graphite [5.246] and 1,4-diketopyrrolopyrroles [5.247].

The CVD process in a fluidized bed (see Section 5.3.1.5) is also used for depositing interference Fe_2O_3 layers on aluminum flakes [5.227] as well as for the formation of titanium nitride on platelet substrates [5.248] or TiO_2 [5.249], [5.250].

Regular scattering of spherical particles (opalescence) of TiO_2 is also now used to generate eye-catching color effects [5.251], [5.252].

5.3.1.7. Uses

Nacreous and interference pigments are used as colorants or part of color formulations for all applications where traditional pigments are used, but where additional color depth, brilliance, iridescence, color shift (flop), and other spectacular effects are desired [5.253]. Mica-based pigments dominate; their combination of pearl and interference effects, brilliance, stability, and weather resistance is unsurpassed. Furthermore, they are non-toxic [5.221].

Nacreous pigments require transparent or at least translucent binders or other carriers. Formulations with other pigments have to take their transparency and color mixing rules into account. Producers specify certain product lines for specific applications on the basis of national regulations and technical considerations. They also provide handling guidelines and starting formulas.

Color systems with nacreous pigments can be formulated in four ways [5.211]:

1) Blending nacreous pigments with other pigments in a single color formulation
2) Using a two-layer coating system in which an upper coat contains the nacreous pigment and the lower coat contains the hiding absorption or metal effect pigments
3) Using a two-layer system with a transparent absorption color formulation applied on top of a layer containing the nacreous pigment
4) Using a multilayer nacreous pigment consisting of an absorbing and/or iridescent coating applied in single coat to the substrate in the form of platelets (see Section 5.3.1.5)

Special care has to be given to ensure parallel alignment of the platelet pigments during application. In plastics formed by injection molding, parallel alignment is hindered by the high viscosity of the polymer melts. Flow lines occur, but can be avoided by appropriate die/mould, flow and surface pattern design. In coatings [5.254], [5.255] and printing inks [5.256], proper formulation, rheology control, settling and shrinking of the films on drying result in parallel, stacked alignment of the platelets. In automotive coatings [5.257] additional surface modification of the pigments is used to increase long-term weather resistance [5.258]–[5.260].

For ceramic applications, the pigment is coated with an additional SnO_2 layer to stabilize it against the aggressiveness of the frits at high temperatures [5.261].

Cosmetic applications require specific nacreous and interference pigments that are approved for use according to national regulations [5.262], [5.263]. Nacreous pigments are not used in cosmetics solely for their optical effects. TiO_2-mica can be used as a sunscreen filter because TiO_2 strongly absorbs light in the near UV region [5.264]. A large variety of surface modifications and additives (e.g., mono-disperse, submicron spheres of SiO_2) are used to improve processability, control oil absorption [5.265] and give a softer skin feel [5.266].

5.3.2. Metal Effect Pigments

Definition. Metal effect pigments consist of flakes or lamellae of aluminum (aluminum bronzes), copper and copper–zinc alloys ("gold bronzes"), zinc, and other metals (DIN 55943, 55944; FSO 4618-1).

Physical Principles. The metallic effect is caused by the reflection of light at the pigment surface (see Fig. 71); luster (remission) decreases when the proportion of light scattered by the edges and corners of the pigment particles increases. As the pigment becomes coarser, the proportion of reflected light becomes larger; brilliance, brightness and, in the case of "nonleafing" aluminum pigments, the flop (color change with viewing angle) therefore increase. With finer pigments, the proportion of scattered light is greater; brilliance, brightness, and flop are therefore lower but the hiding power is better.

The metal effect depends not only on particle size and particle size distribution, but also on the orientation of the metal flakes within the coating film, particle shape, the transparency of the binder matrix, and the presence of colored pigments or dyes.

Particle Size Distribution. The particle size distribution is determined by conventional sieving or other methods such as laser granulometry (see Section 1.2.1: Particle Size).

The required particle size of the pigments depends on the intended use and can vary from a few micrometers (e.g., in offset printing inks), to medium grades of < 25 µm or < 45 µm (e.g., in "metallic" automotive topcoats, gravure and flexographic printing inks), to coarser grades (e.g., in corrosion-inhibiting systems, colorants for plastics), and even up to "spangle" size of several millimeters.

The thickness of the flakes can vary from a hundredth of a micrometer up to approximately one micrometer.

The term shape factor denotes the ratio of the platelet thickness to the platelet diameter.

Production. Metallic pigments were first produced in the Middle Ages when gold leaf was made by hand beating. This was replaced by stamping machines. Metal effect pigments are now usually produced in ball mills, using dry milling (Hametag process) and wet milling (Hall process). Other special production processes are used on a small scale (e.g., production of aluminum foil by vaporization, followed by size reduction).

During the ball milling process, a lubricant is added to prevent cold fusion; it also determines the wetting properties of the metallic pigment, i.e., "leafing" or "nonleafing".

Leafing and Nonleafing. *Leafing pigments* float on the surface of paint or printing ink films as a result of high interfacial tension. They form a coherent surface film whose reflective properties depend on the particle fineness. Stearic acid is the lubricant usually used in the manufacture of these products.

Nonleafing pigments are fully wetted by the medium and are distributed uniformly throughout the coating. Colored metallic effects are produced by combining the pigments with transparent pigments or dyes. The nonleafing properties are obtained by using lubricants that consist of branched-chain or unsaturated fatty acids (e.g., oleic acid) or polar substances (e.g., fatty amines).

Commercial Products. Depending on the production method and area of application, the pigments are supplied commercially in powder form or as solvent-containing pastes. Pastes for the paint industry are normally made with white spirit or a mixture of white spirit and solvent naphtha. Special pastes containing mineral oils are produced for offset printing inks, polar solvents are used for gravure and flexographic inks, and plasticizers are used to make pastes for coloring plastics. Stabilized aluminum pastes with water or water-miscible solvents are available for waterborne coating systems [5.267]. Pigments specially coated with organic (e.g., acrylics) or inorganic materials (e.g., silica) are available for powder coatings.

Uses. Metallic pigments are used to produce metallic or chrome effects and also as functional pigments (e.g., for corrosion protection, as conductive paints, or for protection against electromagnetic radiation).

They are employed in paints and coatings; graphics and printing inks (offset, gravure, flexographic and screen printing, and bronzing); and in plastics (coloration by, e.g., masterbatches, fillers, gel coating techniques, and other coatings).

Other industrial uses of aluminum flake include the building industry (production of aerated concrete) and the chemical industry (e.g., production of titanium dioxide, pyrotechnics, and explosives). Copper powder in flake form is used in the chemical industry (e.g., for phthalocyanine production, for lubricants).

Aluminum pigments are produced by grinding aluminum powder, usually obtained by atomizing molten aluminum. The starting material is mainly aluminum ingots with a purity of 99.5% (DIN 1712), or pure aluminum (> 99.95%) for special outdoor applications (acid-resistant grades). For standard specifications, see Table 1 ("Aluminum pigments and pastes").

Aluminum powder forms explosive mixtures with air; the lower explosive limit being ca. 35–50 g/cm^3. Dust-free aluminum paste is therefore used in most applications.

Aluminum pigments are mainly produced by wet milling in white spirit. The resulting pigment suspension is usually fractionated, sieved, and filtered on a filter press. The filter cake is mixed with solvents to give the usual commercial consistency with 65 wt% solids. Other solvents may be used instead of white spirit, the choice being governed by the intended application. The white spirit can be removed by vacuum drying and replaced by another solvent if necessary (e.g., for printing inks or coloring plastics).

Aluminum behaves as an amphoteric metal, liberating hydrogen from both alkaline and acid aqueous media. In waterborne paints and coloring systems, special stabilized aluminum pigments are therefore required [5.267]. Aluminum pigments must not be used with halogen-containing solvents because they react very violently to release hydrogen halides (Friedel–Crafts reaction).

The leafing properties of aluminum pigments can be adversely affected by polar solvents or binders. "Leafing-stabilized" aluminum pigments are available for these applications.

Along with special pigments for modern environmentally-friendly coating systems (e.g., waterborne paints and powder coatings) described above, aluminum pigments coated with colored metal oxides (e.g., iron oxide) have appeared on the market which produce not only reflection but also interference effects [5.268]. Novel color effects are obtained by combining the metal-oxide-coated pigments with transparent colored pigments.

Aluminum has a very low toxicity and is permitted as a coloring agent (EEC no. E 173, C.I. 77 000). It is also approved by the FDA, § 175 300 (USA).

Copper and Gold Bronze Pigments. Copper and gold bronze pigments (powdered copper–zinc alloys) are usually produced by dry milling. Depending on the alloy composition, the following "natural shades" are produced:

Copper	100% copper
Pale gold	ca. 90% copper and 10% zinc
Rich pale gold	ca. 85% copper and 15% zinc
Rich gold	ca. 75% copper and 25% zinc

Copper–zinc alloys with a higher zinc content (brass) cannot be ground or formed into flake pigments on account of their brittleness. Controlled oxidation of "natural" bronze powders converts them into "fired" bronze powders. These shades (e.g., English green, lemon, ducat gold, fire red) are produced as a result of interference effects that depend on the thickness of the oxide coating.

Copper and gold bronze pigments are not as colorfast as aluminum pigments because they decompose to produce colored oxides and corrosion products. However, stabilized pigments (e.g., with a silica coating) are also available for critical applications in binders with high acid values or that react with copper or zinc.

The classical fields of application for the gold bronze pigments are the graphics industry (bronzing); printing (offset, gravure, flexographic, and screen printing); and coloring plastics. They are also used in the paint industry for decorative finishes (e.g., dip coatings for candles).

Gold bronze pigments are nontoxic. In Germany they are permitted in food packaging materials provided that they do not lead to contamination of the food. They are also allowed for printing on cigarette paper and filter tips according to the tobacco regulations if the zinc content does not exceed 15%. Similar regulations apply in the United States and other countries.

Flake Zinc Pigments. Flake zinc pigments are used mainly as high-quality anticorrosive pigments in powder or paste form. Owing to their platelet structure they have a considerably higher surface area than spherical zinc dust particles. They can therefore take up much more binder which produces a more flexible coating film than that obtained with zinc dust. Other advantages include a lower settling tendency, good remixing, and problem-free application onto the precoated surface.

The flake zinc pigments give a considerably brighter, better metallic effect than zinc dust. Their appearance can be further improved by combining them with aluminum pigments or by applying a topcoat based on aluminum pigment.

Another widespread application for flake zinc pigments is the coating of small articles with complex shapes (e.g., screws, steel springs, bolts, rivets) and for special anticorrosive paints (Section 5.2.11.1).

5.4. Transparent Pigments

Pigments become transparent in binders when the difference between the refractive index of the pigment (which depends on wavelength) and that of the binder is low, and when the particle size of the pigment is in the range 2–15 nm. The decisive factor for transparency is, however, the thickness of the pigment particles in the direction in which the light travels.

In practice, pigment transparency is measured by determining the color difference between a sample of the pigmented system on a black background and the black background itself. The transparency measurement method and the definition of the transparency index are described in DIN 55988.

In a pigment of given chemical composition, transparency can only be influenced by particle size (i.e., by the size distribution of the primary particles) and is thus dependent on the synthesis conditions. Owing to the small size of the primary particles, the pigments tend to agglomerate. Since agglomeration can only be remedied with great difficulty by mechanical grinding, attempts are made to suppress agglomeration by using additives (e.g., sodium salts of higher fatty acids or alkylamines) during synthesis prior to the separation of the pigment from the suspension. In spite of this, almost all transparent pigments are difficult to disperse. They all have a specific surface area (BET) of greater than 100 m^2/g.

5.4.1. Transparent Iron Oxides

Transparent yellow iron oxide has the α-FeO(OH) (goethite) structure; on heating it is converted into transparent red iron oxide with the α-Fe_2O_3 (hematite) structure. Differential thermogravimetric analysis shows a weight loss at 275 °C. Orange hues develop after brief thermal treatment of yellow iron oxide and can also be obtained by blending directly the yellow and red iron oxide powders.

Production. *Transparent yellow iron oxide* [*51274-00-1*], C.I. Pigment Yellow 42:77492, is obtained by the precipitation of iron(II) hydroxide or carbonate with alkali from iron(II) salt solutions and subsequent oxidation to FeO(OH). On an industrial scale, oxidation is usually carried out by introducing atmospheric oxygen into the reaction vessel. Important factors for good transparency are high dilution during precipitation, a temperature of $< 25\,^{\circ}C$ during oxidation, and short oxidation times (< 5 h). Oxidation can be carried out under acidic or basic conditions [5.269], [5.270]. The best results are obtained by using 6-wt% iron(II) sulfate solutions and precipitation with ca. 85% sodium carbonate as a 10-wt% solution. The starting material is usually crystalline $FeSO_4 \cdot 7\,H_2O$, obtained as a byproduct from ilmenite in the pickling of iron or in the production of titanium dioxide according to the sulfate procedure (see Section 2.1.3.1). In order to improve pigment properties, the suspension is matured for about a day before filtration. It is then filtered, dried, and carefully ground to a powder. The primary particles are needle-shaped and have an average length of 50–100 nm, a width of 10–20 nm, and a thickness of 2–5 nm.

Transparent red iron oxide [*1309-37-1*], C.I. Pigment Red 101:77491, is obtained by heating the yellow pigment (e.g., in a cylindrical rotary kiln) at 400–500 °C (Fig. 82).

Transparent red iron oxides containing iron oxide hydrate can also be produced directly by precipitation. A hematite content of $> 85\%$ can be obtained when iron(II) hydroxide or iron(II) carbonate is precipitated from iron(II) salt solutions at ca. 30 °C and when oxidation is carried out to completion with aeration and seeding additives (e.g., chlorides of magnesium, calcium, or aluminum) [5.271]. Transparent iron oxides can also be synthesized by heating finely atomized liquid pentacarbonyl iron in the presence of excess air at 580–800 °C [5.272], [5.273]. The products have a primary particle size of ca. 10 nm, are X-ray amorphous, and have an isometric particle form. Hues ranging from red to orange can be obtained with this procedure, however, it is not suitable for yellow hues.

Transparent brown iron oxides are produced by precipitating iron(II) salt solutions with dilute alkali (sodium hydroxide or sodium carbonate) and oxidizing with air. Only two-thirds of the precipitated iron hydroxide, oxide hydrate, or carbonate is oxidized. Alternatively, the iron oxides can be produced by complete oxidation of

Figure 82. Electron micrograph of a transparent red iron oxide pigment (Sicotrans Red 2815)

the precipitate iron compounds and subsequent addition of half the amount of the initial iron salt solution and precipitation as hydroxide [5.274], [5.275]. However, these products have the composition $FeO \cdot Fe_2O_3$ and are ferromagnetic; they are not of practical importance.

Properties and Uses. As far as resistance to light, weather, and chemicals is concerned, transparent iron oxides behave in a similar manner to the opaque iron oxides (see Section 3.1.1). In addition, they show a high UV absorption, which is exploited in applications such as the coloring of plastic bottles and films used in the packaging of UV-sensitive foods [5.276], [5.277].

Worldwide consumption of transparent iron oxides is 2000 t/a. They are mainly used in the production of metallic paint in combination with flaky aluminum pigments and in the coloring of plastics for bottles and fibers.

Toxicology. Special toxicological studies on transparent iron oxides have not yet been carried out. The results of opaque iron oxides are applicable, see Section 3.1.1.3.

Trade names and producers include Capelle (Gebroeders Cappelle N.V., Belgium), Fastona Transparent Iron Oxide (Blythe Colours Ltd., UK), Sicotrans (BASF and BASF Lacke + Farben, Germany), and Trans Oxide (Hilton Davis, USA).

5.4.2. Transparent Iron Blue

Transparent Iron Blue. Iron blue [*14038-43-8*], C.I. Pigment Blue 27:77520 (Milori blue), also occurs in finely dispersed forms (primary particle size < 20 nm, specific surface area 100 m^2/g) that are more transparent than the conventional iron blue pigments. They are generally produced by the same procedure as the less transparent iron blue pigments (see Section 3.6) but a higher dilution factor is used. The transparency of these pigments is exploited solely in the production of printing inks (illustration gravure). All other properties are described in Section 3.6.

Trade names include Manox Iron Blue (Manox, UK), and Vossen-Blue 2000 (Degussa, Germany). Transparent iron blue pigments are also produced by Dainichiseika (Japan).

5.4.3. Transparent Cobalt Blue and Green

Transparent Cobalt Blue. Cobalt blue [*1345-16-0*] C.I. Pigment Blue 28:77346, is also produced as a transparent pigment by precipitating cobalt and then aluminum as hydroxides or carbonates from salt solutions using alkali. The hydroxides or carbonates are then filtered, washed, dried, and calcined at 1000 °C [5.278]. It is important to carry out the precipitation with high dilution and to distribute the alkali uniformly throughout the entire reaction volume.

Transparent cobalt blue pigments have the form of small tiles (primary particle size 20–100 nm, thickness ca. 5 nm, specific surface area (BET) ca. 100 m^2/g). They are resistant to light and weather and are used in metallic paint, but are of little importance.

Transparent Cobalt Green. Analogous to cobalt blue a transparent cobalt green spinel C.I. Pigment Green 19: 77335 can be manufactured.

The green as well as the blue transparent cobalt pigments are used as filter materials for cathode ray tubes (CRT) [5.279].

5.4.4. Transparent Titanium Dioxide

Titanium dioxide can also be produced with a primary particle size of 10–30 nm, and hence shows transparent properties. Microcrystalline titanium dioxide was first mentioned in japanese patents in 1978.

The use of micro titanium dioxide as a white pigment is limited because its light scattering is very low due to the very fine crystal size, which means that the coloring effect of conventional TiO_2 is lost. The physical properties changed significantly; there is strong UV absorption. Consequently, such fine particle TiO_2 is used as a thermal stable, non-toxic and non-migratory UV light absorbing additive:

- for cosmetics, mainly for sunscreen formulations because of its effective UV protection over the UVC, UVB and UVA spectrum [5.280];
- for automotive paints, especially in combination with aluminum flakes giving a perlescent like appearance; the color flop depends on the concentration of the micro titanium dioxide in the formulation [5.281], [5.282];
- for clear coats and wood varnishes to protect the base due to its transparency and its property to absorbe UV light;
- for plastics to improve the UV durability of the polymer films itself as well as to protect UV sensible foodstuff in plastic wrapping [5.283];
- as heat stabilizers in silicon rubber, as catalysts for hydrogenation [5.284] and oxidation [5.285] as well as for surface protective films for furniture and optical material [5.286].

Microfine TiO_2 can be obtained mainly with rutile structure by different manufacturing routes which depend on starting materials used:

$$TiCl_4 + 4\,NaOH \Rightarrow TiO_2 + 4\,NaCl + 2\,H_2O$$

$$Na_2TiO_3 + 2\,HCl \Rightarrow TiO_2 + 2\,NaCl + H_2O$$

$$Ti(OC_3H_7)_4 + 2\,H_2O \Rightarrow TiO_2 + 4\,C_3H_7OH$$

The process steps include precipitation, neutralisation, filtration and washing, drying and micronisation (see Section 2.1 for details of TiO_2 production). Due to its small particle size the transparent titanium dioxide has a high photoactivity. To reduce this and in order to get a better weatherfastness the fine particles are coated with various combinations of inorganic oxides (e.g. silica, alumina, zirconia, iron) before drying in a similar way as that used for conventional titanium dioxide.

Microfine TiO_2 with predominant anatase structure can also be manufactured by reductive flame hydrolysis of $TiCl_4$ at $< 700\,^{\circ}C$ [5.287]:

$$TiCl_4 + 2H_2 + O_2 \Rightarrow TiO_2 + 4HCl$$

Worldwide consumption is increasing and estimated at 1300 t/a.

Trade names include Titanium Dioxide P 25 (Degussa, Germany), Hombitec RM Series (Sachtleben Chemie, Germany) and Micro Titanium Dioxide MT-Series (Tayca, Japan). Transparent titanium dioxide is also produced by Ishihara (Japan) and Kemira Oy (Finnland).

5.4.5. Transparent Zinc Oxide

Similar to transparent titanium dioxide microparticles of zinc oxide are manufactured by using sol/gel processes or precipitation in the presence of protective colloids to limit particle growth [5.288].

An industrial process [5.289] operates with solutions of zinc sulfate and zinc chloride in the ratio 1:2. Basic zinc carbonate is precipitated by feeding simultaneously the zinc salt solution and a mixed solution of sodium hydroxide and sodium carbonate into a reactor charged with water. The precipitated product is intensively washed several times and then spraydried.

It is used for cosmetics and paints as transparent UV-light shielding chemical.

Trade names include Sachtotec Micro-Zinkoxid (Sachtleben, Germany) and Z-cote HP1 (SunSmart Inc., USA).

5.5. Luminescent Pigments [5.290]–[5.296]

5.5.1. Introduction

Inorganic luminescent materials are synthetically produced crystalline compounds that absorb energy acting on them and subsequently emit this absorbed energy as light, either immediately or over a longer period, in excess of thermal radiation. This light emission, for which, in contrast to thermal radiation, the exciting energy first does not contribute to the thermal energy of the compound, is known as luminescence. It arises from excited states in atoms or molecules which have a lifetime of at least 10^{-9} s. This means that between excitation and light emission there is a time span of 10^{-9} s.

Luminescent inorganic compounds are also known as luminophors and phosphors. Depending on the nature of the exciting energy, the resulting luminescence is named as follows:

Low energy photons	photoluminescence
Cathode rays	cathodoluminescence
X rays	X-ray luminescence
Ions (particles)	ionoluminescence
Mechanical forces	triboluminescence
Electric field strength	electroluminescence
Chemical reactions	chemiluminescence
Biochemical reactions	bioluminescence

Light emission occurring during excitation and up to ca. 10^{-8} s afterwards is called fluorescence, while the glow continuing longer than 10^{-8} s is known as afterglow or phosphorescence. The decay time is the time in which the brightness decreases to 1/10 or 1/e of the initial intensity, for hyperbolic and exponential decay, respectively.

Certain compounds store the absorbed energy over a longer period and only emit it as light under the influence of heat or IR radiation. This process is called stimulation. The opposite phenomenon, in which the light emission is decreased by the action of heat or IR radiation on luminescent materials, is known as quenching.

History. Various reports indicate that luminescent inorganic compounds were already known in antiquity. TITUS LIVIUS (Livy) reports in his book "History of the Romans" that in 186 B.C. the maenads in the bacchanalia had "glowing torches" in their hair, which they then threw into the Tiber. JORRISSON assumes that this was luminescence from calcium sulfide [5.297].

One of the first reports of the synthesis of a luminescent material dates from 1603, when CASCIAVOLUS attempted to obtain gold by heating certain minerals, but instead obtained a rock that emitted reddish light in the dark. Here barium sulfide was formed by reduction of rock containing barium sulfate.

In 1609 BRAND discovered a substance that glows brightly in air and named it phosphorus. Since then inorganic luminescent pigments have been named phosphors and their luminescence, phosphorescence. In the 1600s and 1700s, many other researchers investigated this light phenomenon, including GOETHE.

The systematic scientific investigation of inorganic phosphors only began, however, in the 1800s with BECQUEREL, VERNEUIL, LENARD, and STOKES. STOKES published the first definition of fluorescence in 1852, and drew up the law stating that the exciting light always has a shorter wavelength than the emitted light.

In 1866 SIDOT discovered the luminescence of zinc sulfide, which has been the basis of many important industrial phosphors.

In 1867 BEQUEREL postulated two types of afterglow based on his investigations of uranyl salts: one with exponential and the other with hyperbolic decay.

In 1895 RÖNTGEN used "barium platinocyanate" to make X rays visible. This discovery led to the first industrial breakthrough for inorganic phosphors. In 1887 VERNEUIL recognized the significance of heavy-metal impurities for the luminescence of inorganic compounds.

In the 1900s inorganic phosphors gained increasing scientific and industrial interest. In 1903 and 1904 LENARD and KLATT reported the connection between activators and light emission [5.298].

They were also the first to use fluxes for crystallization and to discuss how they function. The first experiments on producing luminescence solely by the action of an applied electric field were carried out by GUDDEN and POHL in 1920 [5.299], LOSSEW in 1923 [5.300], and DESTRIAU in 1936 [5.301]. URBACH (1926) [5.302] and RANDALL and WILKEN (1945) [5.303] traced the different time constants for the decay of the afterglow back to the differing depth of traps.

In 1938/1939 SEITZ [5.304] and MOTT [5.305] developed the electron configuration diagram, on the basis of which, phenomena such as Stokes' Law became understandable.

Using quantum theory, RIEHL and SCHÖN, among others, developed in 1939 a model for the luminescence of crystalline phosphors which is essentially still valid today [5.306].

5.5.2. Luminescence of Crystalline Inorganic Phosphors

5.5.2.1. Luminescence Processes

The luminescence of inorganic phosphors is comprised of the following processes: 1) absorption and excitation, 2) energy transfer, and 3) emission.

Excitation requires that energy acts on the phosphor that can be absorbed by it. The excitation mechanism differs according to the type of energy involved.

Excitation by Photons. The dependence of the absorptivity of a phosphor upon the photon energy is given by the absorption spectrum. This can be divided into various regions such as basic lattice absorption, edge absorption, and defect absorption [5.307].

Often, excitation only takes place with photons of higher energy. The absorption that is effective for excitation of luminescence is given by the excitation spectrum, which shows the dependence of the luminescence intensity on the photon energy. Strong luminescence occurs when the exciting photon has sufficient energy to promote an electron from the valence band to the conduction band.

Energy Configuration Diagram. This model, based on the energy level diagrams of atoms and molecules, is applicable to luminescence processes in which excitation and emission take place at the same luminescence center.

The energy relationships in a luminescence process are presented in a configurational coordinate diagram (Fig. 83). This illustrates the relationship between the potential energy E of the luminescence center (ordinate) and a space coordinate (abscissa), which gives the representative separation between the atom involved and its nearest neighbors or the deflection from its spatial equilibrium position.

The lower curve represents the ground state of the luminescence center with the vibrational levels, and the upper curve the excited state.

By absorbing excited radiation the electrons are raised from the ground state to the excited state. These transitions take place so rapidly that no displacement of the atomic nuclei occurs (Franck–Condon principle). The space coordinate thus remains unchanged and the transitions can be represented by vertical lines. Because the excited system is not immediately in a state of equilibrium after absorption of energy, it first moves towards the lowest vibrational level with loss of energy to the lattice

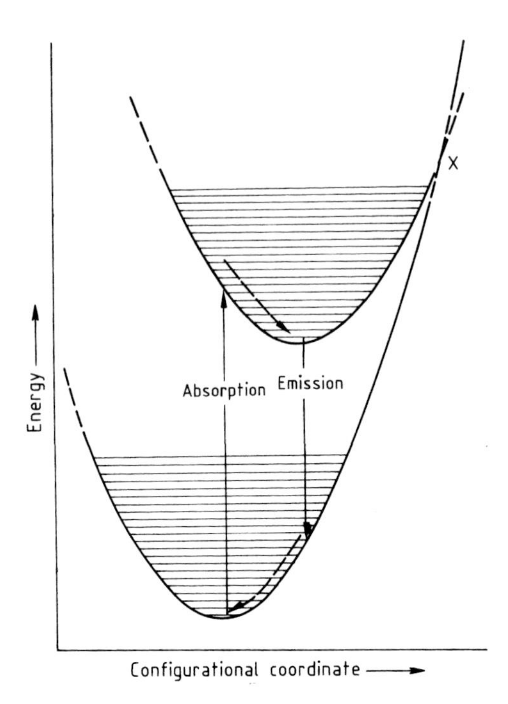

Figure 83. Configurational-coordinate diagram

(lattice relaxation). From there the electrons can return to one of the vibrational levels of the ground state with emission of light and finally relax to the equilibrium position.

Phosphors that do not effectively absorb photons of a specific energy can often be sensitized for this radiation by incorporation of certain ions with higher absorption for this radiation in the host crystal.

Excitation by Electrons. A portion of the electrons impinging on a phosphor are backscattered elastically with loss of energy. The backscattering factor η and the averaged atomic number $\bar{Z}$ of the atoms are related by [5.308]

$$\eta = (\ln\bar{Z} - 5.5)/6$$

For phosphors based on zinc sulfide ($Z_{Zn} = 30$, $Z_S = 16$; hence, $\bar{Z}_{ZnS} = 23$) this gives a backscattering coefficient $\eta \approx 0.27$. This means that 27% of the energy supplied is not available for luminescence. The effect of the electron energy on backscattering is described in [5.309].

The electrons that are not scattered elastically lose their energy on penetrating a phosphor crystal by inelastic scattering and formation of secondary electrons.

The inelastically scattered electrons and the secondary electrons produce charge carriers in the phosphor, which then recombine with emission of luminescent radiation, either directly or after traveling in the lattice.

Excitation by X Rays. As forms of electromagnetic radiation, X and γ rays follow the same absorption rules as photons. Because of the high energy of the secondary electrons formed on absorption, however, the excitation mechanism more closely resembles electron excitation than photon excitation. The subsequent processes then continue as in electron excitation.

Energy Transfer. The excitation of phosphors does not always take place at the luminescence center. A recombination and the associated emission require transfer of the absorbed energy to the luminescence center. This can take place by the following mechanisms:

1) Migration of excited electrons and electron holes
2) Migration of excitons (electron–hole pairs)
3) Resonance between atoms with overlapping electric fields
4) Reabsorption of a light quantum emitted from one activator center by another emission center

The recombination processes are to a great extent independent of the nature of the excitation. The only parameters that are influenced by the type of excitation are the relative intensities of the various emission bands, the quantum yields, and the decay behavior. For photon excitation, the quantum yield can be up to 70 %, whereas it reaches a maximum of 25 % for electron excitation. The afterglow time and afterglow intensity are lower for electron excitation than for photon excitation.

The simplest case of reaction kinetics occurs when the excitation and emission occur in the same atom, molecule, or luminescence center. The recombination can then be treated as a first-order unimolecular reaction. The decay time is independent of the number of other similarly excited atoms or molecules.

If the recombination is delayed, e.g., by migration of excited electrons, luminescence takes place by a second-order bimolecular reaction. The probability of a luminescent recombination of the excited electron with the holes is then proportional to the product of the concentration of electrons and the concentration of holes. The lower the initial intensity is, and the further the decay has progressed, the slower the decay to the half value is. This hyperbolic decay law is only of limited validity. If the excited electron is momentarily trapped before recombination, very complex interactions can arise.

5.5.3. Preparation and Properties of Inorganic Phosphors

5.5.3.1. Sulfides and Selenides

Zinc and Cadmium Sulfides and Sulfoselenides. The raw materials for the production of these sulfide phosphors are high-purity zinc and cadmium sulfides, which are precipitated from purified salt solutions by hydrogen sulfide or ammonium sulfide [5.291], [5.296], [5.307], [5.310]. The concentration of contaminants such as Fe, Co, or Ni must be below 1 % of the activator concentration. The $Zn_{1-x}Cd_xS$ can be produced by mixing precipitated zinc sulfide and cadmium sulfide. However, coprecipitation from mixed zinc–cadmium salt solutions is preferred because of the better homogeneity.

The most important activators for sulfide phosphors are copper and silver, followed by manganese, gold, rare earths, and zinc. The charge compensation of the host lattice is effected by coupled substitution with mono- or trivalent ions (e.g., Cl^- or Al^{3+}). In addition, disorders, such as unoccupied sulfur positions, can also contribute to charge compensation.

For the synthesis of phosphors, the sulfides are homogenized with readily decomposed compounds of the activators and coactivators in the presence of a flux and are fired in quartz crucibles at 800–1200 °C.

The luminescent properties can be influenced by the nature of the activators and coactivators, their concentrations, the composition of the flux, and the firing conditions. In addition, specific substitution of zinc or sulfur in the host lattice by cadmium or selenium is possible, which also influences the luminescent properties. Zinc sulfide is dimorphic and crystallizes below 1020 °C in the cubic zinc-blende structure and above that temperature in the hexagonal wurtzite lattice. When the zinc is replaced by cadmium, the transition temperature is lowered so that the hexagonal modification predominates. Substitution of sulfur by selenium, on the other hand, stabilizes the zinc-blende lattice.

The $Zn_{1-x}Cd_xS$ phosphors have found the greatest industrial use.

Self-Activation. Although pure substances do not normally luminesce, zinc sulfide that has been fired in the presence of a halogen luminesces bright blue [5.311], [5.312]. The luminescence center is assumed to be a cation vacancy. The charge compensation occurs through exchange of S^{2-} by Cl^-.

Cation holes can also be created by coactivation with trivalent metal ions or by incorporation of oxygen [5.313]. The luminescence band of self-activated zinc sulfide with the zinc-blende structure exhibits a maximum at 470 nm. On transition to the wurtzite structure, the maximum shifts to shorter wavelengths. In the mixed crystals zinc sulfide–cadmium sulfide and zinc sulfide–zinc selenide, the maximum shifts to longer wavelengths with increasing cadmium or selenium concentration.

Silver Activation. Doping zinc sulfide with silver leads to the appearance of an intense emission band in the blue region of the spectrum at 440 nm, which has a short decay time. Weak luminescence in the green (520 nm) and red regions can also occur. The blue band is assigned to recombination at substitutionally incorporated silver ions [5.314], [5.315]. The red band is caused by luminescence processes in associates of silver ions occupying zinc positions with neighboring sulfur vacancies [5.316], whereas the nature of the green luminescence center is as yet unknown [5.317].

Industrial silver-activated zinc sulfide phosphors use the intense blue emission exclusively. The ZnS:Ag phosphor for cathode ray tubes is obtained by firing zinc sulfide and silver nitrate at ca. 1000 °C in the presence of sodium chloride (coactivator Cl^-) [5.318]. The afterglow can be further reduced by addition of 10^{-3}–10^{-4}% of nickel ions.

The substitution of zinc by cadmium in the ZnS:Ag phosphor leads to a shift of the emission maximum from the blue over to the green, yellow, red to the IR spectral region [5.307], [5.319]–[5.322].

Copper Activation. Activation with copper causes an emission in zinc sulfide which consists of a blue (460 nm) and a green band (525 nm) in varying ratios, depending on the preparation. The green band is assigned to monovalent copper ions incorporated into the host lattice by substitution in the zinc positions, with halide ions acting as coactivators. The blue band is attributed to associates between Cu^+ ions occupying Zn^{2+} positions and interstitially incorporated Cu^+ ions [5.323]–[5.325]. The phosphor ZnS:Cu can also exhibit luminescence in the red or infrared spectral region, depending on the preparation. The red emission band is assigned to a Cu^+ ion associated with a sulfur vacancy [5.326].

Usually, green-emitting ZnS:Cu phosphors are used for industrial applications. The preparation consists of firing a mixture of ZnS, $Cu(NO_3)_2 \cdot 6H_2O$ (corresponding to 50–1000 ppm Cu in the phosphor), and 5 wt% NaCl at 900–1000 °C in covered fused silica crucibles [5.327].

The decay time increases with the copper concentration and can be increased to several hours by adding small quantities of coactivators. The decay time can be shortened by addition of nickel. It is thus possible to adapt the decay times of copper-doped zinc sulfides to various requirements.

The distance between the valence and conduction bands is altered by partial substitution of Zn^{2+} by Cd^{2+}, and of S^{2-} by Se^{2-}. In this way the emission spectrum can be adjusted (Figs. 84 and 85).

Other Types of Activation. Zinc sulfide forms a wide range of substitutionally mixed crystals with manganese sulfide. Manganese-activated zinc sulfide has an emission band in the yellow spectral region at 580 nm [5.328]. This band can be assigned to the inner transitions of the Mn^{2+} ions incorporated into zinc positions. At higher manganese concentrations, a spectral shift into the orange region takes place as a result of manganese–manganese interactions. At high manganese concentrations, excitation by means of an electric field is possible.

The activation of zinc sulfide with gold leads to luminescence in the yellow-green (550 nm) or blue (480 nm) spectral regions [5.317], [5.329]–[5.332], depending on the

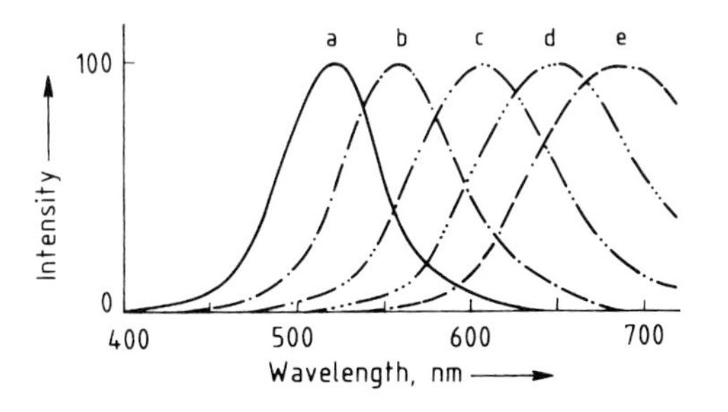

Figure 84. Spectra of copper-activated $Zn_{1-x}Cd_xS$ phosphors (5×10^{-3} wt% Cu) a) $x = 0$; b) $x = 0.1$; c) $x = 0.2$; d) $x = 0.3$; e) $x = 0.4$

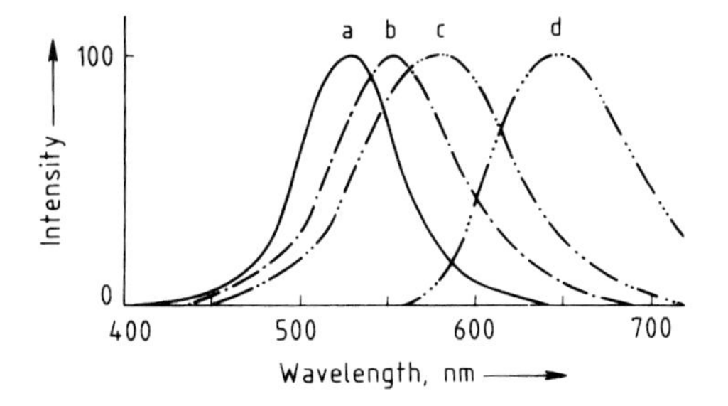

Figure 85. Spectra of copper-activated $ZnS_{1-x}Se_x$ phosphors (5×10^{-3} wt% Cu) a) $x = 0$; b) $x = 0.2$; c) $x = 0.4$; d) $x = 1$

preparation, whereas a blue-white luminescing phosphor is formed on activation with phosphorus [5.333].

Multiple activation of zinc sulfide is also possible. Zinc–cadmium sulfide, doubly activated with silver and gold, which is used as a white-luminescing, one-component phosphor for monochromic cathode-ray tubes [5.334], and the yellow-luminescing ZnS:Cu, Au, Al phosphor, whose emission color corresponds to that of $Zn_{1-x}Cd_xS$:Cu [5.332], are known.

Alkaline-Earth Sulfides and Sulfoselenides. Activated alkaline-earth sulfides have been known for a long time; their luminesence is very varied. Emission bands between the ultraviolet and near infrared can be obtained by varying the activation. They are produced by precipitation of sulfates or selenites from purified solutions, followed by reduction with $Ar-H_2$. The addition of activators, for example, copper nitrate, manganese sulfate, or bismuth nitrate, is followed by firing for 1–2 h. Alkaline-earth halides or alkali-metal sulfates are sometimes added as fluxes.

Among the long afterglow alkaline-earth sulfides, only (Ca, Sr)S:Bi^{3+}, CaS:Bi^{3+}, and CaS:Eu^{2+}, Tm^{2+} still have any real importance because their respective blue, violet, and red luminescence cannot yet be achieved with the less hydrolysis-sensitive zinc sulfide phosphors. The last-mentioned phosphor gives an intensive red afterglow and can substitute the red zinc–cadmium phosphor.

The alkaline-earth sulfides activated with rare earths are of greater importance. These phosphors are considered suitable for use in cathode-ray tubes because of the linear dependence of their brightness on current over a wide range [5.335]–[5.342]. The emission band of MgS with 0.004 mol% europium has a maximum at 600 nm, whose cathode-ray brightness reaches or surpasses that of Y_2O_2S:Eu^{3+}. On substitution of magnesium by calcium, a nonlinear shift of the emission maximum occurs ($Ca_{0.4}Mg_{0.6}S$:Eu^{2+}, $\lambda_{max} = 668$ nm, CaS:Eu^{2+}, $\lambda_{max} = 650$ nm) but at the same time there is a decrease in brightness [5.343].

CaS:Ce^{3+} is a green-emitting phosphor. On activation with 10^{-4} mol% cerium, the emission maximum occurs at 540 nm. Greater activator concentrations lead to a red shift; substitution of calcium by strontium, on the other hand, leads to a blue shift; SrS:Ce^{3+} (10^{-4} mol% cerium) has $\lambda_{max} = 483$ nm [5.344].

MgS:Ce^{3+} (0.1%) has two emission bands in the green and red spectral regions at 525 and 590 nm; MgS:Sm^{3+} (0.1%) has three emission bands at 575 nm (green), 610 (red), and 660 nm (red) [5.340].

Calcium or strontium sulfides, doubly activated with europium–samarium or cerium–samarium, can be stimulated by IR radiation since Sm^{3+} acts as a trap through the transition $Sm^{3+} + e^- \rightarrow Sm^{2+}$ [5.341]. Emission occurs at europium or cerium and leads to orange-red (SrS:Eu^{2+}, Sm^{3+}) or green (CaS:Ce^{3+}, Sm^{3+}) luminescence.

5.5.3.2. Oxysulfides

Red-emitting Y_2O_2S:Eu^{3+} phosphors were developed especially for excitation with cathode rays [5.345], [5.346]. Production generally starts from mixed rare-earth oxides, which are fired with sodium carbonate and sulfur. The polysulfide formed acts as a flux and sulfurizing agent [5.347]. The addition of a small amount of terbium sensitizes the europium luminescence [5.348].

The main emission lines of $Y_2O_2S{:}Eu^{3+}$ occur at 565 and 627 nm. The intensity of the long-wavelength emission increases with the europium concentration, whereby the color of the emission shifts from orange to deep red. For use in color picture tubes, $Y_2O_2S{:}Eu^{3+}$ contains ca. 4 wt% europium [5.349]; $Y_2O_2S{:}Eu^{3+}$ is unsuitable for use in lamps because of extensive thermal quenching of the luminescence above room temperature. Terbium in Y_2O_2S has main emission bands in the blue (489 nm) and green spectral regions (545 and 587 nm), whose intensity ratio depends on the terbium concentration. At low doping levels, $Y_2O_2S{:}Tb^{3+}$ luminesces blue-white, while at higher levels the color tends towards green. Although $Y_2O_2S{:}Tb^{3+}$ has been used for X-ray intensifying screens, $Gd_2O_2S{:}Tb^{3+}$, which also exhibits green luminescence, is predominantly used in X-ray technology because of its higher absorption. White luminescence for use in special cathode-ray tubes can be achieved with blends of $Y_2O_2S{:}Tb$ and $Y_2O_2S{:}Eu$ [5.350] or by double activation of Y_2O_2S with terbium and europium [5.351].

5.5.3.3. Oxygen-Dominant Phosphors

Oxygen-dominant phosphors are generally produced by solid-state reactions in which the components of the phosphor (mostly oxides) are intimately mixed and fired at 500–1500 °C. The desired phosphor is formed by solid-state reaction below the melting point of the compound during a firing period of 3–16 h.

The Tamman rule, which states that a measurable rate of change of position of the lattice components begins at two-thirds of the absolute melting point, gives an indication of the necessary firing temperature [5.352].

The activators are added in the form of oxides, oxalates, carbonates, or other compounds which readily decompose at higher temperatures.

The oxidation state of the activators can be influenced by the choice of the furnace atmosphere. As in the case of the sulfide phosphors, contamination by Fe, Co, or Ni greatly reduces the quantum yield.

As well as higher quantum yields, oxygen-dominant phosphors stand out because of their good temperature stability. For most substances the quantum yield only decreases noticeably above 250 °C.

Borates. Strontium fluoroborate $Sr_3B_{12}O_{20}F_2{:}\ Eu^{2+}$ is used in photocopier lamps, black light lamps, and lamps for medical applications ($\lambda_{max} = 366$ nm).

It is produced by the addition of aqueous ammonia to an aqueous solution of $Eu(NO_3)_3$, $Sr(NO_3)_2$, and H_3BO_3 at 95 °C and pH 8 to precipitate europium-containing $SrB_6O_{10} \cdot 5H_2O$. The latter is mixed with SrF_2 and, after a preliminary reaction at 400 °C, is fired at 820 °C in air and then at 900 °C under N_2–H_2 (95/5) [5.353].

Aluminates. *Yttrium aluminate* $Y_3Al_5O_{12}{:}Ce^{3+}$ (YAG) is produced by precipitation of the hydroxides with NH_4OH from a solution of the nitrates and subsequent firing at 1450 °C [5.354]. To enable a lower reaction temperature of 1300 °C, a freeze-dried powder is also used, which is produced from a solution of $(NH_4)_2Ce(NO_3)_6$, $Al_2(SO_4)_3 \cdot 18\,H_2O$, and $Y_2(SO_4)_3$ [5.355].

The solubility of Ce^{3+} in $Y_3Al_5O_{12}$ depends on the atmosphere in which the mixture is fired. At 1450 °C in a hydrogen atmosphere, up to 6 mol% Ce^{3+} can be incorporated into the $Y_3Al_5O_{12}$ lattice. At higher concentrations, $CeAlO_3$ is formed as a second phase. At 1450 °C in air the solubility is only 2 mol%; higher concentrations lead to CeO_2 as the second phase. In each case the cerium is incorporated as Ce^{3+} [5.355].

When fired in a reducing atmosphere, $Y_3Al_5O_{12}$ exhibits a pronounced afterglow due to traps formed by oxygen vacancies [5.356]. Subsequent annealing in air diminishes this effect and leads to decay times of 200–300 ns; therefore, $Y_3Al_5O_{12}:Ce^{3+}$ is used in flying-spot scanner tubes. The emission maximum is at 550 nm. This phosphor is classified under P46 (TEPAC) and KG (WTDS) (see Section 5.5.4.3).

Cerium magnesium aluminate (CAT) $Ce_{0.65}Tb_{0.35}MgAl_{11}O_{19}$ is produced by coprecipitation of the metal hydroxides from a solution of the nitrates with NH_4OH and subsequent firing at 700 °C for 2 h followed by 1 h at 1500 °C [5.357]. Another possibility is to fire a mixture of Al_2O_3, $MgCO_3$, CeO_2, and Tb_4O_7 for 5 h at 1500 °C with small quantities of MgF_2 or AlF_3 as a crystallizing agent [5.357]. A strongly reducing atmosphere is necessary to ensure that the rare earths are present as Ce^{3+} and Tb^{3+}.

Cerium magnesium aluminate is a highly efficient green phosphor ($\lambda_{max.}$ = 541 nm), which is used in trichromatic fluorescent lamps. The high quantum yield of 65% is effected by the energy transfer from the sensitizer, Ce^{3+}, to the activator, Tb^{3+}.

Barium magnesium aluminate (BAM) $BaMg_2Al_{16}O_{27}:Eu^{2+}$ is produced by firing a mixture of Al_2O_3, $BaCO_3$, $MgCO_3$, and Eu_2O_3 in the presence of a flux in a weakly reducing atmosphere at 1100–1200 °C [5.358]. It is used as the blue component (λ_{max} = 447 nm) in trichromatic fluorescent lamps.

Yttrium aluminum gallium garnet (YAGG) $Y_2Al_3Ga_2O_{12}:Tb^{3+}$ [5.359] can be produced by mixing a stoichiometric quantity of the oxides (5–7 mol% terbium) with 20% BaF_2 in ethanol. After drying, the mixture is fired for 2 h at 1500 °C. The flux is washed out with 20% nitric acid. Yttrium aluminum gallium garnet is a green phosphor with high brightness for use in projection television tubes. It does not show any saturation even at high current densities of 160 μA/cm^2.

For the mixed aluminate–gallate, the attainable brightness shows a greater dependency on the firing temperature than is the case with pure aluminate or gallate. Differences of 100 °C in the firing temperature reduce the brightness by 25–40%.

Gallates. Strontium thiogallate is produced by firing a mixture of high-purity sulfides in a stream of hydrogen sulfide at 900–950 °C for 2 h; Pb^{2+} and Ce^{3+} (with Na^+ for charge compensation) are used as activators [5.360], [5.561].

$SrGa_2S_4:Ce^{3+}, Na^+$ with 4 mol% cerium ($\lambda_{max.}$ = 455 nm, η = 5.0%) and $SrGa_2S_4:Pb^{2+}$ with 8 mol% lead ($\lambda_{max.}$ = 595 nm, η = 4.6%) are used in flying-spot scanner tubes because of their short decay times ($I_0/100$) of 1.55 μs and 1.24 μs, respectively. Their low ageing upon electron beam bombardment is also noteworthy. Because of the relatively broad emission bands, a blend of 15 wt% $SrGa_2S_4: Ce^{3+}$,

Na^+ with 85 wt% $SrGa_2S_4$: Pb^{2+} gives an extensive, regular emission throughout the visible spectrum [5.362].

Silicates. *Zinc orthosilicate* Zn_2SiO_4:Mn^{2+} can be produced by firing a mixture of ZnO, SiO_2, and $MnCO_3$ at 1200 °C for 2 h. Small quantities of MgF_2 or ZnF_2 accelerate the reaction [5.296].

$ZnSiO_4$:Mn is used as a green phosphor ($\lambda_{max.}$ = 525 nm) in fluorescent lamps and cathode-ray tubes and is cited in the TEPAC list as P 1 and in the WTDS as GJ. Its ability to be excited by low-energy electrons, its linear increase in brightness with electron beam intensity, and its high resistance to burn-in, even at high current density, are advantageous. The resistance to burn-in can be improved still further by employing a sol–gel process during the production. This involves reaction of a $[Zn(NH_3)_4](OH)_2$ solution with $Si(OC_2H_5)_4$ to give a gel, which is subsequently dried and fired [5.294]. The decay time of ca. 30 ms can be increased to the order of minutes by the addition of arsenic (a few ppm to 0.5 wt%) [5.363].

$ZnSiO_4$:Mn, As is standardized as P 39 in TEPAC and GR in WTDS.

Yttrium orthosilicate Y_2SiO_5:Ce^{3+} is used as a highly resistant blue cathode-ray phosphor in projection television tubes (η = 6%) [5.364]. It is classified as BH (WTDS) and P 47 (TEPAC). The emission maximum is between 400 nm (0.1% Ce) and 460 nm (10% Ce). The production involves firing a mixture of Y_2O_3, SiO_2, and $Ce(SO_4)_2 \cdot 4H_2O$ for 16 h at 1250 °C. Further firing for 4 h at 1160 °C in a reducing atmosphere is necessary to bring the cerium into the trivalent state. The addition of NaF or YF_3 improves crystallization. Doping with Tb^{3+} instead of Ce^{3+} leads to an efficient (η = 7%) green phosphor, in which coactivation by Ce^{3+} can increase the luminescence by a further 10% [5.365]. An yttrium orthosilicate doped with Ce, Tb, and Mn, is used in artificial teeth.

The phosphors obtained by solid-state reactions are not sufficiently resistant to aging for use in fluorescent lamps. In a pseudo-sol–gel process an aqueous solution of (Y, Tb) $(NO_3)_3$ is treated with SiO_2 and heated to 1200 °C. After subsequent reductive firing under N_2/H_2 at 1700–1800 °C, a phosphor is obtained with a resistance to aging comparable to that of the aluminates [5.366].

Barium disilicate $BaSi_2O_5$:Pb emits in the long-wavelength ultraviolet (λ_{max} = 350 nm). This phosphor is used in blacklight fluorescent lamps; the quantum yield is high.

The phosphor is produced by stirring $BaCO_3$, BaF_2, $Pb(NO_3)_2$, and SiO_2 (fumed silica) to a slurry in methanol. After drying, the mixture is fired for 30 min at 1050 °C, with the heating up and cooling down between 800 and 1050 °C being carried out at a rate of 200 °C/h [5.367].

Germanates. Magnesium fluorogermanate, $3.5\,MgO \cdot 0.5\,MgF_2 \cdot GeO_2$:$Mn^{4+}$ is a brilliant red phosphor used in fluorescent lamps [5.368]. Because of its yellowish color, it strongly absorbs the blue mercury emission lines in these lamps, so that a saturated red light results. The emission maximum lies at 660 nm. The temperature stability of this phosphor is particularly high; therefore, it is used for high-pressure mercury lamps. With rising temperature the absorption edge of the host lattice (350–400 nm) shifts to longer wavelengths. Therefore, the brightness increases with increasing temperature up to 300 °C because the mercury line at 365 nm is then absorbed more effectively. The phosphor is produced by firing a stoichiometric mixture of MgO, MgF_2, and GeO_2 with 0.01 mol% $MnCl_2$ for 5 h at 1150 °C in an oxidizing atmosphere [5.368].

Halophosphates and Phosphates. The halophosphates are doubly activated phosphors [5.369], in which Sb^{3+} and Mn^{2+} function as sensitizer and activator, giving rise to two corresponding maxima in the emission spectrum. The antimony acts equally as sensitizer and activator. It transfers part of its excitation energy to manganese. By changing the activator concentration, the blue antimony emission can be adjusted in favor of the orange-yellow manganese emission in such a way that halophosphates can emit white light of varying color temperature.

Halophosphates crystallize in the apatite structure. The chemical composition can be expressed most clearly as $3\,Ca_3(PO_4)_2 \cdot Ca(F,Cl)_2{:}Sb^{3+}, Mn^{2+}$.

For many years halophosphates were quantitatively by far the most important group of phosphors for low-pressure mercury discharge lamps (fluorescent lamps). The number of studies on the optimization of the quantum yield of halophosphates is correspondingly large. The most important work on the properties and production technology has been reported, for example in [5.345], [5.370].

Halophosphates are produced by the solid-state reaction of high-purity $CaHPO_4$, $CaCO_3$, CaF_2, NH_4Cl, $MnCO_3$, and Sb_2O_3. Because of the volatility of antimony trioxide and to avoid the oxidation of Mn^{2+} and Sb^{3+}, the mixture is usually fired under nitrogen at 1150–1200 °C in covered crucibles.

The substitution of a small proportion of the calcium by cadmium gives a considerable improvement in the radiation stability and an improved quantum yield for halophosphates in fluorescent lamps. The formation of undesirable color centers is strongly suppressed by addition of cadmium [5.371]. The cadmium presumably accelerates the reaction [5.372], thus minimizing side reactions during the firing process.

Because of the toxicity of cadmium, its use was abandoned in the 1980s. Nevertheless, by improving the quality of the raw materials and changing the reaction stoichiometry, it was possible to reproduce the quantum yield of the halophosphates, even without the addition of cadmium [5.373]. The absorption of the exciting UV radiation is strongly dependent on the particle size of halophosphate phosphors and decreases rapidly for particles larger than 3 μm. The particle size is influenced to some extent by the particle size of raw materials (calcium hydrogen phosphate) in the firing mixture. Mechanical separation processes (sieves) are used to free the fired product from undesired coarse and fine particles.

$(Sr,Mg)_3(PO_4)_2{:}Sn^{2+}$ is used in high-pressure mercury lamps. Absorption takes place by Sn^{2+}. The emission band, with a half-band width of 150 nm, has a maximum at 630 nm. Since only the high temperature form ($\geq$ 1300 °C) exhibits good luminescence, it must be cooled rapidly in a reducing atmosphere after firing. The incorporation of small quantities of Al^{3+}, Mg^{2+}, Zn^{2+}, Ca^{2+}, or Cd^{2+} stabilizes the luminescent high-temperature form. In high-pressure mercury lamps, $(Sr,Mg)_3(PO_4)_2{:}Sn^{2+}$ has been replaced by the more effective $YVO_4{:}Eu^{3+}$, Tb^{3+}.

The phosphor $LaPO_4{:}Ce^{3+}$, Tb^{3+} has assumed importance as the green component in trichromatic fluorescent lamps. The emission maximum is at 544 nm and the excitation maximum at 290 nm. It can be produced by solid-state reaction of La_2O_3, Tb_4O_7, $NH_4H_2PO_4$, and CeO_2 or CeF_4 in a reducing atmosphere at 1000–1250 °C. Although $LaPO_4{:}Ce^{3+}$, Tb^{3+} has a high quantum yield comparable with that of CAT, it is insufficiently stable to oxidation in air. Stabilization by large tetravalent cations (Th^{4+}, U^{4+}) [5.374] and by Sc^{3+} [5.375] has been claimed.

$Zn_3(PO_4)_2{:}Mn$, which has an emission maximum at 635 nm, is used as a cathode-ray tube phosphor for data displays under the TEPAC number P27 and WTDS specification RE. It is produced by solid-state reaction between ZnO, $MnCO_3$, and H_3PO_4 in an aqueous slurry. After drying, firing is carried out at 900–1150 °C for several hours in an open crucible [5.376].

$Cd_5Cl(PO_4)_3{:}Mn$ is a yellow-orange-emitting CRT-phosphor for data displays with the WTDS designation LA. It is produced by solid-state reaction at 750 °C after the components $CdSO_4$, CdO, $MnCl_2$ and $NH_4H_2PO_4$ have undergone preliminary reaction in an aqueous medium [5.377].

$3\,Sr_3(PO_4)_2 \cdot SrCl_2{:}Eu^{2+}$ can be excited by radiation from the entire UV range. The excitation maximum lies at 375 nm and the emission maximum at 447 nm. Upon successive substitution of Sr^{2+} by Ca^{2+} and Ba^{2+}, the emission maximum shifts to 450 nm. In the preparation, $SrCl_2$ is used in excess to improve the crystallization [5.378]. The phosphor was used as the blue component in trichromatic fluorescent lamps [5.379] and in the first generation of compact lamps.

Blue-green emitting $Ba_2P_2O_7{:}Ti$ ($\lambda_{max} = 493$ nm) is used as the blue component in mixtures with halophosphates for the adjustment of some cool-white light colors. This phosphor is prepared by firing $BaHPO_4$, TiO_2, and BaF_2 at 1050 °C [5.379]–[5.381].

Oxides. Industrially, the most important of this group is $Y_2O_3{:}Eu^{3+}$. The preparation is generally carried out by precipitating mixed oxalates from purified solutions of yttrium and europium nitrates. Firing the dried oxalates at ca. 600 °C is followed by crystallization firing at 1400–1500 °C for several hours [5.382].

Yttrium oxide and europium oxide form a series of mixed crystals without vacancies because of the small difference between their ionic radii [5.383], [5.384]. $Y_2O_3{:}Eu^{3+}$ shows an intense emission line at 611.5 nm in the red region, corresponding to the transition ${}^5D_0 \rightarrow {}^7F_2$ in europium. The luminescence of this red emission line increases with increasing Eu concentration up to ca. 10 mol%. The concentration in industrial lamp phosphors is ca. 3 mol%. Small traces of Tb enhance the Eu fluorescence of $Y_2O_3{:}Eu^{3+}$. Furthermore, this coactivation leads to a shift of the short-wavelength absorption edge of the host lattice from 220 (Y_2O_3) to 280 nm [5.385]. This allows excitation with short wavelength UV; excitation by electrons is also possible (Section 5.5.4.3). This phosphor for cathode-ray tubes is designated as RE (WTDS) and P56 (TEPAC). $Y_2O_3{:}Eu^{3+}$ shows no significant decrease in luminescence brightness up to 500 °C.

ZnO:Zn is a typical example of a self-activated phosphor. In the case of zinc oxide, it is an excess of zinc which enables the phosphor to luminesce. The production is carried out by thermal oxidation of crystallized zinc sulfide in air at ca. 400 °C. The green luminescence, with a broad maximum at 505 nm, has a very short decay time of 10^{-6} s. As a phosphor for cathode-ray tubes, ZnO:Zn is classified in the TEPAC list as P24 and in the WTDS system as GE.

The compound β-Ga_2O_3:Dy, which has photoluminescence properties but no commercial significance, is described in [5.386].

Arsenates. Magnesium arsenate $6MgO \cdot As_2O_5:Mn^{4+}$ is a very brilliant red phosphor used in fluorescent lamps. Because of its yellow color, it absorbs the blue mercury lines from which the strong red of the fluorescent lamp light results. The emission maximum is at 655 nm. The phosphor is best produced from solutions of $MgCl_2$ and $MnCl_2$ by precipitation with pyroarsenic acid. The dried precipitate is fired for at least 8 h at 900 °C. A second firing at 1100 °C for 5 h is carried out in the presence of boric acid to improve the crystallization [5.387].

Magnesium arsenate has lost its industrial importance because of the toxicity of arsenic.

Vanadates. Of the vanadates activated with rare earths, $YVO_4:Eu^{3+}$ [5.388]–[5.390] has found the greatest industrial application. Vanadates with other activators (YVO_4 with Tm, Tb, Ho, Er, Dy, Sm, or In; $GdVO_4$:Eu; $LuVO_4$:Eu) have attracted only scientific interest [5.391]–[5.393].

In $YVO_4:Eu^{3+}$, the europium emission lines at 619 and 614 nm, due to the transition ${}^5D_0 \rightarrow {}^7D_0$, are not quenched up to 300 °C. Although quenching occurs at higher temperatures, the main lines still retain 50% of their intensity at 500 °C. Small quantities of terbium activate the red Eu^{3+} lines at higher temperatures [5.394].

The long-wavelength absorption edge of $YVO_4:Eu^{3+}$ at 340 nm shifts to longer wavelength with rising temperature. The long-wavelength mercury lines of high-pressure mercury lamps can thus be used for excitation. The incorporation of Bi^{3+} sensitizes the Eu^{3+} emission [5.395] and results in a shift of the luminescence color towards orange, because the emission line at 565 nm, corresponding to the transition ${}^5D_0 \rightarrow {}^7F_1$, increases in intensity. YVO_4 forms a series of mixed crystals with the isostructural YPO_4. Successive substitution of VO_4^{3-} by PO_4^{3-} shifts the absorption edge of $Y(VO_4, PO_4)$ at 340 nm to shorter wavelengths and increases the temperature at which thermal quenching occurs.

For the production of YVO_4:Eu, the mixed oxides of yttrium and europium are thoroughly ground with an alkali-metal hydroxide as flux and high-purity NH_4VO_3. In a two-stage firing process with intermediate grinding, the mixture is fired at 600–650 °C and then at 1100–1200 °C for several hours. Excess alkali-metal vanadate is washed out with dilute sodium hydroxide solution. The effect of the production conditions on the particle size of $YVO_4:Eu^{3+}$ has been studied [5.396].

Niobates and Tantalates. The luminescence of rare-earth tantalates is described in [5.397]. Tantalates crystallize in three different structures, depending on the cation [5.398]. Yttrium tantalate and yttrium niobate–tantalate have good X-ray absorption and are used in X-ray intensifying screens because of their high conversion factor (Section 5.5.4.2.). Substitution of tantalum by small quantities of niobium considerably increases the blue fluorescence when excited by X rays.

The advantage of the X-ray excitable tantalates lies in the fact that the emission spectrum extends into the near UV. The films used with X-ray intensifying screens are particularly sensitive in this region.

The tantalates are produced by a solid-state reaction between Ta_2O_5, Nb_2O_5, and Y_2O_3 in a 14-hour preliminary firing at 1000 °C and a crystallization firing with lithium chloride or lithium sulfate as flux.

Sulfates. The basic lattice of sulfate phosphors absorbs very short wavelength UV radiation. On excitation with X rays or radiation from radioactive elements, a large proportion of the energy is stored in deep traps. For this reason, $CaSO_4$:Mn is used in solid-state dosimeters. Of the glowpeaks which can be selected by thermoluminescence, more than 50 % fail to appear at room temperature because of a self selection of the shallow traps. Other activators, such as lead or rare-earth ions (Dy^{3+}, Tm^{3+}, Sm^{3+}), stabilize the trapped electrons [5.399]–[5.401].

Photoluminescent sulfates are obtained by activation with ions that absorb short-wavelength radiation, for example, Ce^{3+}. Alkali-metal and alkaline-earth sulfates with Ce^{3+} emit between 300 and 400 nm. On additional manganese activation, the energy absorbed by Ce^{3+} is transferred to manganese with a shift of the emission into the green-to-red region.

Sulfate phosphors can be produced relatively easily. Water-insoluble sulfates are thoroughly mixed with the activators and fired in covered crucibles below the melting point. In the case of activation by Ce^{3+} and Mn^{2+} the activator concentration is at least 0.5 mol % [5.402].

Water-soluble sulfates are obtained by evaporating a solution of the mixed salts to dryness.

Tungstates and Molybdates. Magnesium tungstate $MgWO_4$ and calcium tungstate $CaWO_4$ are commercially the most important self-activated phosphors. Magnesium tungstate is often used as a reference material for UV-excited phosphors because of its high quantum yield of 84 % for the conversion of the 50–270-nm radiation into visible light; it is cited as standard phosphor 1027 for quantum output by the NBS [5.403]. The temperature dependence is relatively low; the maximum yield is at 240 K, but is still as high as 50 % at 400 K.

The emission maximum after excitation in a fluorescent lamp is at 482 nm with a half-intensity width of 140 nm. Magnesium tungstate crystallizes in the wolframite structure [5.404]. On additional activation with rare-earth ions their typical emission also occurs. Calcium tungstate crystallizes with the scheelite structure. Magnesium tungstate is produced by heating an excess of MgO or basic magnesium carbonate with high-purity WO_3 or H_2WO_4 at 1000–1100 °C. Excess MgO is washed out with 10 % hydrochloric acid.

Today, $MgWO_4$ is used in blends with other phosphors in high-voltage advertisement illumination tubes and as a fluorescent material in TLC plates. Because of its high effective absorption of X rays, $CaWO_4$ is used in large quantities for the production of X-ray intensifying screens in combination with blue-sensitive films. $CaWO_4$:Pb is still used as the blue phosphor in fluorescent lamps.

The production of self-activated $CaWO_4$ is more difficult than that of $MgWO_4$. Small deviations from stoichiometry in the composition have a marked effect on the efficiency. Also, traces of Pb^{2+} shift the emission peak from 415 nm to 439 nm (0.0006 mol % Pb^{2+}). To bind excess CaO in the production process, a small proportion of sulfate is used. The precipitated $CaWO_4$ is washed until chloride-free, dried, and subjected to crystallization firing at 800 °C with 30% flux. The formation of luminescent $CaWO_4$ from the precipitate is described in [5.405].

Self-activated molybdate phosphorus are described in [5.406]. The energy transfer that occurs in molybdates activated with Eu^{3+} is described in [5.388], [5.407].

5.5.3.4. Halide Phosphors

Alkali-Metal Halides. Luminescent alkali-metal halides can be produced easily in high-purity and as large single crystals. They are therefore often used as model substances for the investigation of luminescence processes. Their luminescence processes can be divided into: 1) the self-luminescence of the undoped crystals, 2) luminescence by lattice defects, and 3) sensitized luminescence.

Self-Luminescence. The action of UV light or ionizing radiation on pure alkali-metal halide crystals causes intense luminescence particularly at low temperature. The emission spectrum is characteristic for each individual compound. This fluorescence is comparable with the recombination luminescence which occurs upon capture of an electron by a V_K center (defect electron).

Luminescence of Lattice Defects. Many defect centers are known in the case of the alkali-metal halides, which are derived from electrons in anion vacancies (F-centers, or color centers). Association of two or more F-centers gives new defect centers, which can each also take up an electron. These lattice defects act as luminescence centers, the emission spectra of which sometimes exhibit a large number of lines.

At certain temperatures phosphorescence can also be observed, which indicates strong relaxation in the excited state.

Sensitized Luminescence. Through the incorporation of foreign ions (e.g., Tl^+, Ga^+, In^+) into the crystal lattice, further luminescence centers are formed. The emission spectra are characteristic for the individual foreign ions. The complicated luminescence mechanism is described in [5.408].

The alkali-metal halide phosphors are produced by firing the corresponding alkali-metal halide and the activator in platinum or fused-silica crucibles under an inert atmosphere.

Some industrially important alkali-metal halide phosphors are listed in Table 56. NaI:Tl and CsI:Tl are used as detectors for X and γ rays because of their high

Table 56. Alkali-metal halide phosphors

Host crystal	Activator		ϱ,g/cm^3	*mp*, °C	Emission peak, nm	Use*
	Type	Conc., mol%				
LiI	Eu	$5-6\times10^{-2}$	3.49	449	440	S
NaI	Tl	10^{-1}	3.67	651	410	S
CsI	Tl	10^{-1}	4.51	624	565–600	S
CsI	Na	$4-6\times10^{-4}$	4.51	624	430	S
LiF	Mg	$10^{-8}-10^{-2}$	2.63	842	400	TLD
LiF	Mg, Ti	$10^{-3}-10^{-2}$	2.63	842	400	TLD
LiF	Mg, Na	$10^{-3}-10^{-2}$	2.63	842	400	TLD

* S = scintillator (see Section 5.5.4.7); TLD = thermoluminescent dosimeter (see Section 5.5.4.7).

absorption capacity for these types of radiation. Europium-activated lithium iodide is preferred for neutron detection because of its favorable capture cross section. The phosphor CsI:Na is particularly important industrially as an X-ray converter in optoelectronic image intensifiers (Section 5.5.4.2). It has a high effective quantum absorption for X rays of energy 20–80 keV, which are used in medicine. Since the vapor pressure curves for CsI and NaI differ only slightly, it is possible to produce luminescent screens with good resolution by evaporation. The emission spectrum of CsI:Na also corresponds well to the spectral sensitivity of the photocathodes usually used for image intensifiers.

Alkaline-Earth Halides. Of the alkaline-earth halide phosphors, those doped with manganese or rare earths have been used industrially (Section 5.5.4.7) (e.g., CaF_2:Mn; CaF_2:Dy).

On excitation with high-energy radiation these compounds store part of the energy, which can subsequently be "called up" by stimulation. If the stimulation is achieved by heating, the resulting luminescence is termed thermoluminescence.

Thermoluminescent phosphors are used industrially in dosimeters. They are produced by coprecipitation of CaF_2 and an activator from a solution of the corresponding cations, or by the reaction of oxides or carbonates of the cations with hydrofluoric acid followed by firing. Moldings with high transparency for use in dosimeters can be produced by high-temperature pressing.

A further industrially important class of phosphors are the alkaline-earth fluorochlorides. The compound $Ba_{0.95}Eu_{0.05}FCl$ exhibits strong emission, with a peak at 375 nm, on X-ray excitation [5.409]. Substitution of Cl^- by Br^- shifts the peak to 385 nm. The Eu^{3+} acts as the luminescence center. The production is carried out by grinding BaF_2 with $BaCl_2$ or $BaBr_2$ in the molar ratio 1:1 with 0.04–0.07 mol% of a europium trihalide. After drying, the mixture is fired in a reducing atmosphere. The Ba^{2+} can be completely or partially substituted by Sr^{2+} or Ca^{2+}, and Cl^- by Br^- or I^-. Information on this and other production processes is given in [5.410]–[5.412]. These phosphors are used in X-ray intensifying screens (Section 5.5.4.2).

The BaFBr:Eu^{2+} phosphors are gaining increasing industrial importance because of their storage effect. On irradiation with X rays (or UV light) some of the Eu^{2+} ions are ionized to Eu^{3+} with loss of electrons to the conduction band. These electrons are captured by F^+ centers (anion vacancies) and can be liberated from these traps by the action of light (stimulation). The electrons return to the Eu^{3+} ions via the conduction band, converting them to Eu^{2+} [5.413]–[5.417].

The production of such storage phosphors is carried out by the method described for BaFCl:Eu^{2+}. These phosphors are used for storage panels in X-ray diagnostics (see Section 5.5.4.2).

Other Halide Phosphors. Information on the most important manganese-activated phosphors is summarized in Table 57.

These fluorides are produced by the following method for (Zn, Mg)F_2:Mn^{2+}. MgF_2 and $ZnSO_4$ are made into a paste in the desired molar ratio and a solution of the activator is added. After drying, the mixture is fired in quartz trays above 1000 °C.

Table 57. Manganese-activated halide phosphors

Composition	WTDS specification	Color location		Decay
		x	y	
$(Zn, Mg)F_2:Mn^{2+}$	LB	0.557	0.442	long
$KMg\ F_3:Mn^{2+}$	LC	0.573	0.426	very long
$MgF_2:Mn^{2+}$	LD	0.559	0.440	very long
$(Zn, Mg)F_2:Mn^{2+}$	LK	0.591	0.407	very long

These orange-emitting phosphors have a long afterglow time and are therefore still used in special radar tubes and oscilloscopes [5.296], [5.307], [5.418], despite their low stability towards burn-in compared with other cathode-ray phosphors.

Oxyhalides. The oxyhalides of yttrium, lanthanum, and gadolinium are good host lattices for activation with other rare-earth ions such as terbium, cerium, and thulium. The use of $LaOCl:Tb^{3+}$ as the green component in projection-television tubes has been discussed [5.419]. $LaOBr:Tb^{3+}$ and $LaOBr:Tm^{3+}$ exhibit high X-ray absorption, and they are used in X-ray intensifying screens [5.420].

The activator concentration (Tb, Tm) is 0.01–0.15 mol%. The phosphors are produced by mixing the corresponding oxides with an excess of ammonium bromide and subsequently firing at 400–500 °C. The reaction product is then subjected to crystallization firing with an alkali-metal bromide at 600–1100 °C [5.421]. The oxybromides are also formed by firing the oxides in a hydrogen bromide atmosphere. The afterglow, which causes problems in medical applications, can be reduced by coactivation with ytterbium. Partial substitution of lanthanum by gadolinium in $LaOBr:Ce^{3+}$ increases the quantum yield upon electron excitation and the quenching temperature [5.422].

5.5.4. Uses of Luminescent Pigments

5.5.4.1. Lighting

Most inorganic phosphors are used in mercury vapor lamps in which a plasma is produced from an inert gas and mercury vapor by electrical discharge. The plasma emits intense UV radiation, whose wavelength depends on the vapor pressure. The conversion of the UV radiation into visible light is effected by phosphors applied as a thin coating on the inside of the glass of the lamp. The numerous phosphors with various emission colors in industrial use today make it possible to produce practically any desired light color or wavelength in the spectrum by means of the UV radiation from the gas discharge.

In *low-pressure (fluorescent)* lamps the phosphor is excited by energy-rich mercury resonance lines at 254 and 185 nm. For many years the most important phosphors were the halophosphates. In the search for more efficient phosphors with greater radiation-stability, blends of $Y_2O_3:Eu^{3+}$, $Ce_{0.65}Tb_{0.35}MgAl_{11}O_{19}$, and $BaMg_2Al_{16}O_{27}:Eu^{2+}$ were used from the 1970s onwards.

Phosphors for *high-pressure mercury vapor lamps* have to fill up the red part of the spectrum of these lamps, which otherwise emit intense radiation in the blue and green regions. The lamp can reach an operating temperature of 150–300 °C. The phosphors must therefore have a very high quenching temperature. In addition they must be readily excitable in the long-wavelength UV. $3.5\ MgO \cdot 0.5\ MgF_2 \cdot GeO_2:Mn^{4+}$ and $YVO_4:Eu^{3+}$ have proved to be suitable.

Special applications of UV phosphors include fluorescent *UV lamps* for tanning (sun studios), photocopier lamps, and lamps for photochemistry, medicine, and varnish hardening. Examples are $Ce(Mg, Ba)Al_{11}O_{19}$ ($\lambda_{max.} = 344$ nm), $BaSi_2O_5:Pb^{2+}$ (351 nm), and $(Ba, Zn, Mg)_3\ Si_2O_7:Pb^{2+}$ (303 nm).

High-voltage tubes with cold cathodes are widely used in *illuminated advertisements*. The light colors can be adjusted by colored glass, the filling gas (neon gives red; helium, white-pink; argon–mercury, blue), and varied precisely by the phosphors (e.g., $CaWO_4:Sm^{3+}$, pink; $MgWO_4$, blue; $Zn_2SiO_4:Mn^{2+}$ green) [5.423], [5.424].

5.5.4.2. X-Ray Technology

Phosphors convert the invisible shadow image which is formed on penetration of an object by X rays into a visible image. They must have a high effective quantum absorption for the energy-rich X rays (20–60 keV). Thus, compounds of elements with high atomic numbers are principally used. The emission spectrum should correspond to the spectral sensitivity of the detector. A compromise between high brightness (coarse particles) and good image-transfer properties (fine particles) must be sought. The afterglow should be so low that on taking pictures of moving parts the image is not blurred. Different quality requirements exist for the various X-ray techniques, the phosphors are therefore adapted to suit the particular application.

Fluoroscopic screens allow direct examination of X-ray images by the eye. (Zn, Cd)S:Ag is used almost exclusively (molar ratio of ZnS:CdS, 60:40; particle size 20–40 μm). The emission spectrum corresponds well to the spectral sensitivity of the eye.

The fluoroscopic screen consists of a substrate (cardboard, plastic sheet), which is first covered with a reflective layer of MgO or TiO_2. The phosphor layer is applied by pouring, after dispersion in a binding agent based on acetylcellulose.

Intensifying screens for X-ray photons are used in medical diagnosis for taking X-ray photographs. The X rays are first converted into light which then irradiates the film. In most cases, two intensifying screens are used, the phosphor screen placed in front of the film (front screen) very often having a thinner coating than the rear screen. The most important phosphors are $CaWO_4$, $Y_2O_2S:Tb^{3+}$, $Gd_2O_2S:Tb^{3+}BaFCl:Eu^{2+}$, $LaOBr:T^{3+}$, and $YTaO_4:Nb^{5+}$ [5.398], [5.425].

Intensifying screens consist of cardboard or plastic sheets (e.g., polyester) as substrates. A reflective layer (e.g., TiO_2) is first applied to this and then the phosphor–binder layer.

Storage panels are similar to intensifying screens. However, they differ in that the phosphor, on excitation with X rays, stores part of the absorbed energy in addition to producing spontaneous emission. The stored energy can later be read out by stimulation with light of a suitable wavelength, for example, with a finely focused laser beam. The possibility thus exists for computer digital radiography, in which information stored simultaneously (in parallel) is then read out serially [5.426].

Storage phosphors include $Y_2O_2S{:}Eu^{3+}$, Bi^{3+}; $GdOBr{:}Sm^{3+}$; $LaOBr{:}Tb^{3+}$, Bi^{3+}; and the barium fluorohalides.

In the optoelectronic X-ray image intensifier (Fig. 86), [5.427], the X-ray phosphor screen (input screen) is in direct optical contact with a photocathode that converts the luminance distribution of the X-ray screen into an electron-density distribution. The liberated electrons are accelerated in an electric field between the photocathode and an anode (20–30 kV) and are focused by electron lenses onto a second phosphor screen (output screen), where conversion of the electron image to a visible image takes place.

The two phosphor screens differ in their construction. Since ca. 1970, the input screens have mostly been made by condensing $CsI{:}Na^+$ onto thin aluminum sheets in vacuum.

The output screen consists of a glass base material (sometimes also fiber optics), onto which fine particles of $(Zn, Cd)S{:}Ag^+$ or $Gd_2O_2S\ Tb^{3+}$ are applied, usually by sedimentation, but also by brushing or spraying. Phosphors with an average particle size of 1.5–3 μm and a narrow particle size distribution are used. The thickness of the layer is ca. 5–10 μm.

In *computer tomography* the patient to be examined lies within a circle of detectors (up to 1200) [5.428]. A continually irradiating X-ray tube rotates around the patient. The detectors measure the intensity of the X rays which have passed through a particular layer of the body. They consist of a phosphor crystal ($CdWO_4$ and $Bi_4Ge_3O_{12}$, λ_{max} 480 nm) which is combined with a photodiode.

5.5.4.3. Cathode-Ray Tubes

Phosphors are an important, quality-determining component of cathode-ray tubes (Fig. 87). In the phosphor screen, modulated electrons are converted into a visible image.

Many methods exist for the production of phosphor screens [5.296]. For monochromatic tubes a suspension of the phosphor in alcohol or water, with addition of water glass solution, is placed in the tube and is deposited by sedimentation on the inner surface of the faceplate. The adhesion is effected by the slowly precipitating silicic acid.

In the coating of color picture tubes, the three components red, green, and blue are applied consecutively as dispersions in a photosensitive resin. This is then exposed to light via a shadow mask so that the resin on the exposed areas becomes water-insoluble. The phosphor on the unexposed areas is washed off with water.

In most cases the phosphor layer is then subjected to an aluminum vapor deposition process. This increases the brightness by reflection of the backward-radiating luminescence and inhibits charge buildup in the phosphor layer.

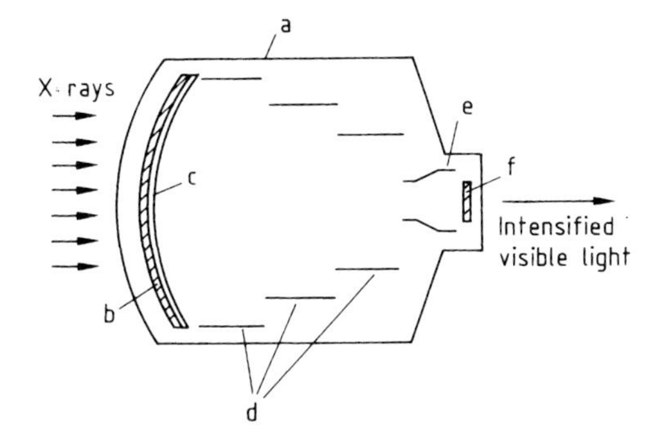

Figure 86. X-ray image intensifier
a) Glass envelope; b) Input conversion screen; c) Photocathode; d) Focusing electrodes; e) Anode; f) Output screen

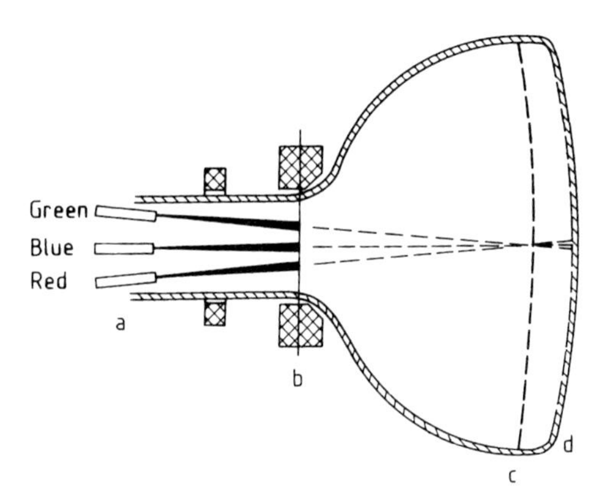

Figure 87. Schematic of a cathode-ray tube
a) Electron guns; b) Optoelectronic system; c) Shadow mask; d) Phosphor screen

Cathode-ray tubes have a very large number of applications which require varying phosphor qualities. Table 58 lists the most important cathode-ray luminescent pigments.

The Worldwide Phosphor Type Designation System (WTDS) the optical data of the specifies phosphors for cathode-ray tubes [5.429]. It replaces several phosphor designation systems (e.g., TEPAC, Pro Electron) previously in use in various countries. The phosphors are characterized by two capital letters, the first giving the position of the emission color in the Kelly Charts of Color Designation:

A: purple-red, purple, violet
B: blue, greenish blue
G: blue-green, green, yellow-green
K: yellow-green
L: orange, yellow-orange
R: red-orange, red, purple-red, pink, purple-pink
W: white
Y: green-yellow, yellow, orange-yellow

The second letter is assigned to a particular product.

The brightness in *television tubes* should generally be as high as possible, the decay time must be adapted to the image frequency to decrease flicker and poor image definition. The white luminescence required for black and white televisions is

Table 58. Phosphors for cathode-ray tubes

Composition	WTDS specification	Emission color	CIE color coordinates		Decay time
			x	y	
Black-and-white television					
$ZnS:Ag^+ + (Zn, Cd)S:Ag^+$	WW	white	0.265	0.285	medium short
$ZnS:Ag^+ + (Zn, Cd)S:Cu^+$	WW	white	0.275	0.311	medium short
Projection tubes					
$Zn_2SiO_4:Mn^{2+}$	GJ	green	0.207	0.713	medium
$(Ca, Mg)SiO_3:Ti^{4+}$	BF	blue	0.164	0.108	medium
$Y_2O_3:Eu^{3+}$	RF	red	0.640	0.335	medium
Color television					
$ZnS:Cu^+, Au^+, Al$	X	green	0.310	0.598	medium
$ZnS:Cu^+$	X	green	0.290	0.611	medium
$ZnS:Ag^+$	X	blue	0.136	0.070	medium
$Y_2O_2:Eu^{3+}$	RF	red	0.640	0.335	medium
$Y_2O_2S:Eu^{3+}$	X	red	0.630	0.340	medium
Monitor tubes					
$ZnS:Cu^+ + Zn_2SiO_4:Mn^{2+}, As^{3+}$	GW	yellow-green	0.238	0.568	medium to long
$(Zn, Mg)F_2:Mn^{2+} + (Ca, Mg)SiO_3:Ce^{3+}$	YD	yellow-orange	0.541	0.456	long
$ZnS:Ag^+ + (Zn, Cd)S:Cu^+$	YC	white	0.330	0.330	medium to long
$(Zn, Cd)S:Cu^+$	KC	yellow	0.423	0.553	medium to long
$Y_2Si_2O_5:Ce^{3+}, Tb^{3+}$		white			very short
Penetration tubes					
$(Zn, Cd)S:Ag^+, Ni^{2+} +$	VB	green (15 kV)	0.398	0.546	medium short
$Y_2O_3:Eu^{3+}$		red (8 kV)	0.655	0.340	medium
$(Zn, Cd)S:Ag^+,Ni^{2+} +$	VC	green (12 kV)	0.414	0.515	medium short
$Y_2O_3:Eu^{3+}$		red (6 kV)	0.672	0.325	medium
$YVO_4:Eu^{3+} +$	VE	yellow-green (17 kV)	0.355	0.510	very long
$ZnS:Cu^+$		red-orange (8 kV)	0.604	0.360	medium
Oscilloscope tubes					
$ZnS:Ag^+$	BE	blue	0.136	0.148	medium
$ZnS:Cu^+$	GH	green	0.226	0.528	medium
$Zn_2SiO_4:Mn^{2+}$	GJ	yellow-green	0.218	0.712	medium
$ZnS:Cu^+$	GL	yellow-green	0.279	0.536	medium short

Table 58. (Continued)

Composition	WTDS specification	Emission color	CIE color coordinates		Decay time
			x	y	
$(Ca,Mg)SiO_3:Ce^{3+}$	GJ	yellow-green	0.280	0.530	medium
$Zn_2SiO_4:Mn^{2+}, As^{3+}$	GR	yellow-green	0.223	0.668	medium short
$Gd_2O_2S:Tb^{3+}$	GY	yellow-green	0.333	0.556	medium
$Y_2O_2S:Tb^{3+}$	WB	white	0.253	0.312	medium
$KMgF_3:Mn^{2+}$	LC	orange	0.573	0.426	very long
$ZnS:Ag^+$	GB	blue	0.151	0.032	medium short
$(Zn, Cd)S:Cu^+$	GB	yellow-green	0.357	0.537	very long
$(Zn, Cd)S:Cu^+$	KB	fluorescence, blue-green phosphorescence, green	0.380	0.520	very long
Radar tubes					
$(Zn, Mg)F_2:Mn^{2+}$	LK	orange	0.591	0.407	very long
$(Ca, Mg)SiO_3:Ce^{3+}$	GV	yellow-green	0.280	0.530	very long
$ZnS:Cu^+$	GF	green	0.240	0.520	long
$Zn_2SiO_4:Mn^{2+}, As^{3+}$	GR	yellow-green	0.223	0.698	long

achieved with blends of $ZnS:Ag^+$ and $(Zn, Cd)S:Cu^+$, which give complementary blue and yellow emission. One-component phosphors which emit white light due to double activation have not been successful [5.333]–[5.334].

In color picture tubes, the final color is formed by additive mixing of red, green, and blue light. Each of the three phosphors (Table 58) is excited by its own modulated electron beam and emits light of intensity corresponding to the particular degree of excitation. The phosphors for color picture tubes have an average particle size of 8–10 μm.

Because of the high resolution required for *monitor tubes*, phosphors with smaller particle sizes (4–6 μm) than for entertainment tubes are often needed. For monitor tubes which reproduce slow movement only, phosphor mixtures with longer decay times are used to diminish image flickering. For the reproduction of faster movement, phosphors with shorter decay times are used. For monochrome monitors with amber as the image color $Cd_5(PO_4)_3Cl:Mn^{2+}$ or a blend of $Y_2O_2S:Eu^{3+}$ and $(Zn, Cd)S:Cu^+$ is used.

The majority of phosphors for cathode-ray tubes are coated with an oxide, silicate, or phosphate before use to improve their processing properties and stability to burn-in. Pigmenting of $Y_2O_2S:Eu^{3+}$ with finely divided Fe_2O_3 and of $ZnS:Ag^+$ with ultramarine or cobalt aluminate is also known. The pigment, which has the same body-color as the emission color of the phosphor, absorbs incident ambient light, effecting an increase in contrast [5.430].

A special case of pigmenting is the coating of one phosphor with another. This type of phosphor (onion-skin) is used in *penetration tubes*. At lower acceleration voltages the luminescence of the outer phosphor dominates, and at higher voltages, that of the inner phosphor. In addition, phosphor blends are used, the components of which exhibit differing brightness–voltage dependencies. By using phosphors with differing decay times or emission colors, monitor tubes with differing afterglow times or colors (multicolor penetration CRT) can be obtained.

Oscilloscopes are electrical measuring instruments for the display and monitoring of rapidly changing electrical processes, for example, changes in voltage and current with time. (Fig. 88). A thermionic cathode emits an electron beam which is directed by the modulator and passes through two pairs of deflector plates placed at right angles to each other and then excites a phosphor screen. For visual examination, green-emitting products are used, and for photographic examination, blue-emitting ones. The phosphor screens are mostly produced by sedimentation, but centrifugal process and cataphoresis are also used.

In construction, a *radar tube* corresponds to the oscilloscope tube. It displays radar echos, i.e., the direction of and the transit time to and from a reflecting object.

Since the pulse duration is of the order of microseconds, the response time of the phosphor must be correspondingly short. On the other hand, the afterglow of the screen should be long enough for the radar image to be regenerated after one rotation of the radar antenna (of the order of seconds). These requirements are achieved by means of a combination of phosphors. In the cascade screen, two superimposed phosphor layers are used. The first consists of a blue, cathodoluminescent, fast-responding phosphor (e.g., $ZnS:Ag^+$). The blue light of this phosphor excites the photoluminescent phosphor with a yellow afterglow in the second layer [e.g., $(Zn, Cd)\,S:Cu^+$].

In penetration tubes, two phosphor layers, separated by a barrier layer, are applied to the phosphor screen. At lower acceleration voltages, only the first phosphor is excited. At higher voltages both phosphors luminesce. Since most electrons are then absorbed in the second layer, this determines the color and afterglow time. It is thus possible with only one electron beam to display simultaneously the inscription in one color with short afterglow and the radar signal in another with long afterglow by changing the voltage, (e.g., phosphor VE in Table 58).

In flying-spot tubes a rapidly moving spot of light, obtained by focusing an electron beam onto the phosphor screen of a cathode-ray tube, scans a photo or a film. The transmitted or reflected light is then converted into signals in a photodiode. A short decay time (< 100 ns) and high spot brightness are important. In monochrome scanning $(Zn, Cd)S:Ag^+, Ni^{2+}$ is used. In color scanning, blends of $Y_3Al_5O_{12}:Ce^{3+}$ with $Ca_2Al_2SiO_7:Ce^{3+}$ or $SrGa_2S_4:Ce^{3+}$ with $SrGa_2S_4:Pb^{2+}$ can be employed.

The *optoelectronic image converter* converts an image invisible to the human eye first into an electron image and the latter into a visible one on a phosphor screen [5.431]. It extends human vision into UV, IR, and X-ray regions. The phosphor

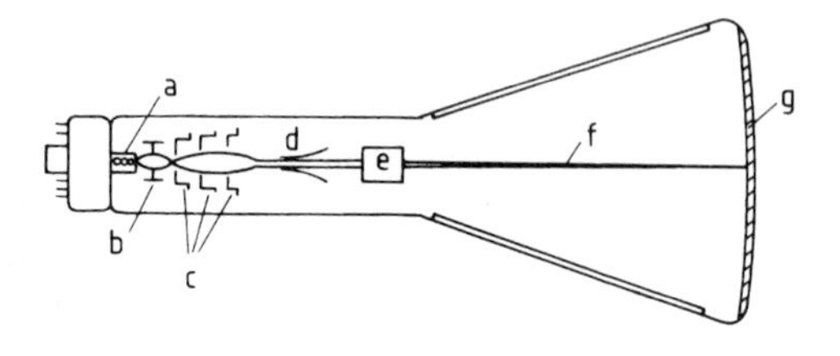

Figure 88. Oscilloscope tube
a) Cathode; b) Modulator; c) Electron lens; d) Y-plates; e) X-plates; f) Electron beam; g) Phosphor screen

should have as high a quantum yield as possible on excitation with 6–20 keV electrons. The requirements for high resolution and good contrast lead to special requirements for the particle size and particle-size distribution of the phosphor. The average particle size should not exceed 4 μm. Some important phosphors are (Zn, Cd) S:Ag^+, ZnS:Cu^+; Gd_2O_2S:Tb^{3+}; and Cd S:Cu^+.

5.5.4.4. Product Coding

Phosphors used for coding mass-produced goods serve to distinguish between particular types or for the control of production batches. Coding of trade marked articles allows rapid recognition of imitations in the fight against product piracy. The phosphor is applied by means of transparent or colorless printing inks in the form of bar codes, numbers, or company emblems. Fluorescent fibers for packaging paper or fluorescent stick-on labels are further forms of coding.

The readily excitable fluorescence or phosphorescence of phosphors can also be made use of in the automation, control, and regulation of automatic sorting processes, in packaging, or in quality control.

Since the 1960s, stamp paper has been treated with fluorescent or phosphorescent phosphors. Since the stamp is stuck to the top right-hand corner of the letter, its recognition by the machine serves to establish the position of the letter and enables machine control of the correct positioning for postmarking and reading. Doped zinc sulfides with average particle sizes of 5–6 μm are used as afterglow phosphors.

The coding of cheques, cheque cards, passports, personal identity cards, securities, gift vouchers, entrance tickets, season tickets, and banknotes with phosphors as invisible security markings is currently widely used. The phosphor acts as a passive protection against forgery and copying. The original, with phosphor, fluoresces under UV light, the copy does not. Phosphors with white body color are generally used.

5.5.4.5. Safety and Accident Prevention

Inorganic luminescent pigments with long afterglows have long been used to make danger areas, escape routes, and switches visible.

For this application, phosphors must have the following properties:

1) Excitability by daylight and artificial light
2) High luminance during the first hour after excitation
3) Long afterglow time, i.e., decrease in luminance to 100 times that still detectable by the human eye (ca. 0.3 mcd/m^2) only after several hours (DIN 67510)
4) An emission spectrum corresponding to the sensitivity curve of the eye
5) High chemical stability

Table 59 lists commercial industrial phosphors. Copper-activated zinc sulfides are the most widely used phosphors for safety purposes.

These zinc sulfides are sufficiently resistant to all media except strong mineral acids. A disadvantage is that they become grey through the action of high intensity

Table 59. Properties of industrially important long-afterglow phosphors for accident prevention

Designation and luminescent color	Composition	Body color	Average particle size (FSSS)	Sieve analysis mesh size, μm	Sieve analysis residue, wt%	Typical luminance, mcd/m²				100-fold visibility limit 0.32 mcd/m² (min.)
						3 min	10 min	30 min	60 min	
LUMILUX green N	ZnS:Cu	yellow-green	40	125	≤1	128	30.4	7.9	3.3	370
LUMILUX green N-ST	ZnS:Cu	yellow-green	45	125	≤2	75	19.0	3.4	1.0	105
LUMILUX green N5	ZnS:Cu, Co	yellow-green	40	125	≤1	123	32.2	9.5	4.4	580
LUMILUX green N10	ZnS:Cu, Co	pale yellow-green	40	125	≤1	80	28.0	10.3	5.2	780
LUMILUX afterglow blend, green	ZnS:Cu, Co	yellow-green	40	125	≤1	126	31.3	8.9	3.9	460
LUMILUX green N-F	ZnS:Cu	yellow-green	25	100	≤0.2	135	29.7	7.2	2.9	310
LUMILUX green N-FF	ZnS:Cu	yellow-green	18	63	≤1	100	24.0	6.2	2.4	260
LUMILUX yellow N	(Zn, Cd)S:Cu	green-yellow	30	125	≤1	93.4	22.0	5.7	2.3	250
LUMILUX orange N	(Zn, Cd)S:Cu	yellow	30	125	≤1	65.0	15.7	4.2	1.8	220
LUMILUX red N100	CaS:Eu, Tm	pink	20	100	≤1	22	9	2	0.9	100
LUMILUX sea green	SrS:Bi	pale green	20	125	≤2	58.9	16.9	5.4	2.6	300
LUMILUX violet N	CaS:Bi	grey	20	125	≤2	15.3	4.1	1.8	0.64	105
LUMILUX blue N	(Sr, Ca)S:Bi	grey	20	125	≤2	31.5	8.9	2.9	1.3	190

UV light in the presence of water, due to photochemical reduction of Zn^{2+} to Zn^0. Outdoor application of these phosphors is limited. The phosphors are used in afterglow paints, plastics, ceramics, and enamels.

The following basic rules should be observed in processing luminescent pigments:

1) The binder system must be colorless and transparent
2) Additives based on heavy metals should not be used especially at high processing temperatures
3) Processes involving the action of large shearing forces on the pigments should be avoided (pressure-induced deterioration of luminescent pigments)

Markings for switches and escape routes, advice and warning signs, and special protective work clothes are produced increasingly from these afterglow phosphors. A low-light escape route system based on afterglowing pigments remains noticeable even in smoke (Fig. 89) [5.432]–[5.434].

5.5.4.6. Dentistry

Natural teeth exhibit blue-white fluorescence in the long-wavelength UV [5.435]–[5.437]. Luminescent pigments are used to imitate this phenomena in artificial teeth. They are added to the ceramic paste at a concentration of 0.3–0.5 wt%. Yttrium silicates doped with cerium, terbium, and manganese give the best results [5.438]. The excitation maximum of these phosphors is in the range 325–370 nm. The fluorescence color of the teeth can be varied by changing the concentration of activators.

5.5.4.7. Other Uses

Green-fluorescing $Zn_2SiO_4{:}Mn^{2+}$ and magnesium tungstate that fluoresces blue under short-wavelength UV are mixed with the sorbents (silica gel, aluminum oxide) of *TLC plates* in proportions of ca. 3 wt% to improve the contrast in the detection of UV-absorbing compounds.

Infrared radiation can be detected by phosphors by various mechanisms. Upon excitation with UV radiation, $ZnS{:}Cu^+, Co^{2+}$ emits an afterglow that is quenched by IR radiation, whereas $ZnS{:}Cu^+, Pb^{2+}$ stores excitation energy and emits it spontaneously under the action of IR radiation. $CaS{:}Sm^{3+}, Ce^{3+}$ and $SrS{:}Sm^{3+}, Eu^{2+}$ only exhibit low fluorescence on excitation by UV radiation and daylight. Through the subsequent action of IR radiation these phosphors can be stimulated to emit green or orange-red light.

Infrared-stimulable phosphors are used as screens for the detection of IR radiation in IR measuring and testing techniques and for the adjustment of IR lasers, and IR light recorders.

Scintillation counters are used to detect energy-rich particles and γ rays. They consist of a combination of a phosphor, usually a single crystal, with a photomultiplier or a photodiode as detector. Alkali-metal iodides (see Table 56), $Bi_4Ge_3O_{12}$, $CaWO_4$, $ZnWO_4$, $CdWO_4$, and $ZnS{:}Ag^+, Ni^+$ are typical phosphors for this application [5.438].

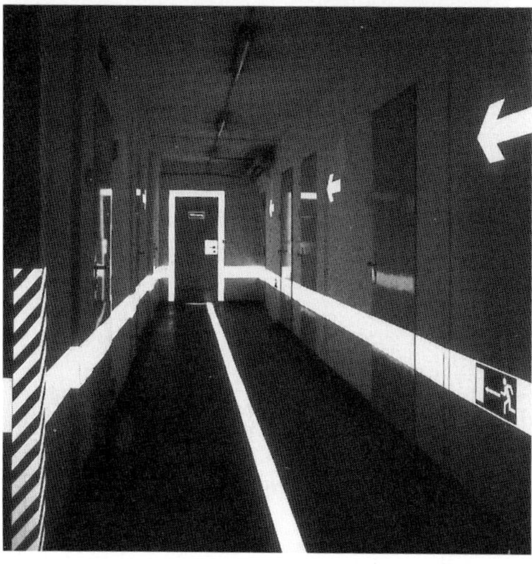

Figure 89. Low-light-level escape route system based on afterglowing LUMILUX[(R)] Green N phosphor

Thermoluminescence dosimeters (TLD) absorb ionizing radiation and accumulate the energy absorbed over a very long period. Heating the TLD results in emission of light proportional to the absorbed energy. $Li_2B_4O_7{:}Mn^{2+}$, LiF (see Table 56), $CaSO_4{:}Dy^{3+}$, and $CaF_2{:}Mn^{2+}$ are used for this application the phosphor is encapsulated in a glass ampoule or is sealed in a heat-resistant synthetic resin.

Phosphors for radio-activation are used as mixtures with a β-radioactive compound (e.g., ^{3}H, ^{14}C, or ^{147}Pm compounds) whose radiation effects continuous luminescence of the phosphor. Tritium-labeled organic binders are used for the production of luminous products and paints. Because of the high cost of tritium, the phosphors must have a high efficiency.

Another radioactive luminescent system is the β-light, which consists of a gas-tight glass vessel that contains gaseous tritium and whose walls are coated with a fine-grained phosphor. The β-lights are mainly used in military applications for the illumination of instruments and scales. Clock faces and hands, escape route signs, and emergency exits are also made visible in darkness through the use of β-lights.

Because of health risks, the use of radioactive substances is declining sharply and they are being replaced by long-afterglow products.

Fluorescent and afterglow products are used to attain special effects in cinema films, on the stage, and in works of art. These *effect pigments* are physiologically harmless phosphors, mainly based on zinc sulfides.

5.5.5. Testing of Industrial Phosphors

Luminescence color is determined by visual comparison, by calculation of the color coordinates x and y from the emission spectrum, or by means of a tristimulus colorimeter under the excitation definitive for the application (UV radiation, cathode rays, etc.).

Luminescence Brightness. A practical brightness measurement for UV-excitable phosphors is the comparison with internal reference phosphors under excitation mainly with short- and long-wavelength (254 nm and 365 nm) UV lamps. The visible parts of the radiation from the mercury discharge must be removed with filters. Integral brightness values in mcd/m^2 are obtained with commercial brightness measuring instruments.

Values of the quantum yield are obtained by comparison against standard phosphors of the NBS. The dependence of the quantum output on the excitation wavelength are obtained by comparison with phosphors whose quantum output is only slightly dependent on the excitation wavelength. Such phosphors are, for the blue emission region, sodium salicylate, for the blue-green, coronene ($C_{24}H_{12}$), and for the orange to orange-red, Lumogen Red T (BASF). Measurements of quantum yield are rarely carried out on industrial phosphors. The absolute measurement of the quantum yields is described in detail in [5.402], [5.439].

For long-afterglow phosphors, the brightness as a function of time after standardized excitation is measured (DIN 67 510).

Phosphors for cathode-ray tubes, television screens, monitor screens, radar screens, and oscilloscopes are tested under electron excitation. Electron energy and density should be similar to the conditions of the tube in which the screen will be used. The phosphors are sedimented or brushed onto light-permeable screens and coated with an evaporated aluminum coating to dissipate charge. The luminescence brightness and color of the emitted light are measured with optical instruments such as photomultipliers or spectrophotometers.

For X-ray excitable phosphors, the intensifier effect is tested as a measure of the luminescence yield. A very simple process uses the exposure of blue- or green-sensitive films. The phosphor is excited by X rays in a cell covered with a light-sensitive film. The blackening of the film is measured with a densitometer and, for given excitation conditions, is a measure of the efficiency of the phosphor.

Decay. The decay time requirements must be adhered to very precisely for cathode-ray tube phosphors. The measuring devices consist of fast excitation sources (flash lamps, lasers), photomultipliers with very low time constants, and an oscilloscope [5.440].

Sedimentation Behavior. The coating of cathode-ray tubes is usually carried out by sedimentation. The sedimentation characteristics of CRT phosphors and their blends under typical application conditions are an important criterion for their suitability. The sedimentation density allows conclusions to be made on the degree of agglomeration of individual crystals.

6. References

References for Chapter 1

[1.1] *Colour Index International:* Society of Dyers and Colourists, 3rd ed., vol. 5, Bradford 1987.

[1.2] *Kirk-Othmer Encyclopedia of Chemical Technology*, 3rd ed., **17**, J. Wiley & Sons, New York 1978–1984, pp. 788–838. H. G. Völz in H. Kittel (ed.): *Lehrbuch der Lacke und Beschichtungen*, vol. II, Verlag W. A. Colomb, Berlin 1974, pp. 27–47.

[1.3] H. Rechmann, G. Sutter, *Farbe + Lack* **82** (1976) 793–796.

[1.4] L. Gall, *Farbe + Lack* **81** (1975) 1015–1024.

[1.5] E. Fritzsche, *Farbe + Lack* **80** (1974) 731–740.

[1.6] H. G. Völz, *DEFAZET – Dtsch. Farben Z.* **28** (1974) 559–563.

[1.7] F. Hund, *Farbe + Lack* **73** (1967) 111–120.

[1.8] G. Kämpf, *Farbe + Lack* **71** (1965) 353–365.

[1.9] B. Honigmann, J. Stabenow in: *Congr. FATIPEC VI*, 1962, Verlag Chemie, Weinheim 1963, pp. 89–92.

[1.10] H. G. Völz, *Farbe + Lack* **90** (1984) 642, 752.

[1.11] F. Kindervater in: *Congr. FATIPEC XII*, 1974, Verlag Chemie, Weinheim 1974, pp. 399–403.

[1.12] S. Keifer, A. Wingen, *Farbe + Lack* **79** (1973) 886–873.

[1.13] D. B. Judd: *Color in Business, Science and Industry*, J. Wiley & Sons, New York 1969.

[1.14] H. G. Völz, *Progr. Org. Coat.* **1** (1973) 1.

[1.15] H. G. Völz, *Angew. Chem., Int. Ed. Eng.* **14** (1975) 655.

[1.16] H. G. Völz: *Industrielle Farbprüfung*, VCH Verlagsgesellschaft, Weinheim 1990.

[1.17] A. Berger, A. Brockes: *Bayer-Farben-Revue*, special issue 3/1, Leverkusen 1971.

[1.18] F. W. Bittmeyer Jr, in: *Kirk-Othmer*, 3rd ed., **6**, p. 523 ff.

[1.19] M. Richter: *Einführung in die Farbmetrik*, Walter de Gruyter, Berlin-New York 1976.

[1.20] P. Kubelka, F. Munk, *Z. Tech. Phys.* **12** (1931) 539–601.

[1.21] P. Kubelka, *J. Opt. Soc. Am.* **38** (1948) 448–457.

[1.22] L. Gall, *Farbe + Lack* **80** (1974) 297–306.

[1.23] K. Hoffmann, *Farbe + Lack* **80** (1974) 118–125, 221–229.

[1.24] F. W. Billmeyer, Jr., R. L. Abrams, D. G. Phillips, *J. Paint Technol.* **45** (1973) 23–30, 31–38; **46** (1974) 36–39; **48** (1976) 30–36.

[1.25] H. G. Völz, *Ber. Bunsen-Ges. Phys. Chem.* **71** (1967) 326–339.

[1.26] G. Mie, *Ann. Phys.* **25** (1908) 377–444.

[1.27] H. C. van de Hulst: *Light Scattering by Small Particles*, J. Wiley & Sons, New York 1957.

[1.28] W. Jaenicke, *Z. Elektrochem.* **60** (1956) 163–174.

[1.29] H. G. Völz, *DEFAZET – Dtsch. Farben Z.* **31** (1977) 469–479.

[1.30] A. Brockes, *Optik (Stuttgart)* **21** (1964) 550–566.

[1.31] R. Brossmann in: *Internationale Farbtagung "Color 69"*, vol. II, Verlag Musterschmidt, Göttingen 1969, pp. 890–900.

[1.31 a] *Landolt-Börnstein*, part 8. H. Wienand, W. Ostertag, *Bull. Bismuth Inst. (Brussels)* (1988) no. 5.

[1.32] H. G. Völz, *Farbe + Lack* **76** (1970) 9–14. S. Keifer, *Farbe + Lack* **81** (1975) 289–294.

[1.33] H. G. Völz, *Farbe + Lack* **88** (1982) 264, 443.

[1.34] W. Schumacher in: *Congr. FATIPEC XIII*. 1976, EREC, Paris 1977, pp. 582–587.

[1.35] F. Finus, *DEFAZET – Dtsch. Farben Z.* **28** (1974) 494–500.
[1.36] Commission International de l'Eclairage, *CIE Technical Report* **116**, CIE Zentralbüro, Wien 1995.
[1.37] S. Keifer, H. G. Völz, *Farbe + Lack* **83** (1977) 180–185, 278–285.
[1.38] H. G. Völz, *Progr. Org. Coat.* **15** (1987) 99.
[1.39] L. Gall in: *Congr. FATIPEC IX,* 1968, Chimie des Peintures, Brussels 1969, pp. 34–39.
[1.40] S. Keifer, *Farbe + Lack* **82** (1976) 811–814.
[1.41] F. Vial in: *Congr. FATIPEC XII.* 1974, Verlag Chemie, Weinheim 1975, pp. 159–167.
[1.42] L. Gall, *Farbe + Lack* **72** (1966) 955–965.
[1.43] L. Gall, *Farbe + Lack* **72** (1966) 1073–1079.
[1.44] H. G. Völz, *DEFAZET – Dtsch. Farben Z.* **30** (1976) 392–396, 454–457.
[1.45] S. Keifer: "Deckvermögen," in: *Anorganische Pigmente für Anstrich- und Beschichtungsstoffe,* Bayer, Leverkusen 1973, Section 3.3.2.
[1.46] H. G. Völz in H. Kittel (ed.): *Lehrbuch der Lacke und Beschichtungen*, vol. II, Verlag W. A. Colomb, Berlin 1974, pp. 22–27.
[1.47] K. Brugger, N. Jehle in: *Congr. FATIPEC XIII.* 1976, EREC, Paris 1977, pp. 15–23.
[1.48] R. Epple, A. Englisch in: *Congr. FATIPEC XII.* 1974, Verlag Chemie, Weinheim 1975, pp. 241–248.
[1.49] G. Kämpf, W. Papenroth, *Farbe + Lack* **83** (1977) 18–29.
[1.50] W. Grassmann, *Farbe + Lack* **66** (1960) 67–74.
[1.51] H. G. Völz, G. Kämpf, H. G. Fitzky, A. Klaeren, 179th National Meeting, American Chemical Society, Houston, Texas, *ACS Symp. Ser.* **151** (1981) 163.
[1.52] H. G. Völz, G. Kämpf, H. G. Fitzky, *Farbe + Lack* **78** (1972) 1037–1049.
[1.53] Th. Bohmann in: *Congr. FATIPEC X* 1970, Verlag Chemie, Weinheim 1971, pp. 123–128.
[1.54] W. Papenroth, P. Koxholt, *Farbe + Lack* **82** (1976) 1011–1016.
[1.55] H. J. Freier, *Farbe + Lack* **72** (1967) 1127–1136.
[1.56] H. Rechmann, et al., in: *Congr. FATIPEC X;* 1970, Verlag Chemie, Weinheim 1971, pp. 3–15.
[1.57] F. Sadowski, *Farbe + Lack* **80** (1974) 29–33.
[1.58] S. Keifer in: *Congr. FATIPEC XII,* 1974, Verlag Chemie, Weinheim 1975, pp. 517–522.
[1.59] P. Kresse in: *Congr. FATIPEC XIII,* 1976, EREC, Paris 1977, pp. 346–353.
[1.60] L. Gall, *DEFAZET – Dtsch. Farben Z.* **30** (1976) 127–128.
[1.61] F. Finus, W. Fleck, *DEFAZET – Dtsch. Farben Z.* **28** (1974) 153–158.
[1.62] G. Lincke: *Einfärben Kunststoffe,* VDI-Verlag, Düsseldorf 1975, pp. 135–142.
[1.63] D. Bialas in H. Kittel (ed.): *Lehrbuch der Lacke und Beschichtungen,* vol. II, Verlag W. A. Colomb, Berlin 1974, pp. 486–514.
[1.64] G. Kämpf, *DEFAZET – Dtsch. Farben Z.* **31** (1977) 347–353.
[1.65] G. R. Joppien, *Farbe + Lack* **81** (1975) 1102–1109.
[1.66] G. Marwedel, *Farbe + Lack* **82** (1976) 789–792.
[1.67] E. Herrmann, *DEFAZET – Dtsch. Farben Z.* **29** (1975) 116–118.
[1.68] S. Keifer, *Farbe + Lack* **79** (1973) 1161–1166.
[1.69] P. Kresse, *DEFAZET – Dtsch. Farben Z.* **28** (1974) 459–466.
[1.70] M. Miele, *Farbe + Lack* **81** (1975) 495–504.
[1.71] A. Zosel in: *Congr. FATIPEC XIII,* 1976, EREC, Paris 1977, pp. 653–660.
[1.72] Th. Bohmann in: *Congr. FATIPEC IX,* 1968, Chimie des Peintures, Brussels 1969, pp. 47–50.
[1.73] O. J. Schmitz in: *Congr. FATIPEC XII,* 1974, Verlag Chemie, Weinheim 1975, pp. 511–515.
[1.74] L. Gall, U. Kaluza, *DEFAZET – Dtsch. Farben Z.* **29** (1975) 102–116.
[1.75] A. Klaeren, H. G. Völz, *Farbe + Lack* **81** (1975) 709–716.
[1.76] H. G. Völz, *Farbe + Lack* **96** (1990) 19.
[1.77] U. Kaluza, *DEFAZET – Dtsch. Farben Z.* **27** (1973) 427–439; **28** (1974) 449–459.
[1.78] H. A. Gardner, G. G. Sward: *Paints, Varnishes, Lacquers and Colors*, Gardner Laboratory, Inc., 12th ed., Bethesda, USA 1962, pp. 251–253.

[1.79] U. Kaluza, *Farbe + Lack* **80** (1974) 404–415.
[1.80] H. A. Gardner, G. G. Sward in [1.78], pp. 501–509.
[1.81] E. Stock: *Analyse der Körperfarben,* Wissenschaftl. Verlags GmbH, Stuttgart 1953.
[1.82] P. Kresse, *Farbe + Lack* **71** (1965) 178–184.
[1.83] W. Papenroth, K. Kox, *DEFAZET – Dtsch. Farben Z.* **28** (1974) 311–313.

References for Chapter 2

[2.1] G. Clarke, *Ind. Min. (London)* **251** (1988) Aug., 17–31.
[2.2] D. Williams, *Eur. Chem. News*, 17–23 June, (1996) 36–39.
[2.3] D. Rohe, *Chemische Industrie*, **10** (1996) 16–18.
[2.4] R. Wyckoff: *Crystal Structure*, J. Wiley & Sons, New York 1965.
[2.5] *Gmelin's Handbuch der anorganischen Chemie*, 8. Auflage, **41** (1951), pp 242.
[2.6] H. P. Boehm, *Chem. Ing. Tech.* **46** (1974) 716.
[2.7] A. R. v. Veen, *Z. Phys. Chem. Suppl.* **162** (1989) 215–219.
[2.8] G. N. Schrauzer, T. D. Guth, *Am. Chem. Soc. Div. Org.* **99** (1977) 7189.
[2.9] R. R. Towner, J. M. Gray, K. M. Porter, US-Geological, Survey Circular 930-G, United States Government Printing Office, Supt. of Docs. no.: I 19.4/2: 930-G (1988).
[2.10] Du Pont, US 2098025, 1935, US 2098055, 1935 (J. E. Booge, J. J. Krchma, R. H. Mc Kinney).
[2.11] K. A. Taylor, *Ind. Min. (London),* (1995) April, 47–63.
[2.12] R. Adams, Environment Matters (GB), February 1990, EM 6/6 (1990).
[2.13] Companhia Vale do Rio Doce, Projecto Titanio, CURO-reviste, vol. 7, no. 23, Mar. 86 (1986).
[2.14] R. Leutz, *Erzmetall* **42** (1989) no. 9, 383.
[2.15] British Titanium Products, DE-OS 2038244, 2038246–248, 1970 (F. R. Williams et al.); 2038245, 1970 (J. Whitehead et al.).
[2.16] Ruthner, DE-OS 1533123, 1966.
[2.17] Oceanic Process Corp., US 3252787, 1966 (C. D. Shiam).
[2.18] Anglo American Corp. of South Africa, DE-AS 1948742, 1969 (T. J. Coyle, H. J. Bovey).
[2.19] Laporte, DE-AS 1184292, 1961.
[2.20] Du Pont, DE-AS 1218734, 1965 (J. W. Reeves).
[2.21] American Cyanamid, DE-OS 2744805, 1977 (P. J. Preston et al.).
[2.22] W. Dunn, DE-OS 2528894, 1975 (W. E. Dunn, Jr.).
[2.23] Cochran, GB 1368564, 1974 (A. G. Starliper, A. A. Cochran).
[2.24] Du Pont, US 3926614, 1974 (H. H. Glaeser).
[2.25] Mitsubishi Chemical, US 3950489, 1974 (S. Fukushima).
[2.26] J. Barksdale: *Titanium*, 2nd ed., Ronald Press Comp., New York 1966, p. 240ff.
[2.27] Bayer, DE 2015155, 1970 (G. Kienast, H. Stütgens, H. G. Zander).
[2.28] FS Ishihara Sangyo Kaisha, Ltd., 30-5166, 1955.
[2.29] Bayer, DE-OS 2951799, 1979 (P. Panek, W. Gutsche, P. Woditsch).
[2.30] Bayer, DE 2454220, 1974 (R. Leiber, J. Leuriclan, J. Renier).
[2.31] J. Barksdale, in [2.7], pp. 264, 276, 278.
[2.32] Du Pont, US 2446181, 1946 (R. B. Kraus); DE 1467357, 1964 (E. L. Larins, O. Kleinfelder); US 3628913, 1969 (K. L. Uhland).
[2.33] Kronos Titan, EP 0265551, 1990 (A. Hartmann, H. Thumm).
[2.34] PPG Industries Inc., DE 1592960, 1967 (H. W. Rahn, K. W. Richardson).
[2.35] Du Pont, DE 1767798, 1968 (J. R. Auld).
[2.36] Cabot Corp., DE 1908747, 1969 (H. Weaver, R. B. Roaper II, Jr.); US 3607049, 1970 (H. Weaver, R. B. Roaper II, Jr.).
[2.37] Du Pont, US 5266108, 1992 (H. M. Hauck).
[2.38] Du Pont, US 5201949, 1992 (A. Allen, G. R. Evers).

[2.39] Du Pont, DE 2342 889, 1973 (A. H. Angerman, C. G. Moore).
[2.40] Du Pont, US 4781761, 1987 (H. W. Jacobson).
[2.41] Ishihara Sangyo Kaisha, J 58134-158 A, 1982.
[2.42] Du Pont, US 2885366, 1956 (R. K. Jeer).
[2.43] Tioxide Group Ltd., GB 1008652, 1961 (A. W. Evans, C. Shon).
[2.44] Kronos Titangesellschaft mbH, DE 1208438, 1960 (H. Rechmann, F. Vial, H. Weber). British Titanium Products, DE 1467412, 1965 (J. R. Moody, G. Lederer).
[2.45] Du Pont, US 4461810, 1984 (H. G. Jacobson).
[2.46] Bayer, DE 2946549, 1977 (K. Köhler, P. Woditsch, H. Rieck, F. Rodi).
[2.47] U. Rothe, K. Vellemann, H. Wagner, *Polymers Paint Coulor J.*, vol. 182, (1992).
[2.48] D. Ainley, *Chem. Rev.* **11** (1993), 18–22.
[2.49] Metallgesellschaft AG, Bayer, DE 2529708, 1977 (K. H. Dörr et al.).
[2.50] Kronos, EP 313715, 1989 (A. Kulling, A. Schinkitz, J. Mauer, J. Steinhausen).
[2.51] Bayer, EP 133505, 1985 (R. Gerken, G. Lailach, E. Bayer, W. Gutsche).
[2.52] Bayer, EP 194544, 1986 (R. Gerken, G. Lailach, A. van Fürden).
[2.53] Bayer, EP 97259, 1984 (R. Gerken, G. Lailach, K. H. Schultz).
[2.54] Bayer, EP 132820, 1985 (G. Lailach, R. Gerken, W. D. Müller, K. Brändle).
[2.55] Bayer AG, EP 393430, 1990 (G. Lailach et al.).
[2.56] W. Schäfer, *Korrespondenz Abwasser*, **2** (1989) 143–147.
[2.57] Du Pont, WO 95/31414, 1995 (P. Hill et al.).
[2.58] I. Fisher, *Ind. Min. (London)* (1996) Aug. 47–55.
[2.59] G. Kämpf et al., *Farbe + Lack* **79** (1973) 9 ff.
[2.60] H. G. Völz et al., *Farbe + Lack* **82** (1976) no. 9, 805 ff.
[2.61] D. V. Boust, *Intertech Conference*, Vancouver, April 1997.
[2.62] K. K. Ricoh, J 58025-363, 1981.
[2.63] Mitchell Market Reports, *Titania and Titanates*, 1990.
[2.64] H. Krüger, *VGB Kraftwerkstechnik* **71** (4) (1991) 371–395.
[2.65] *Energie Spektrum*, April (1991), pp. 20–21.
[2.66] P. S. Indin, *Verfkroniek*, **11** (1994) 17–19.
[2.67] Y. Parent, *Solar Energy*, vol. **56** (5) (1996) 429–437.
[2.68] P. Fournier, *C. R. Hebd. Seances Acad. Sci.* **231** (1950) 1343.
[2.69] H. Mühle, *Am. J. Ind. Med.* **15** (1989) 343–346.
[2.70] H. Clausen: "Zinc-Based Pigments," in P. A. Lewis (ed.): *Pigment Handbook*, 2nd ed., vol. 1, John Wiley & Sons, New York 1988.
[2.71] M. Issel, *Modern Paint and Coatings* **9** (1991) 35–42.
[2.72] M. Cremer: "Non TiO_2 White Pigments with Special References to ZnS Pigments," Industrial Minerals, Pigment & Extenders Supplement, 1985.
[2.73] H. Brown: *Zinc Oxide Rediscovered,* New Jersey Zinc Co., New York 1957.
[2.74] H. Brown: *Zinc Oxide Properties and Applications*, International Lead Zinc Research Organ. New York 1976.
[2.75] K. H. Ulbrich, W. Backhaus: "Zinkoxid in der Gummiindustrie," *Kautsch. Gummi, Kunstst.* **27** (1974) no. 7, 269–272.
[2.76] G. Hänig, K. Ulbrich: "ZnO, Produkt zwischen Pigmentchemie und Hüttenwesen," *Erzmetall* **32** (1979) 140–146.
[2.77] G. Heiland, E. Mollwo, F. Stockmann: *Electronic Processes in Zinc Oxide*, vol. 8, Solid State Physics, Academic Press, New York 1959.
[2.78] Larvik, US 2939783, 1957 (G. Lundevall).
[2.79] G. Meyer: *New Application for Zinc,* Zinc Inst. Inc. New York 1986.
[2.80] E. Ziegler, R. Helbig, *Physik Unserer Zeit* **17** (1986) no. 6, 171–177.
[2.81] Matsushita, DE 3045591, 1980 (Sonoda).
[2.82] Kieffer, in R. Henklin (ed.): *Metalle als Lebensnotwendige Spurenelemente für Pflanzen, Tiere und Menschen, Teil Zink*, Verlag Chemie, Weinheim 1984, pp. 117–123.

References for Chapter 3

[3.1] U. Schwertmann, R. M. Cornell: "*Iron Oxides in the Laboratory, Preparation and Characterisation*", VCH Verlagsgesellschaft, Weinheim 1991. R. M. Cornell, U. Schwertmann: "The Iron Oxides", VCH Verlagsgesellschaft, Weinheim 1996.
[3.2] J. L. W. Jolly, C. T. Collins: "Natural Iron Oxide Pigments," *Iron Oxide Pigments,* part 2, Information Circular-Bureau of Mines 8813, Washington 1980.
[3.3] H. Kittel: *Lehrbuch der Lacke und Beschichtungen,* vol. II, Verlag, W. A. Colomb Berlin 1974, p. 109.
[3.4] M. A. Bouchonnet, *Bull. Soc. Chim. Fr.* **9** (1912) 345.
[3.5] H. Wagner, R. Haug: "Gelbe Eisenoxydfarben," **8d**, *Veröffentlichung des Fachausschusses für Anstrichtechnik bei VDI und VDCh,* VDI-Verlag, Düsseldorf 1934.
[3.6] I. G. Farbenind., US 1813649, 1929 (P. Weise).
[3.7] Minnessota Mining & Manuf. Co., US 2634193, 1947 (G. E. Noponen).
[3.8] Minnessota Mining & Manuf. Co., US 2452608, 1941 (G. B. Smith).
[3.9] Verein Österr. Eisen- und Stahlwerke, OE 176206, 1952 (E. Petzel).
[3.10] The Nitralloy Corp., US 2592580, 1945 (H. Loevenstein).
[3.11] E. V. Carter, R. D. Laundon, *J. Oil Colour Chem. Ass.,* 1990 (1) 7–15.
[3.12] Bayer, DE 2653765, 1976 (B. Stephan, G. Winter).
[3.13] Bayer, DE 3820499, 1988 (B. Kröckert, G. Buxbaum, A. Westerhaus, H. Brunn).
[3.14] Bayer, DE-AS 1191063, 1963 (F. Hund, H. Köller, D. Räde, H. Quast).
[3.15] BASF, DE-OS 2517713, 1975 (W. Ostertag et al.).
[3.16] Magnetic Pigment Co., US 1424635, 1919 (P. Fireman).
[3.17] Ault & Wiberg Co., US 1726851, 1922 (E. H. McLeod).
[3.18] Interchem. Corp., US 2388659, 1943 (L. W. Ryan, H. L. Sanders).
[3.19] C. K. Williams & Co., US 3133267, 1934.
[3.20] Reconstruction Finance Corp., US 2631085, 1947 (L. M. Bennetch).
[3.21] Reymers Holms Gamla Ind., GB 668929, 1950 (T. G. H. Holst, K. A. H. Björned).
[3.22] Glemser, DE 704295, 1937 (O. Glemser).
[3.23] Pfizer, DE-OS 2212435, 1972 (L. M. Bennetch, H. S. Greiner, K. R. Hancock, M. Hoffman).
[3.24] Hollnagel, Kühn, DL 26901, 1960 (M. Hollnagel, E. Kühn).
[3.25] C. K. Williams & Co., US 2620261, 1947 (T. Toxby).
[3.26] West Coast Kalsomine Co., US 1327061, 1917 (R. S. Penniman, N. M. Zoph).
[3.27] National Ferrite Co., US 1368748, 1920 (R. S. Penniman, N. M. Zoph).
[3.28] Frazee, US 1923362, 1927 (V. Frazee).
[3.29] Magnetic Pigment Co., US 2090476, 1936 (P. Fireman).
[3.30] C. K. Williams & Co., US 2111726, 1932 (G. Plews); US 2111727, 1937 (G. Plews).
[3.31] Magnetic Pigment Co., US 2127907, 1937 (P. Fireman).
[3.32] Bayer, DE 902163, 1951 (B. H. Marsh).
[3.33] C. K. Williams & Co., US 2785991, 1952 (L. M. Bennetch).
[3.34] Bayer, FR 1085635, 1953 (F. Hund); DE 1040155, 1954 (F. Hund).
[3.35] Mineral Pigments, US 2633407, 1947 (D. W. Marsh).
[3.36] I. G. Farbenind., DE 463773, 1925 (J. Laux).
[3.37] I. G. Farbenind., DE 515758, 1925 (J. Laux).
[3.38] I. G. Farbenind., DE 551255, 1930 (U. Haberland).
[3.39] Bayer, EP 0249843, 1987 (A. Westerhaus, K. W. Ganter, G. Buxbaum).
[3.40] I. G. Farbenind., DE 466463, 1926 (W. Schubardt, M. Grote).
[3.41] Bayer, EP 0014382, 1980 (G. Franz, F. Hund).
[3.42] R. C. Rowe, *Pharm. Int.* **9** 1988 221–224.
[3.43] Bayer, DE 3326632, 1983 (W. Burow, H. Printzen, H. Brunn, K. Nollen).
[3.43a] *Pigment Handbook*, vol. I, *Properties and Economics*, 2nd ed., John Wiley & Sons, New York 1988, pp. 309–310.

[3.44] S. Keifer, A. Wingen, *Farbe + Lack* **79** (1973) 866–873.
[3.45] J.-L. Lassaigne, *Ann. Chim. Phys., Ser. 2,* **14** (1820) 299–302.
[3.46] H. C. Roth, US 1728510, 1927.
[3.47] PPG Industries, US 4127643, 1977 (W. W. Carlin).
[3.48] BASF, EP-A 0238713, 1986 (N. Müller et al.).
[3.49] Pfizer Inc., EP 0068787, 1982 (V. P. Rao).
[3.50] Bayer, US 4040860, 1976 (M. Mansmann, W. Rambold).
[3.51] Bayer, DE 2635086, 1976 (W. Rambold, H. Heine, B. Raederscheidt, G. Trenczek).
[3.52] British Chrome & Chemicals, US 4296076, 1981 (A. S. Dauvers, M. A. Marshall).
[3.53] C. K. Williams & Co., US 2250789, 1940 (I. W. Ayers).
[3.54] Bayer, DE 1257317, 1963 (F. Hund, D. Räde).
[3.55] Bayer, DE-OS 2358016, 1973 (B. Knickel et al.).
[3.56] Colores Hispañia S. A., ES 438129, 1975.
[3.57] C. K. Williams & Co., US 2560338, 1950 (C. G. Frayne); C. K. Williams & Co., US 2695215, 1950 (W. A. Pollock).
[3.58] Bayer, US 3723611, 1971 (G. Broja, K. Brändle, C. H. Elstermann).
[3.59] Gefahrstoffverordnung – GefstoffV Dec. 16, 1987 (BG Bl. I, 2721).
[3.60] Bayer, DE 3123361, 1981 (J. Rademachers et al.).
[3.61] 37. Abwasser VWV 05. Sept. 1984, GMBL Nr. 22, 344; Water Pollution Control Law, Law No. 138, Dec. 25 1970 (Japan).
[3.62] 178. Mitteilung BGesundhBl. 31 (1988) 363.
[3.63] *Journal Officiel de la République Française* no. 1227 (1983) 75, 97, 109.
[3.64] *Gazzetta Ufficiale della Republica Italiana*, Apr. 20, 1973, 3, 4.
[3.65] *Belgisch Staatsblad* – Moniteur Belge Sept. 24, 1976, 12030.
[3.66] Verpakkingen – en Gebruiksartikelenbesluit (Warenwert) 01. Oct. 1980.
[3.67] EG-Guidelines EEC 88/378, official gazette L 187 16. July 1988.
EEC Council Directive, EEC 76/768, on cosmetic products (18th Suppl.).
[3.68] Japanese Pharmaceutical Affaires 1980.
[3.69] E. Püttbach, *Betonwerk + Fertigteiltechnik* **2** (1987) 124–132.
[3.70] S. Ivankovic, R. Preussmann, *Ed. Cosmet. Toxicol.* **13** (1975) 347–351.
[3.71] U. Korallus, H. Ehrlicher, E. Wüstefeld, *ASP* **3** (1974) 51 ff.
[3.72] EG-Guidelines 67/548/EEC 1967 and Suppl.
[3.73] DCMA Classification and Chemical Description of the Complex Inorganics Color Pigments, 3rd. ed., Dry Color Manufacturers' Association, Alexandria, Virginia, 1991.
[3.73a] V. M. Goldschmidt, *Skr. Nor. Vidensk. Akad. Kl 1: Mat. Naturvidensh. Kl* **1926** Sept., 17.
[3.74] F. Hund, *Angew. Chem.* **74** (1962) 23–27. F. Hund, *Farbe + Lack* **73** (1967) 111–120.
[3.75] Bayer, US 3022186, 1959 (F. Hund).
[3.76] Du Pont, US 2257278, 1939 (H. H. Schaumann).
[3.77] The Harshaw Chemical Comp., US 2251829, 1939 (C. J. Harbert).
[3.78] Bayer, DE-AS 1195913, 1959 (H. Kyri, H. Weber).
[3.79] Ferro Corp., US 3832205, 1973 (H. E. Lowery).
[3.80] Bayer, EP 0318783, 1988 (V. Wilhelm, M. Mansmann).
[3.81] BASF Lacke + Farben AG, EP-A 0233601, 1987 (H. Knittel, R. Bauer, E. Liedek, G. Etzrodt).
[3.82] G. Blasse, *Philips Res. Rep. Suppl.* **3** (1964) 1–139.
[3.83] O. Schmitz-Dumont, B. Raederscheidt, *Naturwissenschaften* **41** (1964) 477–478.
[3.84] Kewanee Oil Company, US 3424551, 1964 (I. E. Owen).
[3.85] R. A. Eppler, *J. Am. Ceram. Soc.* **66** (1983) 794–801.
[3.86] Deutsche Gold- und Silber-Scheideanstalt, formerly Roessler, US 3876441, 1974 (A. Broll, H. Mann).
J. Alarcon, P. Escribano, R. M. Marin, *Br. Ceram. Trans. J.* **84** (1985) 172–175.
[3.87] Bayer, DE-OS 3819626, 1988 (J. Rademachers, V. Wilhelm, S. Keifer, W. Burow).

[3.88] S. H. Murdock, R. A. Eppler, *J. Am. Ceram. Soc.* **71** (1988) C-212–C-214.
[3.89] Bayer, DE-AS 1191063, 1963 (F. Hund, H. Köller, D. Räde, H. Quast).
[3.90] BASF, DE-OS 2517713, 1975 (W. Ostertag et al.).
[3.91] Bayer, DE-OS 1767868, 1968 (W. Holznagel, F. Hund, G. Gerlach).
[3.92] H. B. Krause, H. W. Reamer, J. L. Martin, *Mater. Res. Bull.* **3** (1968) 233–240.
[3.93] BASF Lacke + Farben AG, DE 2936746, 1979 (A. Seitz et al.).
[3.94] Bayer, EP 0075197, 1982 (D. Messer, V. Wilhelm, R. Endres, H. Heine).
[3.95] 178. Mitteilung, Bundesgesundheitsblatt **31** (1988) 363.
[3.96] *Journal Officiel de la République Française* no. 1227 (1983) 75, 97, 109.
[3.97] *Gazzetta Ufficiale della Republica Italiana* April 20, 1973, pp. 3, 4.
[3.98] *Belgisch Staatsblad* – Moniteur Belge Sept. 24, 1976, p. 12030.
[3.99] Verpakkingen – en Gebruiksartikelenbesluit (Warenwert) Oct. 1, 1980.
[3.100] EEC Council Directive, EEC 89/107, on coloring matters authorized for use in foodstuffs intended for human consumption.
[3.101] EEC Council Directive, EEC 76/768, on cosmetic products.
[3.102] EEC Council Directive, EEC 78/25, on coloring matters which may be added to medicinal products.
[3.103] E. Bomhard et al., *Toxicol. Lett.* **14** (1982) 189–194.
[3.104] D. Steinhoff, U. Mohr, *Exp. Pathol.* **41** (1991) 169–174.
[3.105] H. Endriß, D. Räde, *Kunststoffe* **79** (1989) 617–619.
[3.106] K. Weiss, *Z. Naturforsch. A Astrophys. Phys. Phys. Chem.* **2A** (1947) 650–652.
[3.107] S. F. Ravik, *J. Phys. Chem.* **40** (1936) 69.
[3.108] Siegle & Co., DE 1007907, 1955.
[3.109] Bayer, DE-OS 2151234, 1971.
[3.110] H. Endriß, *Kunststoffe* **75** (1985) 10, 758–761.
[3.111] D. Räde, A. Dornemann, *Proceedings of the Third International Cadmium Conference Miami, 1981,* Cadmium Association, London, pp. 37–40.
[3.112] H. Endriß, *Kunststoffe* **69** (1979) 39–43.
[3.113] European Commission: "On the Use of Colourants in Plastic Materials Coming in Contact with Food," Resolution AP (89) 1., Sept. 13, 1989.
[3.114] EEC Council Directive, EEC 83/514, Sept 26, 1983, on the limits for cadmium in waste water.
[3.115] G. Rusch, J. S. O'Grodnick, W. E. Rinehart, *Am. Ind. Hyg. Assoc. J.* **47** (1986) no. 12, 754.
[3.116] H.-J. Klimisch, C. Gembardt, H.-P. Gelbke: *Lung Deposition and Clearance, Lung Pathology and Renal Accumulation of Inhaled Cadmium Chloride and Cadmium Sulfide in Rat*, BASF, Ludwigshafen 1991.
[3.117] S. A. Robinson, O. Cantoni, M. Costa, *Carcinogenesis (London)* **3** (1982) no. 6, 657.
[3.118] M. Costa, J. D. Heck, S. H. Robinson, *Cancer Res.* **42** (1982) 2757.
[3.119] F. Pott et al., *Exp. Pathol.* **32** (1987) 129.
[3.120] H. Oldiges, D. Hochrainer, M. Glaser, *Toxicol. Environn. Chem.* **19** (1988) 217.
[3.121] U. Heinrich et al.: "Investigation of the Carcinogenic Effects of Various Cadmium Compounds after Inhalation Exposure in Hamsters and Mice," *Exp. Pathol.* **38**, in press.
[3.122] EEC Council Directive, EEC 91/338, 1991.
[3.123] E. Zintl, L. Vanino, DE 422947, 1924.
[3.124] Du Pont, US 4026722, 1976 (R. W. Hess); DE 2727864, 1977 (D. H. Piltingsrud); DE 2727863, 1977 (R. W. Hess); US 4063956, 1976 (J. F. Higgins).
[3.125] Montedison, DE 2933778, 1979 (L. Balducci, M. Rustioni); DE 2940185, 1979 (L. Balducci, M. Rustioni); DE 3004083, 1980 (L. Balducci, M. Rustioni); DE 3106625, 1981 (L. Balducci, M. Rustioni).
[3.126] BASF, EP 74049, 1982 (H. Wienand, W. Ostertag, K. Bittler); EP 271813, 1986 (H. Wienand, W. Ostertag, C. Schwidetzky, H. Knittel); DE 3926870, 1989 (H. Wienand, W. Ostertag, C. Schwidetzky).

[3.127] Bayer, DE 3315850, 1983 (P. Köhler, P. Ringe); DE 3315851, 1983 (P. Köhler, P. Ringe, H. Heine); EP 492224, 1991 (F. Schwochow); DE 4119668, 1991 (F. Schwochow, R. Hill); EP 723998, 1996 (H. Schittenhelm, R. Hill).
[3.128] Ciba-Geigy, EP 239526, 1987 (F. Herren); EP 304399, 1988 (R. Sullivan); EP 430888, 1990 (L. Erkens, G. Schmitt, H. Geurts, W. Corvers); DE 4037878, 1990 (F. Herren, L. Erkens); US 5399335, 1993 (R. Sullivan).
[3.129] BASF, EP 551637, 1992 (E. Liedek, H. Knittel, H. Reisacher, N. Mronga, H. Ochmann, H. Wienand); EP 640556, 1994 (G. Etzrodt, H. Knittel, H. Reisacher).
[3.130] H. Endriß, *Farbe + Lack* **100** (1994) no. 6, 397–398; H. Endriß, M. Haid, *Kunststoffe* **86** (1996) no. 4, 538–540.
[3.131] F. Hund, *Farbe + Lack* **73** (1967) 111–120.
[3.132] L. J. H. Erkens et al., *J. Oil Colour Chem. Assoc.* **71** (1988) no. 3, 71–77.
[3.133] H. Wagner et al. *Z. Anorg. Allg. Chem.* **208** (1931) 249; *Farben-Ztg.* **38** (1933) 932.
[3.134] J. F. Clay & Cromford Color, GB 730176, 1951; H. Lesche, *Farbe + Lack* **65** (1959) 79, 80.
[3.135] Du Pont, US 2808339, 1957 (J. J. Jackson).
[3.136] Du Pont, DE-OS 1807891, 1969 (H. R. Linton).
[3.137] Bayer, DE-OS 1952538, 1969 (C. H. Elstermann, F. Hund).
[3.138] ICI, DE-OS 2049519, 1970 (Ch. H. Buckley, G. L. Collier, J. B. Mitchell).
[3.139] Ten Horn Pigment, DE-OS 2600365, 1976 (J. J. Einerhand et al.).
[3.140] BASF, DE 3323247 A1, 1983 (E. Liedek et al.).
[3.141] Heubach, DE 3806214 A1, 1988 (I. Ressler, W. Horn, G. Adrian).
[3.142] Heubach, DE 3906670 A1, 1989 (I. Ressler, W. Horn, G. Adrian).
[3.143] R. Williams, Jr.: "Continuous Chrome Yellow Process," *Chem. Eng. (N.Y.)* 1949, March.
[3.144] Chemokomplex Vegyipari-Gep es Berendezes Export, Import Vallalat, DE-AS 1592848, 1971 (J. Scholtz et al.).
[3.145] H. Schäfer, *Farbe + Lack* **77** (1971) no. 11, 1081–1089.
[3.146] Hoechst, DE-OS 2127279, 1971 (R. Kohlhaas et al.).
[3.147] Hoechst, DE-OS 2062775, 1970 (R. Kohlhaas et al.).
[3.148] Sherwin-Williams, US 2237104, 1938 (N. F. Livingston).
[3.149] E. Renkwitz, *Farben-Ztg.* **28** (1923) 1066; H. Levecke, *Farbe + Lack* **42** (1936) 41–43; H. Berger, *Arbeitsschutz* **1941**, III/44.
[3.150] *Paint Oil Chem. Rev.* **95** (1933), 86–92; A. E. Newkirk, S. C. Horning, *Ind. Eng. Chem. Ind. Ed.* **33** (1941) 1402–1407.
[3.151] *Internationale MAK-Werte-Liste* (1990), Ecomed Verlag; *TRGS 505 Blei* (1988), Carl Heymanns Verlag.
[3.152] EEC Council Directive, EEC 82/605, July 28, 1982.
[3.153] J. M. Davies, *Lancet* **2** (1978) 18; *Br. J. Ind. Med.* **41** (1984) 158–169.
[3.154] W. C. Cooper, Dry Colour Manufacturers' Association, Arlington 1983.
[3.155] TA-Luft (Technische Anleitung zur Reinhaltung der Luft (1986), C. H. Beck'sche Verlagshandlung, München 1987.
[3.156] 37. Anhang zur Rahmen-Abwasser-Verwaltungsvorschrift vom 27.08.1991; § 7a Wasserhaushaltsgesetz.
[3.157] 37. Abwasser VwV (05.09.1984) für Direkteinleiter (Herstellung Anorganischer Pigmente).
[3.158] Gefahrstoffverordnung (GefStoffV) vom 26.10.1993, Stand 10.94, und Chemikalien-Verbotsverordnung (ChemVerbV) vom 26.10.1993, Stand 10.94, Deutscher Bundesverlag GmbH, Bonn.
[3.159] EEC Council Directive, EEC 67/548, June 27, 1967 together with EEC Council; Directive EEC 93/21, April 27, 1993 (18th adaption to EEC 67/548) and EEC Council Directive EEC 93/72, Sept. 1st, 1993 (19th adaption to EEC 67/548).
[3.160] EEC Council Directive EEC 94/96, Dec. 12, 1994 (21th adaption to EEC 67/548).

[3.161] EEC Council Directive EEC 93/18, April 05, 1993 (3rd adaption to EEC 88/379).
[3.162] Erste Verordnung zur Änderung chemikalienrechtlicher Verordnungen vom 12. Juni 1996, Deutscher Bundesverlag GmbH, Bonn.
[3.163] EEC Council Directive EEC 94/60, Dec. 20, 1994 (14th amendment of EEC 76/769).
[3.164] EEC Council Directive, EEC 89/178, Feb. 22, 1989.
[3.165] Lebensmittel- und Bedarfsgegenständegesetz vom 15.08.1974 (BGBl S. 1945).
[3.166] Sicherheit von Spielzeug (Safety of toys), EN 71 Part 3, Dec. 1994.
[3.167] S. E. Tarling, P. Barnes, A. L. Mackay, *J. Appl. Cryst.* **17** (1984) 96–99.
[3.168] S. E. Tarling, P. Barnes, J. Klinowsky, *Acta Cryst.* **B44** (1988) 128–135.
[3.169] R. J. H. Clark, M. L. Franks, *Chem. Phys. Lett.* **34** (1975) 69–72.
[3.170] R. J. H. Clark, D. G. Cobbold, *Inorg. Chem.* **17** (1978) 3169–3174.
[3.171] R. J. H. Clark, T. J. Dines, M. Kurmoo, *Inorg. Chem.* **22** (1983) 2766–2772.
[3.172] Interchemical Corp., US 2535057, 1950 (A. E. Gessler, C. A. Kummins).
[3.173] Toyo Soda, JP-KK 81045513, 1978.
[3.174] DE-OS 4402538.
[3.175] Anonymus, *Farbe & Lack*, **100** (1994) 887.
[3.176] Reckitt's Colours Ltd., *The Cost of Whiteness*, Hull, United Kingdom.
[3.177] *Ullmann Encyklopädie der technischen Chemie*, 4th ed., **18**, Verlag Chemie, Weinheim 1979, pp. 623.
[3.178] C. Clauss, E. Gratzfeld in H. Kittel (ed.): *Pigmente*, 3rd ed., Wissenschaftl. Verlags GmbH, Stuttgart 1960.
[3.179] M. F. Dix, A. D. Rae, *J. Oil Colour Chem. Assoc.* **61** (1978) 69.
[3.180] Degussa *Vossen-Blau-Pigmente*, Frankfurt/M. 1973.
[3.181] G. K. Wertheim et al., *J. Chem. Phys.* **54** (1971) 3235.
H. Buser et al., *J. Chem. Soc. Chem. Commun.* **23** (1972) 1299.
[3.182] M. L. Napijalo, V. Stefancic, *Fizika (Zagreb)* 8 Suppl. (1976) 16.
[3.183] R. J. Emrich et al., *J. Vac. Sci. Techn. A* **5** (1987) 1307.
[3.184] F. Herren et al., *Inorg. Chem.* **19** (1980) 956.
A. Ludi, *Chem. Unser Zeit* **22** (1988) 123.
[3.185] Degussa AG, DE 1188232, 1964 (E. Gratzfeld).
[3.186] Degussa AG, DE 976599, 1952 (H. Verbeek, E. Gratzfeld).
[3.187] Chem. Fabrik Wesseling AG, DE 1061935, 1955 (H. Verbeek, E. Gratzfeld).
M. Sakakibara, S. Teshirogi, JP 51082317, 1976. Toya Soda, JP 55062969, 1980.
[3.188] Degussa AG, OE 233669, 1962 (E. Gratzfeld).
[3.189] Degussa AG, DE-OS 1792418, 1968 (E. Gratzfeld, E. Clausen, E. Ott).
[3.190] Degussa AG, DE 1937832, 1969 (E. Gratzfeld).
[3.191] Degussa AG, DE 1949720, 1969 (E. Gratzfeld, E. Kühn).
[3.192] L. Müller-Fokken, *Farbe + Lack* **84** (1978) 489.
[3.193] H. Ferch, H. Schäfer: *18th AFTPV-Kongreßbuch*, Nice 1989, p. 315.
[3.194] "Fotometrische Messung tiefschwarzer Systeme". Schriftenreihe Pigmente Nr. 24, Degussa AG, 60287 Frankfurt/M. (1989).
[3.195] "Vossen-Blau zur Färbung von Fungiziden." Schriftenreihe Pigments Nr. 50, Degussa AG, Frankfurt/M. 1985.
[3.196] W. Koblet, *Schweiz. Z. Obst Weinbau* **1** (1965) 8.
[3.197] H. Wiedmer et al.: *Agro-Dok* no. D4341, Sandoz AG, Basel 1977.
[3.198] R. Ciferri: "Le 4 Stagioni" (Montecatine) **4** (1963) no. 2, 2.
[3.199] H. Winkler: Schriftenreihe Pigmente Nr. 20, 3rd ed., Degussa AG, Frankfurt 1992.
[3.200] V. Nigrovic, *Int. J. Rad. Biol.* **7** (1963) 307.
[3.201] V. Nigrovic, *Phys. Med. Biol.* **10** (1965) 81.
[3.202] I. V. Tananayev, *Zh. Neorg. Khim.* **1** (1956) 66.
[3.203] P. Dvorak, *Z. Gesamte Exp. Med.* **89** (1969) 151.
[3.204] P. Dvorak, *Arzneim. Forsch.* **20** (1970) 1886.

[3.205] P. Dvorak, *Z. Naturforsch. B* **26** (1971) 277.
[3.206] V. Nigrovic, F. Bohne, K. Madshus, *Strahlentherapie* **130** (1966) 413.
[3.207] P. Dvorak, *Z. ges. exp. Med.* **151** (1969) 89–92.
[3.208] T. A. Shashina et al., Gig. Tr. Prof. Zabol. **1** (1991) 35–36.
[3.209] J. M. Verzijl et al., *Clinical Toxicol.* **31** (1993) 553–562.
[3.210] B. D. Nielsen et al., *Z. Naturforsch.* **45** (1990) 681–690.
[3.211] Degussa AG, unpublished report: Degussa AG US-IT-Nr. 84-0074-DKT (1984) and 88-0083-DKT, 88-0084-DKT (1988a).
[3.212] NPIRI (National Printing Ink Research Institute Raw Materials Data Handbook); Napim, New York, 4 (1983) 21.
[3.213] Degussa AG, unpublished report: Degussa AG US-IT-Nr. 77-0069-FKT (1977).
[3.214] Degussa AG, unpublished report: Degussa AG US-IT-Nr. 85-0081-DKT, 85-0080-DKT (1985a), 87-0038-DKT, 87-0039-DKT, 87-0040-DKT (1987) and 88-0085-DKT (1988b).
[3.215] F. Leuschner, H. Otto, unpublished (1967); in: Kosmetische Färbemittel, Harald Boldt Verlag KG, Boppard (1977).
[3.216] V. Nigrovic et al., *Strahlentherapie* **130** (1966) 413–419.
[3.217] M. Günther, Kernforschungszentrum Karlsruhe, KFK 1326, Gesellschaft für Kernforschung m.b.H., Karlsruhe, unpublished report US-IT-Nr. 70-0001-FKT (1970).
[3.218] P. Dvorak et al., *Naunyn-Schmiedebergs Arch. Pharmak.* **269** (1971) 48–56.
[3.219] K. Madshus et al., *Int. J. Radiation Biol.* **10**, 519–520 (1966).
[3.220] K. Madshus, A. Strömme, *Z. Naturforsch.* **A 23** (1968) 391–392.
[3.221] V. Pai, *West Indian Med. J.* **36** (1987) 256–258.
[3.222] Degussa AG, unpublished report: Degussa AG US-IT-Nr. 85-0082-DGO, 85-0085-DGO, 85-0088-DGO, 85-0078-DGO (1985b).
[3.223] Degussa AG, unpublished report: Degussa AG US-IT-Nr. 79-0046-DKO, 79-0047-DKO, 79-0088-DKO (1979) and 85-0079-DKO, 85-0083-DGO (1985c).

References for Chapter 4

[4.1] A. I. Medalia, D. Rivin, *Carbon* **20** (1982) no. 6, 481–492.
[4.2] H. P. Boehm, *Farbe + Lack* **79** (1973) no. 5, 419–432.
[4.3] R. D. Heidenreich, W. M. Hess, L. L. Ban, *J. Appl. Crystallogr.* **1** (1968) 1; K. A. Burgess, C. E. Scott, W. M. Hess, *Rubber Chem. Technol.* **44** (1971) no. 1, 230–248.
[4.4] W. C. Wake (ed.): *Fillers for Plastics,* Iliffe Books, London 1971.
[4.5] Inter-Governmental Maritime Consultive Organization, London, *International Maritime Dangerous Goods Code,* 11–75, 4081–4082.
[4.6] *Rep. Invest. U.S. Bur. Mines* **6598** (1965).
[4.7] *U.S. Bur. Mines Technical Paper* **610** (1940).
[4.8] H. W. Davidson et al., *Manufactured Carbon,* Pergamon Press, Oxford-New York 1968.
[4.9] Degussa, DE-OS 2 410 565, 1974.
[4.10] Cabot Corp., DE-OS 2 507 021, 1975.
[4.11] Cabot Corp., US 3 010 794, 3 010 795, 1958.
[4.12] Cities Service Co., DE-OS 1 592 853, 1967.
[4.13] Ashland Oil & Refining Co., US 3 649 207, 1969; Phillips Petroleum Co., US 3 986 836, 1974.
[4.14] G. Kühner, G. Dittrich, *Chem. Ing. Tech.* **44** (1972) 11.
[4.15] Cities Service Co., US 3 490 869, 1966.
[4.16] Cabot Corp., DE-OS 2 507 021, 1975.
[4.17] *Ullmann,* 3rd ed., vol. 14, pp. 793–810.
[4.18] Degussa, DE 895 286, 1951; DE 1 129 459, 1960.
[4.19] Cities Service Co., US 3 593 371, 1968; Phillips Petroleum Co., US 3 956 445, 1974; Cabot Corp., DE-OS 2 004 493, 1970; Cabot Corp., GB 1 370 704, 1972.

[4.20] Shawinigan Chem. Ltd., US 2453440, 1944; US 2492481, 1948; J. Wotschke, K. Paasch, *Schweiz. Arch.* **5** (1949) 173.
[4.21] Cabot Corp., US 3342554, 1963.
[4.22] M. S. Jyengar, *Chem. Technol.* **2** (1972) no. 9, 565–569.
[4.23] Degussa, DE 742664, 1940; Cabot Corp., US 2420810, 1941.
[4.24] Degussa, GB 895990, 1958.
[4.25] Cabot Corp., US 2439442, 1943.
[4.26] T. C. Patton (ed.): *Pigment Handbook*, vol. 3, J. Wiley & Sons, New York 1973.
[4.27] Cabot Corp., US 3383232, 1968; Phillips Petroleum Co., US 3870785, 1975.
[4.28] W. M. Hess, L. L. Ban, G. C. McDonald, *Rubber Chem. Technol.* **42** (1969) 1200; J. C. Motte, *Rev. Gen. Caoutch. Plast.* **49** (1971) no. 5, 417–421.
[4.29] S. Brunauer, P. H. Emmett, E. J. Teller, *J. Am. Chem. Soc.* **60** (1938) 309.
[4.30] J. H. de Boer et al., *J. Catal.* **4** (1965) 319–323, 643–648.
[4.31] R. Bode, *Kautsch. Gummi Kunstst.* **36** (1983) no. 8, 660–676.
[4.32] R. Bode, H. Ferch, D. Koth, W. Schumacher, *Farbe + Lack* **85** (1979) 7–13.
[4.33] H. Ferch, *Verfkroniek* **47** (1974) 40–46.
[4.34] H. P. Boehm, *Farbe + Lack* **79** (1973) 419–432.
[4.35] *Printing Ink Handbook*, 2nd ed., National Association of Printing Ink Manufacturers, Elmsford N.Y. 1967.
[4.36] D. Rivin, R. G. Smith, *Rubber Chem. Technol.* **55** (1982) no. 3, 707–761.
[4.37] J. M. Robertson and R. G. Smith, *Patty's Industrial Hygiene and Toxicology, Fourth Edition Volume 2, Part D, Chapter 28*, p. 2395, John Wiley & Sons, Inc. (1994).
[4.38] ECBECB (European Committee for Biological Effects of Carbon Black), Bulletin No. **7** (1992), *Polyaromatic Hydrocarbons in Carbon Black*.
[4.39] J. M. Robertson and T. H. Ingalls, *Am. Ind. Hyg. Assoc. J.* **50** (1989) 510.
[4.40] W. A. Crosbie, *Arch. Environ. Health*, **41** (1986) 346.
[4.41] J. M. Robertson and T. H. Ingalls, *Arch. Environ. Health* **35** (1980) 181.
[4.42] C. A. Nau, J. Neal, and V. A. Stembridge, *Am. Med. Assoc. Arch. Ind. Health* **18** (1958) 511.
[4.43] C. J. Kirwin et al., *J. Toxicol. Environ. Health* **7** (1981) 973.
[4.44] F. Buddingh, M. J. Bailey, B. Wells, and J. Maesemmeyer, *Am. Ind. Hyg. Assoc. J.*, **42** (1981) 503.
[4.45] J. L. Mauderly et al., in: *Health Effect Research Report No. 68*, Health Effect Institute, Cambridge, Massachusetts (1994).
[4.46] U. Heinrich et al., *Inhal. Toxicol.* **7** (1995) 533.
[4.47] G. Oberdorster, in: *Toxic and Carcinogenic Effects of Solid Particles in the Respiratory Tract*, ILSI Press, Washington DC, 335 (1994).
[4.48] American Conference of Governmental Industrial Hygienists, *Documentation of the Threshold Limit Values for Substances in Workroom Air-Carbon Black*, rev. ed., ACGIH, Cincinnati, OH, p. 220, (1991).
[4.49] K. Gardiner, W. N. Trethowan, J. M. Harrington, et al., *Annn. Occup. Hyg.* **5** (1992) 477.
[4.50] IARC (Lyon) "Evaluation of Carcinogenic Risk of Chemicals to Humans", vol. **33** Carbon Black (1984); Confirmed in March, 1987.
[4.51] IARC (Lyon) "Evaluation of Carcinogenic Risk of Chemicals to Humans", vol. **65** Carbon Black (1996).

General References:

[4.52] J. B. Donnet, R. C. Bansal, M.-J. Wang (Ed.) *Carbon Black, 2nd ed.*, Marcel Dekker, New York, 1993.
[4.53] H. Ferch *Pigmentruße* (U. Zorll Ed. *Die Technologie des Beschichtens*), Vincentz Verlag, Hannover, 1995.

References for Chapter 5

[5.1] Armour Research Foundation, US 2694656, 1947 (M. Camras).
[5.2] Bayer, DE 1061760, 1957 (F. Hund).
[5.3] EMI, GB 765464, 1953 (W. Soby).
[5.4] BASF, DE 1204644, 1962 (W. Balz, K. G. Malle).
[5.5] VEB Elektrochemisches Kombinat, Bitterfeld, DD 48590, 1965 (W. Baronius, F. Henneberger, W. Geidel).
[5.6] Agfa-Gevaert, DE 1592214, 1967 (W. Abeck, H. Kober, B. Seidel).
[5.7] Bayer, DE 1803783, 1968 (F. Rodi, H. Zirngibl).
[5.8] Pfizer, US 3498748, 1967 (H. S. Greiner).
[5.9] Sakai Chemical Industries, US 4202871, 1980 (S. Matsumoto, T. Koga, K. Fukai, S. Nakatani).
[5.10] A. R. Corradi et al., *IEEE Trans. Magn.* **MAG-20** (1984) 33–38.
[5.11] Bayer, DE 1266997, 1959 (W. Abeck, F. Hund).
[5.12] Y. Imaoka, S. Umeki, Y. Kubota, Y. Tokuoka, *IEEE Trans. Magn.* **MAG-14** (1978) 649.
[5.13] 3M, US 3573980, 1968 (W. D. Haller, R. M. Colline).
[5.14] Hitachi Maxell, DE 2235383, 1972 (Okazoe, Akira).
[5.15] B. L. Chamberland, *CRC Critical Reviews in Solid State and Material Science*, 1977, 1–31.
[5.16] J. N. Ingram, T. J. Swoboda, US 2923683 (1957).
[5.17] M. Essig et al., *IEEE Trans. Magn.* **MAG-26** (1990) 69.
[5.18] Du Pont, US 3512930, 1968 (W. G. Bottjer et al.).
[5.19] BASF, EP-A 218234, 1985 (W. Steck et al.).
[5.20] BASF, unpublished.
[5.21] M. Kryder et al., *IEEE Trans. Magn.* **MAG-23** (1987) 45.
[5.22] Du Pont, US 2923683, 1957 (J. N. Ingraham et al.).
Bayer, DE-AS 1467328, 1963 (F. Hund et al.).
[5.23] Du Pont, US 3034988, 1958 (J. N. Ingraham et al.).
[5.24] E. Köster in C. D. Mee, E. D. Daniel (eds.): *Magnetic Recording*, vol. I, McGraw Hill, New York 1987.
[5.25] H. Auweter et al., *IEEE Trans. Magn.* **MAG-26** (1990) 66.
[5.26] M. W. Müller et al., *IEEE Trans. Magn.* **MAG-26** (1990) 1897.
[5.27] *Ullmann Encyklopädie der technischen Chemie*, 4th ed., **7**, Verlag Chemie 1979, pp. 77.
[5.28] G. R. Cole et al., *IEEE Trans. Magn.* **MAG-20** (1984) 19.
[5.29] G. R. Cole, *IEEE Trans. Magn.* **MAG-20** (1984) 19.
H. Chen, D. M. Hiller, J.E. Hudson, C. J. A. Westenbroek, *IEEE Trans. Magn.* **MAG-20** (1984) 24.
[5.30] G. Bate, *J. Appl. Phys.* **52** (1981) 2447.
[5.31] M. Kishimoto, S. Kitahata, M. Amemiya, *IEEE Trans. Magn.* **MAG-22** (1986) 732–734.
[5.32] T. Fujiwara, *IEEE Trans. Magn.* **MAG-23** (1987) 3125.
[5.33] H. Yokoyama et al., *IEEE Trans. Magn.* **28** (1992) 2391.
[5.34] H. Hibst, *Angew. Chem.* **94** (1982) 263.
[5.35] H. Stäblein in E. P. Wohlfarth (ed.): *Ferromagnetic Materials*, vol. 3, North Holland Publ., Amsterdam 1982.
[5.36] Toshiba, EP-A 39773, 1980 (H. Endo et al.).
[5.37] Toda, EP-A 150580, 1983 (N. Nagai et al.).
[5.38] Dowa Mining, DE-OS 3527478, 1984 (K. Aoki).
[5.39] Sakai Chemical, DE-OS 3529756, 1984 (S. Jwasaki et al.).
[5.40] Ishihara, EP-A 299332, 1987 (K. Nakata et al.).
[5.41] Ugine Kuhlmann, DE-OS 2003438, 1969 (M. G. de Bellay).
[5.42] Toda, EP-A 164251, 1984 (N. Nagai et al.).
[5.43] Toda, EP-A 232131, 1986 (N. Nagai et al.).

[5.44] Matsushita, EP-A 290263, 1987 (H. Toril et al.).
[5.45] Toshiba, DE 3041960, 1979 (O. Kubo et al.).
[5.46] BASF, DE-OS 3702036, 1987 (G. Mair).
[5.47] R. E. Fayling, *IEEE Trans. Magn.* **MAG-15** (1979) 1567.
[5.48] R. J. Veitch, *IEEE Trans. Magn.* **MAG-26** (1990) 1876.
[5.49] Y. Okazaki et al., *IEEE Trans. Magn.* **MAG-24** (1989) 4057.
[5.50] Sony Magnascale, *High Speed Video Duplicating System*, Sprinter Brochure, 1986.
[5.51] D. E. Speliotis, *IEEE Trans. Magn.* **MAG-26** (1990).
[5.52] O. Kubo et al., *IEEE Trans. Magn.* **MAG-24** (1988) 2859.
[5.53] H. Wienand, W. Ostertag, *Farbe + Lack* **88** (1982) no. 3, 183–188.
[5.54] S. Wiktorek, *Surface Coat. Australia* **24** (1987) no. 5, 11–16.
[5.55] W. Funke, *Farbe + Lack* **86** (1980) no. 8, 730.
[5.56] P. Kresse, *DEFAZET Dtsch. Farben Z.* **32** (1978) no. 5, 216, 217.
[5.57] W. Funke, *Farbe + Lack* **89** (1983) no. 2, 86–91.
[5.58] H. Haagen, R. Konzelmann, *Farbe + Lack* **94** (1988) no. 11, 893, 894.
[5.59] H. Haagen, D. Martinovic, *Farbe + Lack* **95** (1989) no. 12, 892–895.
[5.60] E. V. Carter, R. D. Laundon, *J. Oil Colour Chem. Assoc.* 1990 (1), 7–15.
[5.61] K. A. van Oeteren, *Farbe + Lack* **79** (1973) no. 1, 45–48.
[5.62] V. Cupr, *Maschinenmarkt IMM Ind. J.* **84** (1978) 439–442.
[5.63] K. A. van Oeteren, *Fette Seifen Anstrichm.* **84** (1982) no. 6, 242–245.
[5.64] E. Leitel et al., *Werkst. Korros.* **39** (1988) 1–7.
[5.65] Ch. Nitsche, *Farbe + Lack* **90** (1984) no. 6, 468–473.
[5.66] M. Cohen, F. J. Beck, *Z. Elektrochem.* **61** (1958) 696.
[5.67] BASF: Sicor-Pigmente, company brochure, Ludwigshafen 1987.
[5.68] Dr. Hans Heubach GmbH & Co. KG: Heucophos company information, Langelsheim 1989.
[5.69] Halox-Pigments: *Halox-Pigments*, company information, Pittsburgh 1973.
[5.70] Société nouvelle des Couleurs Zinciques, company information, Beauchamp 1990.
[5.71] Colores Hispañia: *Anticorrosive Pigments*, company information, Barcelona 1990.
[5.72] Ten Horn Pigment bv., company information, Maastricht 1977.
[5.73] Teikoku Kako Co. Ltd.: *K-White*, company information, Osaka 1987.
[5.74] N. R. Whitehouse, *Polym. Paint Colour J.* **178** (1988) no. 4211, 239.
[5.75] Sherwin-Williams Chemicals: *Moly White*, company information, Coffeyville, Kansas, 1986.
[5.76] NL Chemicals: *Nalcin 2*, company information, Hightstown, N.J., 1983
[5.77] PPG Industries Inc., US 4707405, 1987 (E. Evans et al.)
[5.78] Buckman Laboratories Inc.; *Barium metaborate*, company information, Memphis, Tennessee, 1989.
[5.79] Alex R. Vleeshouwers, *Polym. Paint Colour J.* **178** (1988) no. 4224, 788, 790.
[5.80] G. Meyer, *Farbe + Lack* **68** (1962) no. 5, 315; *Dtsch. Farben-Z.* **20** (1966) 8; *Farbe + Lack* **79** (1967) no. 6, 529.
[5.81] J. Ruf, *Werkst. Korros.* **20** (1969) 861–869.
[5.82] J. A. Burkill, J. E. O. Mayne, *J. Oil Colour Chem. Assoc.* 71 (1988) no. 9, 273–275.
[5.83] G. Sziklai, J. Szucs, *Hung. J. Ind. Chem.* **10** (1982) 215–221.
[5.84] P. J. Gardner, I. W. McArn, V. Barton, G. M. Seydt, *J. Oil Colour Chem. Assoc.* **73** (1990) no. 1, 16.
[5.85] G. Rasack, *Farbe + Lack* **84** (1978) no. 7, 497–500.
[5.86] G. Adrian, A. Bittner, *J. Coat. Technol.* **58** (1986) 59–65.
[5.87] P. Kresse, *Farbe + Lack* **83** (1977) no. 2, 85–95.
[5.88] P. Reichle, W. Funke, *Farbe + Lack* **93** (1987) no. 7, 537–538.
[5.89] M. Svoboda, *Farbe + Lack* **92** (1986) no. 8, 701–703.
[5.90] Goslarer Farbenwerke Dr. Hans Heubach, DE 3046698, 1980 (W. Haacke et al.).
[5.91] Bayer, DE-OS 2656779, 1976 (F. Hund, P. Kresse).

[5.92] Herberts GmbH, DE-OS 3720779, 1989 (H. Becker, F. Sadowski, W. Stephan).
[5.93] Hoechst AG, DE-OS 2849712, 1978 (G. Mietens et al.); Hoechst AG, EP 0028290, 1980 (K. Hestermann et al.).
[5.94] Hoechst AG, DE-OS 2916029, 1979 (A. Maurer et al.).
[5.95] Goslarer Farbenwerke Dr. Hans Heubach, EP 0308884, 1988 (A. Bittner); DE-OS 3731737, 1987 (A. Bittner).
[5.96] Mizusawa Kagaku Koayo K.K., DE 2159342, 1971 (Y. Sugahara et al.).
[5.97] Pfizer & Co., US 3443977, 1969 (L. M. Bennetch).
[5.98] Teikoku Kako, JP 55160-059, 1980 (M. Kinugasa).
[5.99] Int. Standard Electric Corp., DE-OS 3042630, 1980 (C. F. Drake et al.).
[5.100] Toy Soda, EP 35-798, 1981 (H. Minamide).
[5.101] Nippon Paint Co, EP 0259748, 1987 (T. Okal et al.).
[5.102] Colores Hispañia, ES 8605-015, 1985.
[5.103] Hooker Chemicals & Plastics Corp.: *Ferrophos*, company information, Niagara Falls, N.Y., and D. E. de Jong bv. Apeldoorn, Holland, 1976.
[5.104] Goslarer Farbenwerke Dr. Hans Heubach, DE 3532806.1, 1985 (M. Gawol, G. Adrian).
[5.105] PPG Industries, US 4837253, 1989 (J. D. Mansell et al.).
[5.106] H. Kossmann, *Farbe + Lack* **91** (1985) no. 7, 588–594.
[5.107] S. Hellbordt, *Farbe + Lack* **85** (1979) no. 6, 485.
[5.108] J. D'Ans, V. Groope, *Dtsch. Farben-Z.* **15** (1961) 51, 69.
[5.109] U. R. Evans, *J. Chem. Soc.* **1927**, 1020.
[5.110] J. E. D. Mayne, *J. Chem. Soc.* **1948**, 1932.
[5.111] J. L. Rosenfeld, *Lakokras. Mater. Ikh. Primen.* **1961**, 50.
[5.112] H. H. Uhlig, *Chem. Eng. News* **24** (1946) 3154.
[5.113] G. H. Cartledge, *Corrosion (Houston)* **18** (1962) 316.
[5.114] J. Ruf: *Korrosionsschutz durch Lacke und Pigmente,* A. W. Colomb, H. Heenemann GmbH, Stuttgart–Berlin 1972.
[5.115] J. Ruf, *Farbe + Lack* **75** (1969) no. 10, 943.
[5.116] GeffStoffV, TRgA 602, part 2, 1987.
[5.117] Verordnung über gefährliche Arbeitsstoffe (Arbeitsstoffverordnung – ArbStoffV) Feb. 11, 1982 (BGBl I, 144).
[5.118] Nippon Paint KK, JP 61276-861, 1985.
[5.119] G. Lincke, O. Schroers, K. D. Nowak, *Farbe + Lack* **82** (1976) no. 11, 1003.
[5.120] G. Meyer, *DEFAZET Dtsch. Farben Z.* **5** (1963) 201–205.
[5.121] W. J. Banke, *Mod. Paint Coat.* **70** (1980) Febr., 45–47.
[5.122] American Metal Climax, CH 602899, 1973 (D. R. Robitaille et al.); DE 2334541, 1973 (D. R. Robitaille et al.)..
[5.123] Noranda Miners Ltd., US 4132667, 1977 (D. Kerfoot); DE 2814454, 1978 (D. Kerfoot).
[5.124] J. Ruf, *Farbe + Lack* **79** (1973) no. 1, 22–27.
[5.125] BASF AG, company information, Ludwigshafen 1990.
[5.126] Horst Michaud et al., *Chem. Ztg.* **112** (1988) 287–294.
[5.127] Amer. Cyanamide Co., US 2213440, 1939; US 2213441, 1939.
[5.128] Duisburger Kupferhütte, DT 956219, 1956.
[5.129] I. G. Farbenind., DT 651684, 1934.
[5.130] N. Lowicki, *Z. Erzbergbau Metallhüttenwes.* **2** (1949) 282, 283.
[5.131] S. A. Miller, B. Bann, *J. Appl. Chem.* **6** (1956) 89–93.
[5.132] V. Kontnik, V. Novak, *Chem. Prum.* **9** (1959) 18–22.
[5.133] Süddeutsche Kalkstickstoffwerke, DT 803351, 1949.
[5.134] Lab. Chim. G. Guainazzi, IT 428147, 1947.
[5.135] Admin. Des. Mines Bouxwiller, FR 966349, 1948.
[5.136] E. Stock, H. Blum, *Dtsch. Farben-Z.* **5** (1951) 317–321.
[5.137] Th. Goldschmidt, DT 940291, 1952.

[5.138] Duisburger Kupferhütte, DT 819 689, 1949.
[5.139] BASF AG, company information, Ludwigshafen 1990.
[5.140] Techn. Jahresbericht 1957 der Berufsgenossenschaft der chem. Industrie der Bundesrepublik Deutschland.
[5.141] Int. Minerals and Chem. Corp., US 4 738 720, 1986 (P. Eckler, L. Ferrara).
[5.142] The British Petroleum Company, EP 0 316 066, 1988 (T. E. Fletcher).
[5.143] Grace & Co., company information, London 1986.
[5.144] B. P. F. Goldie, *J. Oil Colour Chem. Assoc.* **71** (1988) no. 9, 257–270.
[5.145] Grace & Co., *Polym. Paint Colour J.* **177** (1987) no. 4189, 260.
[5.146] A. F. Wells: *Structural Inorganic Chemistry,* Clarendon Press, Oxford 1962.
[5.147] J. D'Ans, W. Breckheimer, H. J. Schuster, *Werkst. Korros.* **8** (1957) 677. J. D'Ans, H. J. Schuster, *Farbe + Lack* **61** (1955) no. 10, 453. J. E. O. Mayne, *Farbe + Lack*, **76** (1970) no. 3, 243.
[5.148] G. Lincke, *Farbe + Lack* **77** (1971) no. 5, 443.
G. Lincke, *Congr. FATIPEC* **12th** 563.
[5.149] H. Berger, *Farbe + Lack* **88** (1982) no. 3, 180–182.
[5.150] P. N. Martini, A. Bianchini, *J. Appl. Chem.* **19** (1969) 147.
[5.151] J. F. H. van Eijnsbergen, *Farbe + Lack* **71** (1965) no. 10, 1005.
G. Meyer, *Dtsch. Farben-Z.* **16** (1962) 347.
[5.152] Bayer AG, DE 2 642 049, 1976 (F. Hund et al.).
Bayer AG, DE 2 625 401, 1976 (F. Hund et al.).
[5.153] Tada Kogyo, US 3 904 421, 1975 (S. Shimizu et al.).
[5.154] P. Kresse, *Farbe + Lack* **84** (1978) no. 3, 156–159.
[5.155] M. Svoboda, *Farbe + Lack* **96** (1990) no. 7, 506–508.
[5.156] Bayer: Anticor 70, product information, Leverkusen 1989.
[5.157] Lindgens, company information, Köln-Mülheim 1991.
[5.158] M. Leclercq, *Farbe + Lack* **97** (1991) no. 3, 207–210.
[5.159] T. Szauer, A. Miszczyk, *ACS Symp. Ser.* **322** (1986) 229–233.
[5.160] K. M. Oesterle, R. Oberholzer, *Dtsch. Farben-Z.* **18** (1964) 151.
[5.161] J. D'Ans, H. J. Schuster, *Farbe + Lack* **63** (1957) no. 9, 430.
[5.162] G. Grillo, *Tech. Rundsch.* **60** (1968) 19.
[5.163] A. Laberenz, *Farbe + Lack* **68** (1962) no. 7, 765.
[5.164] E. V. Schmid, *Farbe + Lack* **84** (1978) no. 1, 16–19.
[5.165] M. A. Newnham, *Paint Technol.* **42** (1968) 16.
[5.166] R. Hawner, *Farbe + Lack* **69** (1963) no. 6, 611.
[5.167] Kärntner Montanindustrie GmbH, company information, 1991.
[5.168] E. Carter, *Paint Resin* **55** (1985) 4, 30.
[5.169] E. V. Schmid, *Farbe + Lack* **90** (1984) no. 9, 759–765.
[5.170] N. Wamser, *Farbe + Lack* **96** (1990) no. 6, 435–441.
[5.171] R. D. Athey, *J. Water Borne Coat.* **8** (1985) 1, 7–10.
[5.172] W. Grass, H. D. Merzbach, D. Skudelny, *Farbe + Lack* **94** (1988) no. 11, 895–897.
[5.173] N. Wamser, E. Urbano, *Farbe + Lack* **95** (1989) no. 2, 109, 110.
[5.174] R. D. Athey, *J. Water Borne Coat.* **8** (1985) 2, 10–20.
[5.175] Eckhard, company information, Fürth 1989.
[5.176] G. Cinti, A. Papo, G. Torriano, *J. Oil Colour Chem. Assoc.* **68** (1985) 29–34.
[5.177] C. H. Hare, M. G. Fernald, *Mod. Paint Coat.* **74** (1984) no. 10, 138–146.
[5.178] R. Besold, *Farbe + Lack* **89** (1983) no. 3, 166–173.
[5.179] W. Funke, *Farbe + Lack* **87** (1981) no. 9, 787, 788.
[5.180] BASF Lacke + Farben AG, DE 2 502 781, 1975.
[5.181] R. Pantzer, *Dtsch. Farben-Z.* **19** (1975) 13.
R. Pantzer, *Oberfläche-Surf.* **17** (1976) 181.
[5.182] J. Ruf, *Congr. FATIPEC* **12th** (1974) 571.

[5.183] Ciba-Geigy AG, EP 0128862, 1984 (G. Berner et al.).
[5.184] Ciba-Geigy, company information, Basel 1986.
[5.185] Henkel KGaA, DE 2824508, 1979 (H. Eschwey et al.).
[5.186] Henkel KGaA, company information, Düsseldorf 1983.
[5.187] BASF Lacke+Farben AG, DE 3616721, 1986; EP 246570, 1987 (E. Liedek, G. Hägele).
[5.188] Laboratorium für Pharmakologie und Toxikologie, Hamburg-Hausbruch 1972.
[5.189] Institut für Toxikologie, Bayer AG, Wuppertal 1972.
[5.190] L. M. Greenstein, in P. R. Lewis (ed.): *Pigment Handbook*, 2nd ed., Vol. 1, Wiley & Sons, New York, 1988, p. 829.
[5.191] W. Bäumer, *Farbe + Lack* **79** (1973) 747.
[5.192] K. D. Franz, R. Emmert, K. Nitta, *Kontakte (Darmstadt)* (1992) (2) 3.
[5.193] K. D. Franz, H. Härtner, R. Emmert, K. Nitta, in *Ullmann's Encyclopedia of Industrial Chemistry*, Vol. 20 A, Verlag Chemie, Weinheim, 1992, p. 347.
[5.194] R. Glausch, M. Kieser, R. Maisch, G. Pfaff, J. Weitzel: *Perlglanzpigmente*, Curt R. Vincentz Verlag, Hannover, 1996.
[5.195] H. Simon: *The Splender of Iridiscence*, Dodd, Meal and CO, New York, 1971.
[5.196] H. F. Taylor, *Drugs Oil Paints* **3** (1937) 106.
[5.197] A. H. Pfund, *J. Franklin Inst.* **183** (1917) 453.
[5.198] A. v. Unruh, *Kunststoffe* **8** (1918) 49.
[5.199] S. J. Williamson, H. Z. Cummins: *Light and Color in Nature and Art*, J. Wiley & Sons, New York, 1983.
[5.200] K. Nassau: *The Physics and Chemistry of Color*, J. Wiley & Sons, New York, 1983.
[5.201] H. R. Linton (Du Pont), US 3087828 (1963), US 3087829 (1963). C. A. Quinn, G. J. Rieger, R. A. Bolomey (Mearl), US 3437515 (1969). H. Kohlschütter et al. (Merck KGaA), US 3553001 (1971).
[5.202] W. Ostertag, *Nachr. Chem. Tech. Lab.* **42** (1994) 849.
[5.203] O. S. Heavens: *Optical Properties of Thin Solid Films*, Dover Publ., New York, 1965.
[5.204] S. Hackisu, K. Furusawa, *Sci. Light (Tokyo)* **12** (1963) 157.
[5.205] H. Becker, H. Rechmann, F. Rosendahl, *Farbe + Lack* **73** (1967) 17.
[5.206] C. Schmidt, M. Fritz, *Kontakte (Darmstadt)* (1992) (2) 15.
[5.207] F. Hofmeister, *Eur. Coat J.* **1** (1987) 400.
[5.208] A. M. Gaudin, *J. Phys. Chem.* **41** (1937) 811.
[5.209] R. Emmert, *Cosm. Toiletries* **104** (1989) 57.
[5.210] J. A. Dobrowolski, F. C. Ho, A. Waldorf, *Appl. Opt.* **28** (1989) 2702.
[5.211] S. Teaney, I. Denne, *Kontakte (Darmstadt)* (192) (2) 46.
[5.212] S. Hachisu, *Sci. Light (Tokyo)* **6** (1957) 20.
[5.213] L. M. Greenstein, *Proc. Sci. Sect. Toilet. Goods Assoc.* **45** (1966) 20.
[5.214] A. J. Krajkeman, *Paint Manuf.* **22** (1952) 371.
[5.215] T. A. Marchmay, *Sci. Amer.* **185** (1921) 533.
[5.216] P. Alexander, *Manuf. Chem.* **57** (1986) 60.
[5.217] Y. Morita, *J. Chem. Educ.* **62** (1985) 1072.
[5.218] S. Hachisu, *Sci Light (Tokyo)* **8** (1959) 19 and 26.
S. Hachisu, S. Okamoto, *Sci. Light (Tokyo)* **11** (1962) 157.
[5.219] H. A. Miller, *Bull. Bismuth Inst. (Brussels)* **52** (1987) 1.
[5.220] K. R. Skulberg et al., *Scan. J. Work Environ. Health* **11** (1985) 65.
[5.221] B. K. Bernhard et al., *J. Toxicol. Environ. Health* **29** (1985) 65.
[5.222] R. Esselsborn, K. D. Franz: *XVIIth Congress AFTPV*, Nice, 1987, p. 148.
[5.223] W. Ostertag (BASF), US 4552593 (1985).
[5.224] S. Roesch, *Z. Wiss. Mikrosk. Tech.* **64** (1959) 236 and **66** (1966) 136.
[5.225] H. Kubota, *Prog. Opt.* **1** (1961) 212.
[5.226] G. Pfaff, R. Maisch, *Farbe + Lack* **101** (1995) 89.
[5.227] W. Ostertag, N. Mronga, P. Hauser, *Farbe + Lack* **93** (1987) 973.

[5.228] G. Gehrenkamper, F. Hofmeister, R. Maisch, *Eur. Coat. J.* **3** (1990) 80.
[5.229] R. Emmert, M. Weigand (Merck KGaA), EP 307747 (1989).
[5.230] K. D. Franz, K. Ambrosius, S. Wilhelm, K. Nitta (Merck KGaA), WO 93/19131.
[5.231] K. D. Franz, K. Ambrosius, C. Prengel (Merck KGaA), EP 354374 (1990).
[5.232] K. D. Franz, K. Ambrosius, A. Knapp, D. Brückner (Merck KGaA), US 4867793 (1989).
[5.233] G. Pfaff, P. Reynders, *Chem. Rundschau Jahrbuch* (1993) 31.
[5.234] T. Watanabe, T. Noguchi (Merck KGaA), US 4603047 (1986).
[5.235] D. Brückner, R. Glausch, R. Maisch, *Farbe + Lack* **96** (1990) 411.
[5.236] T. Daponte, P. Verschaeren, M. Kieser, G. Edler (Merck KGaA/Hyplast), WO 94/05727.
[5.237] F. Hund, G. Franz (Bayer), EP 14382 (1982).
[5.238] E. V. Carter, R. D. Laundon, *J. Oil Colour Chem. Assoc.* (1990) (1) 7.
[5.239] W. Ostertag, N. Mronga (BASF), EP 283852 (1988).
[5.240] W. Ostertag, N. Mronga, *Macromol. Symp.* **100** (1995) 163.
[5.241] G. Ash (Optical Coating Laboratory), US 4434010 (1981).
[5.242] R. W. Phillips, A. F. Bleikolm, *Appl. Opt.* **35** (1996) 5529.
[5.243] R. Glausch, G. Pfaff, R. Maisch, *XXIst FATIPEC Congr.*, Amsterdam, Vol. 2, 1992, p. 33.
[5.244] G. Bauer, K. Osterried, C. Schmidt, R. Vogt, H. Kniess, M. Uhlig, N. Schül, G. Brenner (Merck KGaA), WO 93/08237.
[5.245] W. Ostertag: *XIXth, FATIPEC Congr.*, Aachen, Vol. 1, 1988, p. 103.
[5.246] F. Bäbler, A. Weissmüller (Ciba-Geigy), US 4517320 (1985).
[5.247] P. Bugnon, F. Herren, B. Medinger (Ciba-Geigy), EP 408498 (1991).
[5.248] B. R. Swallow (Mearl), EP 401141 (1990).
[5.249] P. Eskelinen: *Prepr. IFSCCJ Congress*, Yokohama, Vol. 1, 1992, p. 196.
[5.250] V. P. S. Judin, *Polym. Paint Coat. J.* **182** (1992) 4312.
[5.251] S. Panush (BASF), US 4753829 (1988).
[5.252] A. P. Philipse, *J. Mater. Sci. Lett.* **8** (1989) 1371.
[5.253] O. Mayborodoff, *Double Liaison – Chim. Peint* **308** (1981) 36.
[5.254] R. Maisch, G. Houseman, K. D. Franz, *Polym. Paint Color J.* **178** (1988) 100.
[5.255] K. Dorfner, J. Ohngemach, *Plast. Eng.* **43** (1987) 33.
[5.256] G. Houseman, *J. Oil Colour Chem. Assoc.* (1987) (11) 329.
[5.257] S. Panush, *J. Paint Technol.* **45** (1981) 39 and **57** (1985) 67.
[5.258] C. J. Rieger, L. Armanini (Mearl), US 4134776 (1979).
[5.259] K. D. Franz et al. (Merck KGaA), US 4482389 (1984).
[5.260] K. Nitta, T. Watanabe, I. Suzuki (Merck KGaA), US 4828623 (1989).
[5.261] A. Rau, K. D. Franz, K. Ambrosius (Merck KGaA), US 5022923 (1991).
[5.262] A. Thurn-Müller, J. Hollenberg, J. Liston, *Kontakte (Darmstadt)* (1992) (2) 35.
[5.263] G. Möschl, F. K. Soehngen, M. Kieser, *Seifen Öle Fette Wachse* **106** (1980) 93, **111** (1985) 643 und **112** (1986) 45.
[5.264] L. Armanini, *Cosm. Toiletries* **106** (1991) 53.
[5.265] M. L. Schlossmann, *Cosm. Toiletries* **105** (1990) 53.
[5.266] A. Thurn-Müller, J. Dietz, C. Prengel (Merck KGaA), EP 406657 (1990).
[5.267] R. Besold, R. Reißer, E. Roth, *Farbe + Lack* **97** (1991) 311.
[5.268] W. Ostertag, *Congr. FATIPEC XIX*, vol II/103, 1988.
[5.269] US 2558302, 1951 (G. C. Marcot et al.).
[5.270] US 2558304, 1951 (G. C. Marcot et al.).
[5.271] Bayer, DE-OS 2508932, 1975 (F. Hund, G. Linde).
[5.272] Bayer, DE 2210279, 1973 (F. L. Ebenhöch, K.-P. Hansen, H. Stark).
[5.273] BASF, DE 2344196, 1973 (F. L. Ebenhöch, D. Werner, G. Bock).
[5.274] BASF F+F AG, DE-OS 2228555, 1972 (H. Gaedecke, R. Bauer).
[5.275] Magnetic Pigment Comp., US 1424635, 1922 (P. Fireman) Reichard-Coulston Inc., US 2574459, 1947 (L. H. Bennetch)
[5.276] F. Finus, *Farbe + Lack* **7** (1975) 604–607.

[5.277] G. Narvuglio, R. F. Sharrock, R. J. Kennedy, *Oil J., Col. Chem. Assoc.* **61** (1978) 79–85.
[5.278] BASF, DE 2840870, 1978 (A. Seitz).
[5.279] TOKYO SHIBAURA DENKIKBUSHIKI KAISHA EP 0019710, 1980 (Wakatsuki).
[5.280] V. P. S. Judin, V. T. Salonen, *Seifen–Oele–Fette–Wachse-Journal*, **119** (1993) no. 8, 491.
[5.281] W. H. Kettler, G. Richter, *Farbe + Lack* **98** (2/1992) 93.
[5.282] BASF US 932741, 1986 (S. Punush).
[5.283] D. R. Robertson, F. Gaw, *Congr.* Add '95, Paper 12.
[5.284] M. A. Vannice, R. L. Garten, *J. Catal.* **63** (1980) 255.
[5.285] Huels, DE-OS 3010710, 1984 (K. Neubold, K.-D. Gollner).
[5.286] Mitsubishi Materials Corp. J 07062326.
[5.287] Degussa EP 609533, 1994 (W. Hartmann, D. Kerner).
[5.288] Nippon Shokubai Co. J 07232919, 1994.
[5.289] L. Brüggemann, DE 3900243, 1993 (G. Walde, A. Rudy).

General references for Chapter 5.5

[5.290] E. Newton Harvey: *A History of Luminescence (until 1900)*, Academic Press, New York 1966.
[5.291] N. Riehl: *Einführung in die Lumineszenz*, Verlag Karl Thiemig KG, München 1970.
[5.292] E. J. Adirowitsch: *Einige Fragen zur Theorie der Lumineszenz der Kristalle*, Akademie Verlag, Berlin, 1953.
[5.293] D. Curie: *Luminescence in Crystals*, Methuen & Co. Ltd., London 1963.
[5.294] Felix Fritz: *Leuchtfarben: Geschichte, Herstellung, Eigenschaften*, Chem. Technischer Verlag Dr. G. Bodenbender, Berlin 1940.
[5.295] P. Goldberg (ed.): *Luminescence of Inorganic Solids*, Academic Press, New York 1966.
[5.296] W. Espe: *Werkstoffkunde der Hochvakuumtechnik*, vol. III, VEB Deutscher Verlag der Wissenschaften, Berlin 1961.

Specific references for Chapter 5.5

[5.297] W. P. Jorrissen, *J. Chem. Educ.* **25** (1948) 685–686.
[5.298] P. Lenard, V. Klatt, *Ann. Phys.* **15** (1904) 225–282, 425–484, 633–673.
[5.299] B. Gudden, R. W. Pohl, *Z. Phys.* **2** (1920) 192.
[5.300] O. W. Lossew, *Telegrafia i Telefonia* **18** (1923) 61.
[5.301] G. Destriau, *J. Chem. Phys.* **33** (1936) 587.
[5.302] F. Urbach, *Wien. Ber.* IIa **139** (1926) 473.
[5.303] J. T. Randall, M. F. H. Wilkens, *Proc. R. Soc. London* **A184** (1945) 347, 366, 390.
[5.304] F. Seitz, *Trans. Faraday Soc.* **35** (1939) 74.
[5.305] N. F. Mott, R. W. Gurney, *Trans. Faraday Soc.* **35** (1939) 69.
[5.306] N. Riehl, M. Schön, *Z. Phys.* **114** (1939) 682–704.
[5.307] H. W. Leverenz: *An Introduction to Luminescence of Solids*, John Wiley & Sons, New York 1950.
[5.308] S. G. Tomlin, *Proc. Phys. Soc. London* **82** (1963) 465–466.
[5.309] J. E. Holliday, E. J. Sternglass, *J. Appl. Phys.* **28** (1957) 1189.
[5.310] R. Ward: *Preparation and Characteristics of Solid Luminescence Materials*, Cornell Symp. 1946, Wiley and Sons, New York 1948.
[5.311] A. Schleede, *Angew. Chem.* **50** (1937) 908.
[5.312] B. J. Verwey, F. A. Kröger, *Philips Tech. Rev.* **13** (1951) 90–95.
[5.313] N. Riehl, H. Ortmann, Angew. Chem. u. Chem. Ingenieurtechnik, Monogr. 72, 1957.
[5.314] K. Era, S. Shionoya, Y. Washizawa, *J. Phys. Chem. Solids* **29** (1968) 1827–1841.
[5.315] Y. Uehara, *J. Chem. Phys.* **62** (1975) 2982–2994.
[5.316] J. Dieleman, S. H. de Bruin, C. S. van Doorn, J. H. Haanstra, *Philips Res. Rep.* **19** (1964) 311–318.

[5.317] N. R. J. Poolton, *J. Phys. Chem: Solid State Phys.* **20** (1987) 5867–5876.
[5.318] Kasei Optonix, JP (6295378) 8 795 378, 1987 (T. Hase).
[5.319] S. Asano, *Sci. Light (Tokyo)* **4** (1955) 32.
[5.320] W. van Gool, *Philips Res. Rep.* **13** (1958) 157–166.
[5.321] A. Bril, H. A. Klasens, *Philips Res. Rep.* **10** (1955) 305
[5.322] W. Lehmann, *J. Electrochem. Soc.* **110** (1963) 754–758.
[5.323] S. Rothschild, *Trans. Faraday Soc.* **42** (1946) 635.
[5.324] F. A. Kröger, J. E. Hellingman, N. W. Smit, *Physica (Amsterdam)* **15** (1949) 990–1018.
[5.325] H. Blicks, N. Riehl, R. Sizmann, *Z. Phys.* **163** (1961) 594–604.
[5.326] S. Shionoya in N. Riehl, H. Kallmann (eds.): *Int. Lumineszenzsymposium über die Physik und Chemie der Szintillatoren,* Thiemig-Verlag, München 1966.
[5.327] H. Kawai, S. Kuboniwa, T. Hoshina, *Jap. J. Appl. Phys.* **13** (1974) 1593–1603.
[5.328] F. A. Kröger, *Physica (Amsterdam)* **6** (1939) 369–379.
[5.329] F. A. Kröger, J. A. M. Dikhoff, *Physica (Amsterdam)* **16** (1950) 297–316.
[5.330] F. A. Kröger, N. W. Smit, *Physica (Amsterdam)* **16** (1950) 317–328.
[5.331] K. Kynev, V. Kuk, *Z. Naturforsch. A* **34 A** (1979) 262–264.
[5.332] M. Tamatani, *Gold Bull.* **13** (1980) 98.
[5.333] A. H. McKeag, P. W. Ranby, *J. Electrochem. Soc.* **96** (1949) 85–89.
[5.334] F. A. Kröger, A. Bril, J. A. M. Dikhoff, *Philips Res. Rep.* **7** (1952) 241.
[5.335] W. Lehmann, *J. Electrochem. Soc.* **117** (1970) 1389–1393.
[5.336] W. Lehmann, *J. Electrochem. Soc.* **118** (1971) 1164–1166.
[5.337] W. Lehmann, F. M. Ryan, *J. Electrochem. Soc.* **119** (1972) 275–277.
[5.338] W. Lehmann, *J. Lumin.* **5** (1972) 87–107.
[5.339] W. Lehmann, F. M. Ryan, *J. Electrochem. Soc.* **118** (1971) 477–482.
[5.340] R. P. Rao, M. de Murcía, J. Gasiot, *Radiat. Proc. Dosim.* **6** (1984) 64.
[5.341] K. Chakrabarti et al., *Symp. Lumin. Sci. Tech.* **88** (1988) 133.
[5.342] R. P. Rao, *J. Electrochem. Soc.* **132** (1982) 2033–2034.
[5.343] H. Kasano, K. Megumi, H. Yamamoto, *J. Electrochem. Soc.* **131** (1984) 1953–1960.
[5.344] F. Okamoto, K. Kato, *J. Electrochem. Soc.* **130** (1983) 432–437.
[5.345] RCA, US 3 418 246, 1968 (M. R. Royce).
[5.346] R. A. Brown, *J. Chem. Educ.* **56** (1979) 732–733.
[5.347] O. Kanehisa, T. Kano, H. Yamamoto, *J. Electrochem. Soc.* **132** (1985) 2023–2027.
[5.348] C. X. Guo, W. P. Zhang, C. S. Shi, *J. Lumin.* **24/25** (1981) 297–300.
[5.349] H. Forest, *J. Electrochem. Soc.* **120** (1973) 695–697.
[5.350] Sony Corp., JP 6 222 887, 1987 (S. Hashimoto).
[5.351] Sony Corp., JP 6 222 888, 1987 (S. Hashimoto).
[5.352] G. Tamman: *Lehrbuch der Metallkunde,* Voss, Leipzig, Hamburg 1932.
[5.353] K. H. Butler: *Fluorescent Lamp Phosphors,* The Pennsylvania State University Press, London 980.
[5.354] T. Y. Tien et al., *J. Electrochem. Soc.* **120** (1973) 278–281.
[5.355] M. F. Yan, T. C. D. Huo, H. C. Ling, *J. Electrochem. Soc.* **134** (1987) 493–498.
[5.356] E. F. Gibbons et al., *J. Electrochem. Soc.* **120** (1973) 835–837.
[5.357] J. M. P. J. Verstegen, J. L. Sommerdijk, J. G. Verriet, *J. Lumin.* **6** (1973) 425–431.
[5.358] J. M. P. J. Verstegen, A. L. N. Stevels, *J. Lumin.* **9** (1974) 406–414.
[5.359] K. Ohno, T. Abe, *J. Electrochem. Soc.* **134** (1987) 2072–2076.
[5.360] T. E. Peters, J. A. Baglio, *J. Electrochem. Soc.* **119** (1972) 230–236.
[5.361] T. E. Peters, *J. Electrochem. Soc.* **119** (1972) 1720–1723.
[5.362] T. E. Peters, *J. Electrochem. Soc.* **122** (1975) 98–101.
[5.363] in [5.298] p. 68.
[5.364] A. H. Gomes de Mesquita, A. Bril, *Mater. Res. Bull.* **4** (1969) 643–650.
[5.365] J. Shmulovich, G. W. Berkstresser, C. D. Brandle, A. Valentino, *J. Electrochem. Soc.* **135** (1988) 3141–3151.

[5.366] T. E. Peters, R. B. Hunt, R. G. Pappalardo, F. T. Taubner, *Proc. Electrochem. Soc.* **88** (1988) 199–203.
[5.367] J. W. Gilliland, *J. Lumin.* **4** (1971) 345–356.
[5.368] L. Thorington, *J. Opt. Soc. Am.* **40** (1950) 579.
[5.369] H. G. Jenkins, A. H. McKeag, P. W. Ranby, *J. Electrochem. Soc.* **96** (1949) 1–12.
[5.370] P. D. Johnson in P. Goldberg (ed.): *Luminescence of Inorganic Solids*, Academic Press New York, London 1966, pp. 287–336.
[5.371] E. F. Apple, *J. Electrochem. Soc.* **110** (1963) 374–380.
[5.372] R. Knütter, *Techn.-wiss. Abh. Osram-Ges.* **11** (1973) 269–276.
[5.373] U. Müller, R. Schirmer, *Techn.-wiss. Abh. Osram-Ges.* **12** (1986) 444.
[5.374] F. Zwaschka, *Techn.-wiss. Abh. Osram-Ges.* **12** (1986) 470.
[5.375] Kasei Optonix JP 63 264 694, 1988 (S. Fujino, F. Takahashi).
[5.376] J. J. Brown, F. A. Hummel, *J. Electrochem. Soc.* **111** (1964) 1052–1057.
[5.377] R. W. Wollentin, C. K. Lui Wei, R. Nagy, *J. Electrochem. Soc.* **99** (1952) 131–136.
[5.378] F. C. Palilla, B. E. O'Reilly, *J. Electrochem. Soc.* **115** (1968) 1076–1081.
[5.379] F. M. Ryan, *J. Lumin.* **24/25** (1981) 827–834.
[5.380] S. T. Henderson, P. W. Ranby, *J. Electrochem. Soc.* **98** (1951) 479–482.
[5.381] D. E. Harrison, *J. Electrochem. Soc.* **107** (1960) 217–221.
[5.382] Y. Kotera, T. Higashi, M. Sugai, A. Ueno, *J. Lumin.* **31/32** (1984) 709–711.
[5.383] R. S. Roth, S. J. Schneider, *J. Res. Nat. Bur. Stand., Sect. A.* **64** (1960) 309.
[5.384] H. Wulff, W. Kempfert, G. Herzog, *Wiss. Z. Ernst-Moritz-Arndt Univ. Greifsw., Math. Naturwiss. Reihe* **36** (1987) 30.
[5.385] N. C. Chang, *J. Appl. Phys.* **34** (1963) 3500–3504.
[5.386] W. C. Herbert, H. B. Minnier, J. J. Brown, *J. Electrochem. Soc.* **115** (1968) 104–105.
[5.387] H. A. Klasens, *Philips Res. Rep.* **9** (1954) 377.
[5.388] L. G. van Uitert, R. R. Soden, R. C. Linares, *J. Chem. Phys.* **36** (1962) 1793–1796.
[5.389] L. G. van Uitert, R. C. Linares, R. R. Soden, A. A. Ballmann, *J. Chem. Phys.* **36** (1962) 702.
[5.390] A. K. Levine, F. C. Palilla, *Appl. Phys. Lett.* **5** (1964) 118.
[5.391] S. Faria, D. T. Palumbo, *J. Electrochem. Soc.* **116** (1969) 157–158.
[5.392] S. Faria, C. W. Fritsch, *J. Electrochem. Soc.* **116** (1969) 155–157.
[5.393] L. H. Brixner, E. Abramson, *J. Electrochem. Soc.* **112** (1965) 70–74.
[5.394] S. Faria, E. J. Mehalchick, *J. Electrochem. Soc.* **121** (1974) 305–307.
[5.395] R. K. Datta, *J. Electrochem. Soc.* **114** (1967) 1057–1063.
[5.396] P. Wacher, D. J. Bracco, F. C. Palilla, *Electrochem. Technol.* **5** (1967) 358–363.
[5.397] G. Blasse, A. Bril, *J. Lumin.* **3** (1970) 109–131.
[5.398] L. H. Brixner, H.-Y. Chen, *J. Electrochem. Soc.* **130** (1983) 2435–2443.
[5.399] K. H. Becker, A. Scharmann: *Einführung in die Festkörperdosimetrie,* Thieme Pocket Books, vol. 56, Verlag Karl Thiemig KG, München 1975.
[5.400] J. R. Danby, K. Holliday, N. B. Manson, *J. Lumin.* **42** (1988) 83–88.
[5.401] M. R. Mulla, S. H. Pawar, *J. Lumin.* **40/41** (1988) 179–180.
[5.402] J. Rudolph, H. Ruffler, *Tech. Wiss. Abh. Osram Ges.* **7** (1958) 223.
[5.403] A. Bril, W. Hoekstra, *Philips Res. Rep.* **16** (1961) 356.
[5.404] D. Ülkü, *Z. Krystallogr.* **124** (1967) 192–219.
[5.405] P. R. Wonka in H. Ortmann (ed.): *Zur Physik und Chemie der Kristallphosphore*, vol. 2, Akademie-Verlag, Berlin (DDR) 1962.
[5.406] F. A. Kröger: *Some Aspects of the Luminescence of Solids,* Elsevier, Amsterdam 1948.
[5.407] J. J. Borchardt, *J. Chem. Phys.* **42** (1965) 3743.
[5.408] E. Fischer in [5.293], pp. 151–181.
[5.409] A. L. N. Stevels, F. Pingault, *Philips Res. Rep.* **30** (1975) 227–290.
[5.410] Siemens, DE-OS 2 642 226, 1978 (H. Degenhardt).
[5.411] Siemens, DE-OS 2 542 481, 1976 (A. L. N. Stevels).
[5.412] Kasei Optonix, JP 1 576 241, 1977 (E. Mori).

[5.413] K. Takahashi, K. Kohda, J. Miyahara, *J. Lumin.* **31/32** (1984) 266–268.

[5.414] K. Takahashi, J. Miyahara, *J. Electrochem. Soc.* **132** (1985) 1492–1494.

[5.415] H. v. Seggern, T. Voigt, K. Schwarzmichel, *Siemens Forsch. Entwicklungsber.* **17** (1988) 125–130.

[5.416] D. M. de Leeuw, T. Kovats, S. P. Herko, *J. Electrochem. Soc.* **134** (1987) 491–493.

[5.417] Y. Amemiya et al., International Conference on Biophysics and Synchrotron Radiation, Frascati, July 14–16, 1986.

[5.418] R. J. Kurtz, *J. Electrochem. Soc.* **109** (1962) 18–24.

[5.419] G. Blasse, A. Bril, *Philips Res. Rep.* **22** (1967) 481–504.

[5.420] General Electric, US 3 795 814 1974 (J. G. Rabatin); US 4 070 583, 1978 (J. G. Rabatin).

[5.421] General Electric, US 3 591 516, 1971 (J. G. Rabatin); US 3 607 770, 1971 (J. G. Rabatin).

[5.422] D. M. de Leeuw, C. A. H. A. Mutsaers, H. Mulder, D. B. M. Klaassen, *J. Electrochem. Soc.* **135** (1988) 1009–1014.

[5.423] H. Spangenberg: *Leuchtröhrenanlagen für Lichtreklame und moderne Beleuchtung*, Helios-Verlag, Berlin 1955.

[5.424] H. Kullack: *Hochspannungs-Leuchtröhrenanlagen*, Verlag Technik, Berlin 1951.

[5.425] H. Degenhardt, *Electromedica* **3** (1981) 154–158.

[5.426] M. Sonoda, M. Takano, J. Miyahara, H. Kato, *Radiology* **148** (1983) 833–838.

[5.427] C. W. Bates in D. A. Garrett, D. A. Bracher (eds.): *American Society for Testing and Materials*, Philadelphia, STP-716 1980, p. 45–65.

[5.428] F. Roland, B. Ramm: *Das Röntgenbild: Theorie, Methode, Technik*, Thieme Verlag, Stuttgart 1980.

[5.429] P. Keller, *Proc. SID* **24** (1983) 323–328.

[5.430] K. Carl, J. A. M. Dickoff, W. Eckenbach, *Philips tech. Rev.* **40** (1982) 48.

[5.431] F. Eckart: *Elektronenoptische Bildwandler und Röntgenbildverstärker*, Verlag Johann Ambrosius Barth, Leipzig 1962.

[5.432] G. Krokeide in J. D. Sime (ed.): *Safety in the Built Environment*, E & FN Spon Ltd., London-New York 1988, pp. 134–146.

[5.433] K. P. Günther, J. Mirow, H. E. Dolle, H. Dörre, *Brandschutz* **3** (1988).

[5.434] G. M. B. Webber, P. J. Hallman in J. D. Sime (ed.): *Safety in the Built Environment*, E & FN Spon Ltd., London-New York 1988, pp. 147–157.

[5.435] H. Stübel, *Pflügers Arch. Gesamte Physiol. Menschen Tiere* **141** (1911) 1–14.

[5.436] D. Fujimoto, *Biochem. Biophys. Res. Commun.* **84** (1978) 52–57.

[5.437] Y. Fukushima, T. Araki, M.-O. Yamada, *Cell. Mol. Biol.* **33** (1987) 725–736.

[5.438] H. Grassmann, H.-G. Moser, E. Lorenz, *J. Lumin.* **33** (1985) 109–113.

[5.439] A. Bril, W. van Meurs-Hoekstra, *Philips Res. Rep.* **19** (1964) 296–306.

[5.440] N. Riehl: *Physik. und technische Anwendungen der Lumineszenz*, Springer Verlag, Berlin 1941.

Index

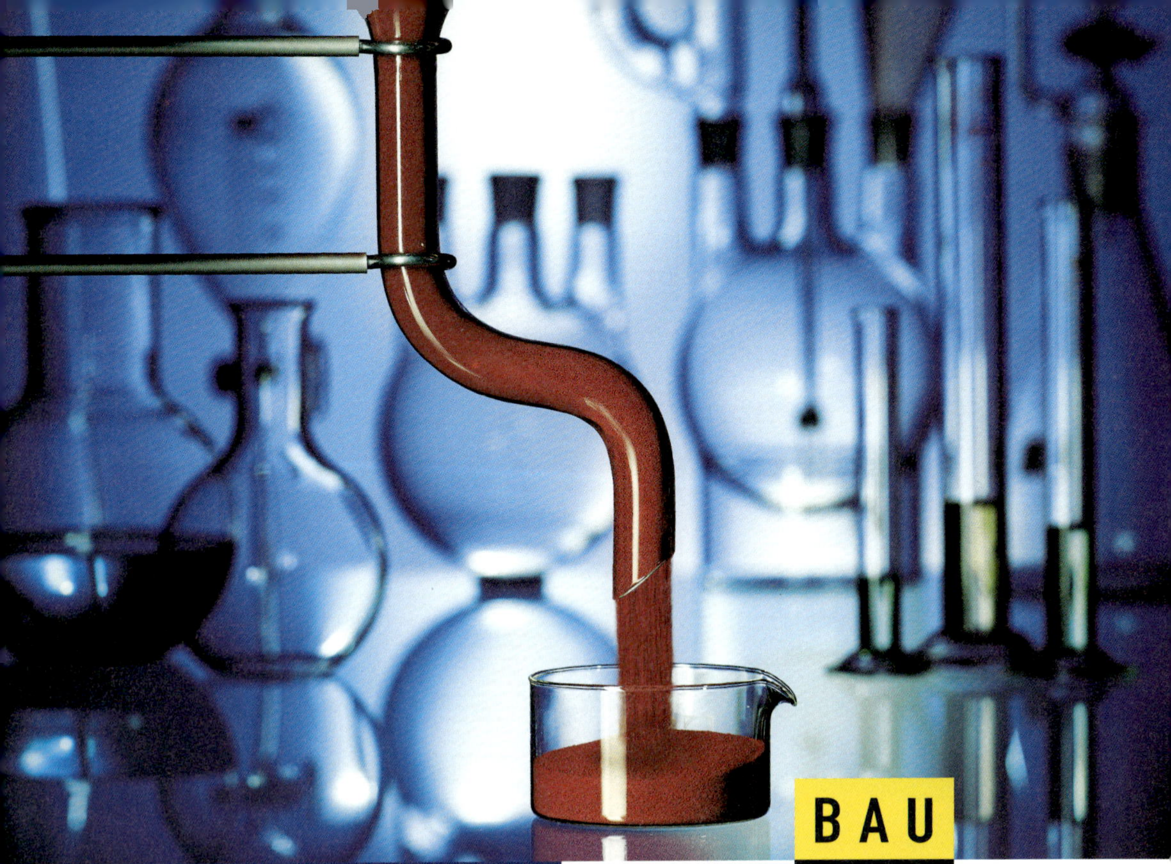